FIRST INTERNATIONAL COURSE ON CONDENSED MATTER

ACIF Series — Vol. 8

Bogota,
July 7-18 1986

Editors
DANIELE PROSPERI
University of Rome,
La Sapienza
SERGIO ROSATI
University of Pisa
GALILEO VIOLINI
Centro International de Fisica
and
University of Calabria

World Scientific
Singapore • New Jersey • Hong Kong

Published by

World Scientific Publishing Co. Pte. Ltd.
5 Toh Tuck Link, Singapore 596224
USA office: 27 Warren Street, Suite 401-402, Hackensack, NJ 07601
UK office: 57 Shelton Street, Covent Garden, London WC2H 9HE

British Library Cataloguing-in-Publication Data
A catalogue record for this book is available from the British Library.

FIRST INTERNATIONAL COURSE ON CONDENSED MATTER

ISBN-13 978-9971-5-0743-5
ISBN-10 9971-5-0743-9

FIRST INTERNATIONAL COURSE ON CONDENSED MATTER

Other ACIF Publications in this Series

1. *Stochastic Processes Applied to Physics and Other Related Fields,* edited by B. Gomez, S. M. Moore, A. M. Rodriguez Vargas, and A. Rueda (1983).

2. *Workshop on the Search for Gravitational Waves,* edited by E. Posada and G. Violini (1983).

3. *Non Conventional Energy Sources,* edited by G. Furlan, H. Rodriguez, and G. Violini (1984).

4. *Galaxies, Quasars and Cosmology,* edited by L. Z. Fang and R. Ruffini (1985).

5. *Reactor Physics for Developing Countries and Nuclear Spectroscopy Research,* edited by K. Lieb and G. Medrano (1986).

6. *Proceedings of the First Latin American School on Biophysics,* edited by R. Fayad, A. M. Rodriguez-Vargas, and G. Violini (1987).

7. *Methods and Applications of Nonlinear Dynamics,* edited by A. W. Saenz (1988).

EDITORS' FOREWORD

It is a great pleasure for us to present the Proceedings of the First International Course on Condensed Matter as a volume of the ACIF series. The Course was organized by the Centro Internacional de Física and held in Bogota from July 7 to July 18, 1986.

The purpose of the organizers was to include an activity in the area of the physics of strongly interacting systems in the programs of the Center, in order to promote horizontal cooperation between Latin-American scientists in this area of physics. Two problems are of particular importance in the study of such systems, namely the knowledge of their interactions and the calculation of their structure and properties.

The nucleon-nucleon interaction remains a fascinating puzzle. The picture of nucleons as point-like particles interacting by two-body potentials can be adequate for many problems of nuclear structure, but it encounters severe difficulties as the energy increases. The quark models for hadrons are now becoming an important tool also for nuclear physics and for such a reason this class of models has been widely examined in the Course.

The structure of quantum systems, as nuclei, nuclear or neutron matter, liquid Helium, has been studied in the framwork of different many-body theories. In the last few years the techniques for detailed calculations on strongly interacting systems have received a great impulse also thanks to the increased computational possibilities now available. These techniques have been extensively discussed in the Course and illustrated in this volume, in the hope that this material will allow young people to perform explicit calculations on such systems.

Some of the authors are grateful to the Istituto Nazionale di Fisica Nucleare (Italy) for financial support.

Daniele Prosperi

Sergio Rosati

Galileo Violini

CONTENTS

CORRELATED BASIS FUNCTION THEORY FOR FERMION SYSTEMS
by A. Fabrocini and S. Fantoni

MONTECARLO TECHNIQUES IN THE MANY BODY PROBLEM
by R. Guardiola

CONSTITUENT QUARK MODELS FOR HADRONS AND NUCLEI

by
R.Mignani(*), D.Prosperi(*)
Physics Department, I University of Rome "La Sapiemza"
and
Istituto Nazionale di Fisica Nucleare - Sezione di Roma (Italy)

(*)Dipartimento di Fisica and INFN, Sezione di Roma,
P.le Aldo Moro 2, I-00185 Roma (Italia)

CONSTITUENT QUARK MODELS FOR HADRONS AND NUCLEI

by R. Mignani and D. Prosperi

1 – Introduction.

It is a well known and experimentally established fact that nuclei are essentially made out of protons and neutrons (which are nothing but the two components, in isospin space, of a single particle, the nucleon). The word *essentially* is here used to mean that, actually, a full understanding of the nuclear structure requires more than a simple description in terms of nucleons.

The picture of nuclei as compounds of pointlike nucleons, interacting via a short range force which can be well approximated by a two-body potential, works sufficiently well for the ground state and the low-lying excited states ([1]). For instance, the electromagnetic properties of nuclei (like magnetic and quadrupole moments) are quite satisfactorily accouted for by this simple model. But things change drastically when one probes the nuclear structure at higher energies. Then, two kinds of effects come into play. Firstly, one has to take account explicitly of the finite size of nucleons, that are not pointlike particles but extended objects with dimensions comparable to the internuclear distance. Secondly, the interaction between nucleons is to be described by the exchange of particles (bosons) which mediate the strong interaction. In first approximation, it is sufficient to assume (according to the old Yukawa's hypothesis) that only the pion is the carrier of the strong force. Indeed, pion exchange accounts well for the long-range part of the nucleon-nucleon (NN) interaction. Howewer, a reproduction of the NN interaction involves a relatively large number of different particles, like the Δ-resonance (which appears in the description of the intermediate-range part of the interaction), the ρ and ω vector mesons, and so on. The corresponding contributions to the NN interaction just represent the first terms of a practically infinite series.

The proposal by Gell-Mann and Zweig in 1964 of quarks as constituents of hadrons ([2]), and the subsequent large success of the quark model in high energy physics ([3]), force us to a radical change in the present picture of nuclear structure. Hadrons (and therefore nucleons, pions, and so on) are no more fundamental objects. The true, fundamental strong force is not the nuclear one, but the *colour* force among quarks, that is mediated by the *gluon* field. The corresponding quantum field theory, based on quarks and colour, and called Quantum Chromodynamics (QCD) is today generally accepted as being the theory of the strong interactions ([4]). From the QCD standpoint, nuclei are to be considered as *molecules* of quarks, and the nuclear force is just a residual of the strong colour force, very much as the molecular force is a residual of the Coulomb force. The nucleon-nucleon interaction is thus described by quark interchange and gluon exchange processes.

Howewer, this new picture of hadrons and strong interactions based on quarks and QCD has had very little impact in nuclear physics until about a decade ago. The reason is twofold. From the experimental point of view, there was no evidence at all for effects due to quarks degrees of freedom in nuclei. From the theoretical standpoint, perturbative methods in QCD can be applied only at very short distances; at large distances (of the order of the range of the nuclear force), a non-perturbative approach is needed, which is presently beyond technical calculation capability.

In the last decade, the state of affairs in the field of quarks and nuclei is rapidly changed, and there is now an increasing deal of interest ([5]) on the subject. The so-called EMC effect (recently observed at $CERN$ and apparently showing a different behaviour of deep inelastic lepton scattering on free nucleons and on nuclei) seems to provide the first experimental evidence for a role of quarks in nuclei. On the theoretical side, a lot of work on the applications of perturbative *QCD* to nuclear physics has been done by Brodsky and coworkers [6], who have been able to derive rigorously a number of properties of short-distance nuclear properties, like nucleon and deuteron asymptotic form factors, factorization theorems, amplitudes for meson photoproduction and deuteron photo-and electro-disintegration, and so on. Moreover, a number of phenomenological models, which incorporate the main features expected from *QCD*, have been developed in order to deal with quarks in nuclei at intermediate and large distances (where perturbative *QCD* fails). They include non-relativistic constituent quark models [7], bag models in their various versions [8], Skyrme [9] and soliton [10] models. All these models have been successfull (partially, at least) in describing some aspects of nuclear physics for few-body systems at large distances, for instance the main features of the nucleon-nucleon interaction.

Clearly, a number of questions remain open. The most serious problem is to fully understand the relationship between the two alternative pictures of the nuclear structure: that valid at short distances, in terms of gluon and quark exchanges, and the classical description based on meson exchange, adopted for large distances. A possible solution to this problem is provided by the so-called *hybrid models*, which try to incorporate the good features of both the *QCD* based and the classical nuclear approach.

These lectures are aimed to provide an introduction to the subject of the quark approach to nuclear physics. We assume that the reader does posses a sound knowledge of standard nuclear physics, and try to acquaint him with the basic concepts and methods of quark physics from the nuclear standpoint. For reasons of time and space, we have not made any attempt to give an exhaustive review of all possible topics in the field, and focused our attention on the non-relativistic, phenomenological quark models (basically, because a number of excellent review articles and lectures already exist on relativistic models, like the bag [8] and so on). Furthermore, these lectures do not presume at all to be self-contained. In this sense, they may be better regarded as a guide to a critical reading of the existing literature [11].

The content of these lectures is as follows. In chapt. 2, we give the essential concepts of the quark model, colour and *QCD*. Chapt. 3 is concerned with the experimental evidence of the quark structure of nucleons, including also a brief section on the *EMC* effect. Sect. 3.6 discusses the main tools and results of perturbative *QCD* applied to nuclear physics. Phenomenological, non-relativistic constituent quark models are considered in chapt. 4, with special emphasis on their nuclear applications. Chapt. 5 is concerned with a special approach to nuclear calculations in the framework of the quark model that is revealed itself very fruitful in the last years, i.e. the "resonating group method". Chapt. 6 is devoted to final remarks.

2 – Elements of Quantum Chromodynamics.

In this chapter, we will try to give an answer as simple as possible to the question: what is Quantum Chromodinamics (QCD)([4])? Clearly, a complete answer would require a book (or a number of books!) by itself. Therefore, we shall confine ourselves to report only the basic concepts and features of QCD, together with a few topics directly concerning applications to nuclear physics.

QCD is today generally believed to be the solution to the old problem of a theory for the strong interactions. It is a quantum field theory, based on the group of *colour* $SU(3)_c$, which had as early as 1964 been suggested to solve the problem of Pauli principle for hadron spectroscopy, in the context of the *quark model*([3]). As such, QCD belongs to the class of field theories known as *non-abelian gauge theories*. There is a similarity between QCD and the quantum field theory of electromagnetic interactions, quantum electrodynamics (QED). Colour is the analogous of electrical charge (then, in a sense, it is a *strong charge*); strong interactions are mediated by massless particles analogous of photons (*gluons*) (which, howewer, carry themselves colour). By this analogy, hadrons correspond to atoms, and nuclei to molecules.

Let us discuss in more details, and on a more sound theoretical basis, the structure of QCD.

2.1 – Foundations of QCD: quarks and colour.

As we have briefly mentioned above, the idea of a new quantum number, *colour*, basic for the development of QCD, was firstly suggested in the framework of the quark model of hadrons. Let us spend a few words about quarks.

When the word *quark* was first introduced in elementary particle physics by Gell-Mann([2]), it was considered rather an useful synonimy for *basis vector of the fundamental* SU(3) *representation*, than the name of a physical entity. Actually, in 1964 all strongly-interacting particles could be attributed to representations of the unitary group SU(3), of which the three quarks u, d, s, form the fundamental representation. The SU(3) symmetry scheme permitted to account for hadron spectroscopy, including masses, magnetic moments and so on. As it is well known, one has to assume that quarks are spin 1/2 particles, endowed with fractional ($-1/3$ or $2/3$) electric charge. In the standard terminology, the different kinds of quarks are distinguished by the quantum number of *flavour* †.

In the 70's, the discovery of new particles has forced to change the standard SU(3) scheme to SU(6), thus introducing new kinds of quarks and new flavours (see Table I). In the meantime, experimental evidence was accumulating (mainly due to the results from deep inelastic scattering and e^+e^- experiments) for the existence of quarks as point-like constituents of hadrons.

Today quarks are generally considered *almost* as real as all other particles (the only

---

† The names of colour and flavour for quark quantum numbers were inspired by the following words of the ancient Greek philosopher Anaxagoras: "...And since these things are so, it is necessary to think that in all the objects that are compound there existed many things of all sorts, and germs of all objects, having all sorts of forms and *colours* and *flavours*."

difference being their permanent confinement inside hadrons, whereby they escape actual detection: see sect.2.5), and the building blocks of baryons and mesons. Therefore, the old, rather abstract quark model has evolved into a *constituent quark model.* In other words, once one assumes quarks to be physical particles, the subsequent attempt was to overcome the mere classification scheme and try to give it a true dynamical basis. QCD is just such an effort.

TABLE I – Quark quantum numbers (flavours)

quark	charge (q/e)	i-spin T	T_3	baryon number	strangeness	charm
u	+2/3	1/2	+1/2	1/3	0	0
d	-1/3	1/2	-1/2	1/3	0	0
s	-1/3	0	0	1/3	1	0
c	+2/3	0	0	1/3	0	1
b,t,...	...	...	...	...	...	...

In the quark model, all baryons are supposed to be built up by three quarks (i.e. they have essentially the structure q^3), while mesons are composed by a quark and an antiquark $(q\bar{q})$ †. For instance, the nucleons have essentially the following structure: proton: $|uud\rangle$; neutron: $|udd\rangle$. (Incidentally, this implies that, in nuclear physics, we are mainly interested to aggregates of quarks u,d. They constitute a — practically degenerate — doublet of the standard isospin).

It follows immediately by the three-quark structure of baryons that it is impossible to apply to quarks the usual Fermi statistics, unless one endowes them with a further quantum number([12]). Indeed, for instance, in the quark model the wave function ψ_0 of a nucleon turns out to be completely symmetric under the exchange of the usual indices (t_3, s_3, l, l_3) for any quark pair, contrarily to what demanded by the Pauli principle. This problem was solved by introducing a new, hidden degree of freedom, named *colour*, thus enlarging the set of quantum numbers needed to characterize a state. In order that a three-quark wavefunction can be antisymmetrized in the colour space (thus yielding a *white* state ψ_c, i.e. a state with zero colour), it is necessary that each quark flavour comes in three distinguishable colours: let us denote them, as usual, by red (r), green (g) and blue (b). Thus the total wavefunction of the nucleon, $\psi_0 \times \psi_c$, acquires the required symmetry properties. Similar arguments lead to conclude that the existence of the colour degree of freedom is essential to our understanding of the whole baryon spectroscopy ([13]). Further and conclusive evidence for the need of such new degree of freedom have come from the

† From a mathematical point of view, this amounts to say that baryons belong to a SU(3) representation contained in the product of three fundamental representations, whereas meson representations are contained in the product of the fundamental representation by its conjugate. See sect. 4.2.

experimental results on various observables which are sensitive to the number of distinct species of quarks, like the rate of the decay $\pi^0 \to 2\gamma$ ([14]), and the cross-section ratio $R_{e^+e^-} = \sigma(e^+e^- \to hadrons)/\sigma(e^+e^- \to \mu^+\mu^-)$ ([15]) ‡.

The colour hypothesis has permitted not only to solve the problem of statistics in the quark model, thus giving a better insight into hadron spectroscopy; but it provided a dynamical basis to the quark model, by playing a fundamental role in building up QCD. Indeed, as we shall see, QCD is basically a field theory invariant under local rotations in colour $SU(3)_c$ space.

2.2 – The colour group $SU(3)_c$.

Let ψ_c^f be the wave function of a quark with flavour f and colour c. The set of the three different colour components can be represented by the column matrix

$$\psi^f = (\psi_c^f) = \begin{pmatrix} \psi_r^f \\ \psi_g^f \\ \psi_b^f \end{pmatrix}. \tag{2.1}$$

Therefore, the most general trasformation acting on the colour variables of a quark will be obviously given by a 3×3 matrix $\widehat{M}$ (with elements M_{ab}). Explicitly, one has

$$\psi^f \to (\psi^f)' \tag{2.2}$$

with

$$(\psi^f)' = (\widehat{M}\psi^f) = (\sum_c M_{ac}\psi_c^f) \,. \tag{2.3}$$

If $\widehat{M} = \widehat{U}$, where $\widehat{U}$ is an unitary matrix, we shall refer to the trasformation (2.2) as a *rotation of ψ^f in colour space.*

It is easy to check that the operator

$$\widehat{U}(\alpha) = exp(i\frac{\alpha}{2}\widehat{\lambda}) \tag{2.4}$$

is unitary, provided that the parameter α is real and the 3×3 matrix $\widehat{\lambda}$ is hermitian. Operator (2.4) will induce a rotation, whose amount is determined by α; the matrix $\widehat{\lambda}$ is called the *generator* of the trasformation †. If α is an infinitesimal parameter, we get the *infinitesimal rotation* operator:

$$\widehat{U}(\alpha) = \widehat{1} + i\frac{\alpha}{2}\widehat{\lambda} \tag{2.5}$$

---

‡ The theoretical value of $R_{e^+e^-}$, calculated by means of the quark-parton model, is $R_{e^+e^-} = N_c \sum_f e_f^2$ (N_c = number of coloured quarks; e_f = charge of quark with flavour f) and, for $N_c = 3$, agrees well with the experimental values $R_{e^+e^-} = 2$ below the charm threshold, $R_{e^+e^-} = 10/3$ above the charm threshold and $R_{e^+e^-} = 11/3$ above the beauty threshold; see ref. ([15]).

† Clearly, eq.(2.4) is the analogous of a rotation in spin space, generated by the Pauli matrices $\widehat{\sigma}$: $\widehat{U} = exp(i\frac{\alpha}{2}\widehat{\sigma})$.

(unitary up to the first order in α), where $\widehat{1}$ is the 3×3 unity matrix.

In the case we are considering, all rotations in colour space are generated by eight ‡ linearly independent, traceless matrices $\widehat{\lambda}_i (i = 1, \ldots, 8)$, which can be put in the following canonical form:

$$\widehat{\lambda}_1 = \begin{pmatrix} 0 & 1 & 0 \\ 1 & 0 & 0 \\ 0 & 0 & 0 \end{pmatrix} \qquad \widehat{\lambda}_2 = \begin{pmatrix} 0 & -i & 0 \\ i & 0 & 0 \\ 0 & 0 & 0 \end{pmatrix}$$
$$\widehat{\lambda}_3 = \begin{pmatrix} 1 & 0 & 0 \\ 0 & -1 & 0 \\ 0 & 0 & 0 \end{pmatrix} \qquad \widehat{\lambda}_4 = \begin{pmatrix} 0 & 0 & 1 \\ 0 & 0 & 0 \\ 1 & 0 & 0 \end{pmatrix}$$
$$\widehat{\lambda}_5 = \begin{pmatrix} 0 & 0 & -i \\ 0 & 0 & 0 \\ i & 0 & 0 \end{pmatrix} \qquad \widehat{\lambda}_6 = \begin{pmatrix} 0 & 0 & 0 \\ 0 & 0 & 1 \\ 0 & 1 & 0 \end{pmatrix} \tag{2.6}$$
$$\widehat{\lambda}_7 = \begin{pmatrix} 0 & 0 & 0 \\ 0 & 0 & -i \\ 0 & i & 0 \end{pmatrix} \qquad \widehat{\lambda}_8 = \frac{1}{\sqrt{3}} \begin{pmatrix} 1 & 0 & 0 \\ 0 & 1 & 0 \\ 0 & 0 & -2 \end{pmatrix}$$

(Gell-Mann matrices), whose elements will be denoted in the following by $(\lambda_i)_{cc'}$.

The group of all unitary trasformations $\widehat{U}(\alpha)$ of the kind (2.4), with generators the Gell-Mann matrices, is the unitary group SU(3). In the present case, we shall call it the *colour group* $\mathrm{SU}(3)_c$. From a mathematical point of view, it is a Lie group of order eight (the number of generators) and rank two (the number of diagonal generators).

Let us note that the algebraic structure of the first three generators $\widehat{\lambda}_1$, $\widehat{\lambda}_2$, $\widehat{\lambda}_3$ generalizes that of the Pauli matrices (which, as it is well known, are used in describing the spin and isospin components). Indeed, we can write

$$\widehat{\lambda}_i = \begin{pmatrix} \widehat{\sigma}_i & 0 \\ 0 & 0 \end{pmatrix} \qquad i = 1, 2, 3 \tag{2.7}$$

with the $\widehat{\sigma}_i$'s being the Pauli matrices. In particular, the matrix $\widehat{\lambda}_3$ is diagonal and has three eigenvalues, $+1, 0, -1$. The second diagonal matrix is $\widehat{\lambda}_8$, with eigenvalues $+1/\sqrt{3}, +1/\sqrt{3}, -2/\sqrt{3}$.

Sometimes it is useful to work with suitable linear combinations of the diagonal matrices, like the *(colour) hypercharge*

$$\widehat{\lambda}_Y = \frac{1}{\sqrt{3}} \widehat{\lambda}_8 = \frac{1}{3} \begin{pmatrix} 1 & 0 & 0 \\ 0 & 1 & 0 \\ 0 & 0 & -2 \end{pmatrix} \tag{2.8}$$

and the *(colour) charge*

$$\widehat{\lambda}_Q = \frac{\widehat{\lambda}_3}{2} + \frac{\widehat{\lambda}_Y}{2} = \begin{pmatrix} 2/3 & 0 & 0 \\ 0 & -1/3 & 0 \\ 0 & 0 & -1/3 \end{pmatrix} . \tag{2.9}$$

‡ Indeed, the number of parameters needed to specify a $N \times N$ traceless matrix is $N^2 - 1$.

The three possible colour states are usually represented by the *weight diagram* of Fig.1, where Y_c and $I_3^{(c)}$ are the eigenstates of $\hat{\lambda}_Y$ and $\hat{\lambda}_3$, respectively.

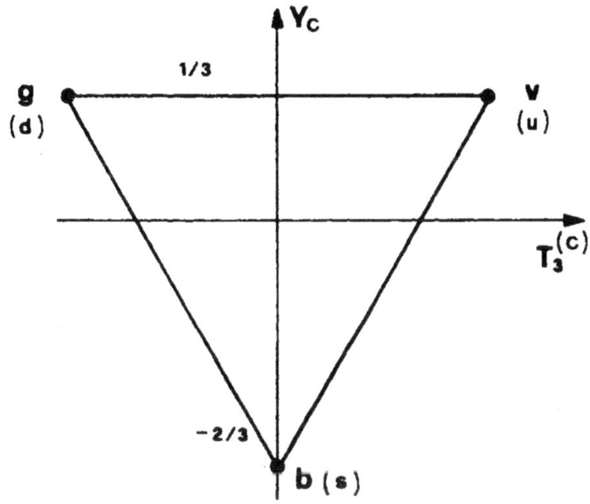

Fig. 1 – Weight diagram of the fundamental 3-dimensional representation of the colour unitary group $SU(3)_c$. The flavours of the fundamental triplet of $SU(3)_f$ are also reported in brackets.

It is easy to check by direct computation that the generators $\hat{\lambda}_i$ satisfy the commutation relations

$$[\hat{\lambda}_j, \hat{\lambda}_k] = 2if_{jkl}\hat{\lambda}_l \tag{2.10}$$

(sum over l understood), where the f_{jkl}'s are the *structure constants* of SU(3) and are antisymmetric under the permutation of any two indices †. The nonzero f_{jkl} are given in Table II. In the following, we shall also use the matrices

$$\hat{T}_i = \hat{\lambda}_i/2 \tag{2.11a}$$

with commutation relations

$$[\hat{T}_j, \hat{T}_k] = if_{jkl}\hat{T}_l \tag{2.11b}.$$

TABLE II – Structure functions of SU(3)

jkl	123	147	156	246	257	345	367	458	678
f_{jkl}	1	1/2	-1/2	1/2	1/2	1/2	-1/2	$\sqrt{3}/2$	$\sqrt{3}/2$

† The antisymmetry of the structure constants is insured by the generator property $Tr(\hat{\lambda}_j\hat{\lambda}_k) = \delta_{jk}$.

Eq. (2.10) tells us that the commutator of two generators can be expressed as a linear combination of the generators — a property far from trivial. All groups which possess such a property — and, therefore, whose generators satisfy eq. (2.10) — are called *Lie groups* (familiar examples of Lie groups are the rotation group and the Lorentz group)([16]) †. It is easy to realize that the structure constants fully specify all algebraic properties of the group ‡. Any explicit realization of a given Lie group $\mathcal{G}$ by means of $N \times N$ matrices is called a *(matrix) representation* of $\mathcal{G}$. We shall come back to the concept of SU(3) representations (and their products) in sect. 4.2. However, we anticipate that — as it is easy to realize from the previous discussion and the explicit forms of the quark wavefunction (2.1) and the Gell-Mann matrices (2.6) — coloured quarks belong to the fundamental 3-dimensional representation of $SU(3)_c$. Gluons (the quanta of the strong interaction), being themselves coloured, belong to the eight-dimensional $SU(3)_c$ representation (it is also customary to say that quarks and gluons are a *triplet* and an *octet* of $SU(3)_c$, respectively). Hadrons are colourless (*white*), and therefore are classified according to 1-dimensional representations of $SU(3)_c$ (i.e. they are *singlets*).

A physical theory will be said *colour invariant*, if all its equations are invariant under rotations in colour space. Let us recall that, under a unitary trasformation $\hat{U}(\alpha)$, an operator $\hat{A}$ changes according to

$$\hat{A} \to \hat{A}' = \hat{U}(\alpha)\hat{A}\hat{U}^{-1}(\alpha) \tag{2.12}$$

(*similarity trasformation*). Then, it is easily seen that $\hat{A}$ is invariant ($\hat{A}' = \hat{A}$) if it commutes with all $SU(3)_c$ generators $\hat{\lambda}_i$:

$$[\hat{\lambda}_i, \hat{A}] = 0 \qquad \forall i \ (i = 1, \ldots, 8) \ . \tag{2.13}$$

Given an operator $\hat{A}$, and any two-quark wavefunctions ψ^g, ψ^f (g,f being the flavour quantum numbers) it is easy to check that the quantity

$$I^{gf} = (\psi^g)^+ \hat{A} \psi^f = \sum_{c,c'} (\psi^g_c)^+ A_{c,c'} \psi^f_{c'} \tag{2.14}$$

(with $c = r, g, b$) is colour invariant. Indeed, by eqs. (2.4), (2.12), one gets

$$(I^{gf})' = (\psi^{g'})^+ \hat{A}' \psi^{f'} = (\psi^g)^+ \hat{U}(-\alpha) \hat{U}(\alpha) \hat{A} \hat{U}(-\alpha) \hat{U}(\alpha) \psi^f = I^{gf} \ .$$

---

† From a more rigorous viewpoint, a Lie group is a continuous group (i.e. its elements depend on continuous parameters) such that the parameters of the product element are analitic functions of those of the factors. A Lie group is compact if the domain of variation of all the parameters is compact (i.e. bounded and closed). It is easy to see that the rotation group is compact, whereas the Lorentz group does not.

‡ However, let us notice that the converse is not true. The explicit form of the structure constants depends on the basis chosen for the generators. In general, there is a continuum of different sets of structure constants for a given Lie group. This is connected to the fact that to any Lie group corresponds a Lie algebra whose product is given by the commutator.

In the following, we shall see that the colour group $SU(3)_c$ plays a fundamental role in building up a dynamical theory of strong interactions. This is just accomplished by assuming $SU(3)_c$ to be an exact local symmetry of the QCD Lagrangian, i. e. postulating local invariance under phase transformations involving coloured fields.

In conclusion, let us note that all the algebraic properties we have discussed for the colour group $SU(3)_c$ hold as well for the unitary group connected to the flavour variables of quarks, $SU(3)_f$.

However, it must be stressed that: (i) today, the flavour group does not coincide with the old $SU(3)_f$ of the *eightfold way*, since at least six flavours exist (cf. table I); then, one has to consider, in general, the group $SU(6)_f$; (ii) $SU(3)_c$ is assumed to be an exact internal symmetry at the quark level, while $SU(6)_f$ is only a broken symmetry (a phenomenological evidence for the breaking of $SU(6)_f$ is provided by the different masses of the various kinds of quarks: see the next section).

2.3 – QCD Lagrangian.

Throughout this paper, we shall make no distinction in notation between quark wave functions and operators: so, ψ^f (ψ^f_c) will denote the (fermionic) field operator for flavour f (and colour c) (see eq. (2.1)). The free lagrangian for massless quarks can then be written as †

$$\mathcal{L}_q(\underset{\sim}{x}) = \overline{\psi}^f(\underset{\sim}{x}) i\gamma^\mu \partial_\mu \psi^f(\underset{\sim}{x}) = \overline{\psi}^f_c(\underset{\sim}{x}) i\gamma^\mu \partial_\mu \psi^f_c(\underset{\sim}{x}), \tag{2.15}$$

and is obviously invariant under global colour rotations.

Writing the corresponding lagrangian for interacting quarks requires the tools of *non-abelian gauge theories*([17]). A local gauge transformation (phase rotation) in colour space is given by

$$\begin{aligned} \psi^f(\underset{\sim}{x}) &\to \hat{U}(\underset{\sim}{x})\psi^f(\underset{\sim}{x}) \\ \overline{\psi}^f(\underset{\sim}{x}) &\to \overline{\psi}^f(\underset{\sim}{x})\hat{U}^+(\underset{\sim}{x}) \end{aligned} \tag{2.16}$$

where the operator $\hat{U}(\underset{\sim}{x})$ explicitly reads

$$\hat{U}(\underset{\sim}{x}) = \exp(ig\alpha_l(\underset{\sim}{x})\hat{T}_l). \tag{2.17}$$

Here, the $\hat{T}_l$ are defined in eq. (2.11a), the $\alpha_l(\underset{\sim}{x})$'s are eight arbitrary real functions and g is a suitable constant, characteristic of the theory (*strong coupling constant*). Such a gauge transformation is said *local*, because the parameters α_l depend on the coordinates, and *non-abelian*, due to the noncommutativity of the generators $\hat{T}_i$.

It is easy to see that the free lagrangian (2.15) is not invariant under transformations of the kind (2.16),(2.17). Indeed, the gradient term transforms as

$$\partial_\mu \psi^f(\underset{\sim}{x}) \to \hat{U}(\underset{\sim}{x})[\partial_\mu \psi^f(\underset{\sim}{x}) + ig\hat{T}_l(\partial_\mu \alpha_l(\underset{\sim}{x}))\psi^f(\underset{\sim}{x})], \tag{2.18}$$

† Hencefort, a sum over repeated indices will be always understood. Greek letters are space-time indices ($\mu = 0,1,2,3$), and $\underset{\sim}{x} = (t, \mathbf{x})$ is the position four-vector.

which involves more than a mere phase change. The additional gradient-of-phase term spoils local phase invariance. It is impossible to build up a locally gauge-invariant theory of free coloured quarks. Local gauge invariance may be attained, however, by the introduction of a vector field $\hat{A}_\mu$, defined as

$$\hat{A}_\mu(\underset{\sim}{x}) = A^i_\mu(\underset{\sim}{x})\hat{T}_i \tag{2.19}$$

with eight components $A^i_\mu(\underset{\sim}{x})$ ($i = 1, \ldots, 8$) in colour space. Under a local gauge transformation, $\hat{A}_\mu(\underset{\sim}{x})$ becomes

$$\hat{A}_\mu \to \hat{U}\hat{A}_\mu\hat{U}^{-1} - \frac{i}{g}(\partial_\mu\hat{U})\hat{U}^{-1}. \tag{2.20}$$

The fields $A^i_\mu(\underset{\sim}{x})$ are called *gauge fields*, and are the analogous of the four vector potential A_μ in QED. To every field component in colour space corresponds a vector boson particle (the *gluon*). The eight (coloured) gluons are the quanta of the strong force (like photon in QED). Let us now define the *gauge-covariant derivative*

$$D_\mu = \partial_\mu - ig\hat{A}_\mu(\underset{\sim}{x}). \tag{2.21}$$

It is not difficult to check that $D_\mu\psi$ has the same transformation properties of ψ:

$$D_\mu\psi \to \hat{U}(\underset{\sim}{x})(D_\mu\psi). \tag{2.22}$$

Then, the lagrangian

$$\mathcal{L}_q(\underset{\sim}{x}) = \overline{\psi}^f(\underset{\sim}{x})i\gamma^\mu D_\mu\psi^f(\underset{\sim}{x}) \tag{2.23}$$

is indeed invariant under local gauge transformations. To get the full QCD lagrangian, we have to construct a gauge-invariant kinetic term for the gauge fields $A^i_\mu(\underset{\sim}{x})$. This is called a *Yang-Mills term* and can be put in the form

$$\mathcal{L}_{YM} = -\frac{1}{4}Tr(\hat{F}_{\mu\nu}\hat{F}^{\mu\nu}) = -\frac{1}{4}F^l_{\mu\nu}F^{l\mu\nu} \tag{2.24}$$

where $\hat{F}_{\mu\nu}$ explicity reads

$$\hat{F}_{\mu\nu} = F^l_{\mu\nu}\hat{T}_l = [D_\mu, D_\nu] \tag{2.25}$$

and therefore

$$F^l_{\mu\nu} = \partial_\mu A^l_\nu - \partial_\nu A^l_\mu - gf^{ljk}A^j_\mu A^k_\nu. \tag{2.25a}$$

For a comparison with QED, let us note that, for $l = 0, f^{ljk} = 0$, $\hat{T}_0 = \hat{I}$ ($\hat{I}$ being the identity matrix) (i.e. the invariance group becomes U(1)), $A_\mu(\underset{\sim}{x})$ reduces to the four-vector potential and $F_{\mu\nu}$ to the usual electromagnetic field tensor. Moreover, as it is easy to verify, transformation (2.20) becomes the standard gauge transformation $A_\mu \to A_\mu - \partial_\mu\alpha$.

The QCD lagrangian is the sum of $\mathcal{L}_{YM}$ and $\mathcal{L}_q$:

$$\mathcal{L}_{QCD} = -\frac{1}{4}Tr(\hat{F}_{\mu\nu}\hat{F}^{\mu\nu}) + \overline{\psi}^f(\underset{\sim}{x})i\gamma^\mu D_\mu\psi^f(\underset{\sim}{x}) \tag{2.26}$$

and describes the interaction between the $SU(3)_c$ quark triplet and the $SU(3)_c$ gluon octet. Clearly, by construction, $\mathcal{L}_{QCD}$ is locally colour invariant. Unlike QED, the Yang-Mills term $\mathcal{L}_{YM}$ describes the interaction of massles gluons among themselves (since, carrying colours, they feel the strong force), and, as it is easy to see, contains terms cubic and quartic in the gauge fields A^i_μ. This amounts to say that QCD is a nonlinear field theory (unlike QED, that is linear), that is a peculiar feature of non-abelian gauge theories.

In a pure colour framework, quarks are massless. However, they acquire mass by their electromagnetic and weak interactions. The quark mass m_f is different for different flavours. In other words, the flavour symmetry group $SU(6)_f$ is badly broken (as already mentioned at the end of the previous subsection). Then, the QCD lagrangian must be modified as follows:

$$\mathcal{L}^{QCD}_{m_f \neq 0} = \overline{\psi}^f(\underset{\sim}{x})(i\gamma^\mu D_\mu - m_f g)\psi^f(\underset{\sim}{x}) - \frac{1}{4}Tr(\widehat{F}_{\mu\nu}\widehat{F}^{\mu\nu}) \tag{2.27}$$

The lagrangian (2.27) is the starting point to build Feynman rules. To this aim, a basic feature of QCD is its being a renormalizable quantum field theory ([18]). This implies, among the others, that one may apply the usual perturbative methods (with some caution).

Clearly, things are far from easy. A number of technical points (like the choice of a gauge for A_μ, etc.) are to be solved before actually applying QCD to a given problem. For brevity's sake we shall not enter into details here, and just refer the reader to the excellent literature on the subject (ref. ([4])).

2.4 – Asymptotic freedom.

It is well known that, in interacting field theory, scattering amplitudes are not given by simple Born diagrams, but higher-order corrections (radiative corrections) must be taken into account. Since the modifications to lowest-order contributions are in general dependent upon kinematical variables, they can be represented in a convenient way by the so-called *running coupling constant*, i.e. an effective coupling that depends on the kinematics of the process.

The invariant kinematic variable that is most frequently used is the four-momentum transfer (squared)

$$\underset{\sim}{q}^2 = (|\underset{\sim}{k} - \underset{\sim}{k}'|)^2 \tag{2.28}$$

where $\underset{\sim}{k}$, $\underset{\sim}{k}'$ are the initial and final four-momentum, respectively. Since, in general, it is $\underset{\sim}{q}^2 < 0$ (spacelike intermediate particle), it is more convenient to use the invariant

$$\underset{\sim}{Q}^2 = -\underset{\sim}{q}^2 = -(mass)^2. \tag{2.29}$$

By the uncertainty principle, larger values of momentum transfers correspond to shorter distances.

In QED, vacuum polarization graphs lead from the fine structure constant α to an effective Coulomb interaction $\alpha(\underset{\sim}{Q}^2)$ increasing with $\underset{\sim}{Q}^2$. The QED vacuum behaves as a polarizable medium: a test charge at large distances (with respect to the molecular scale) will be seen endowed with a smaller effective charge, due to the medium polarization (*screening effect*).

Let us now consider what happens in QCD. The interaction between two quarks is given by the infinite series of Feynman diagrams in Fig. 2. The graphs of the first row are the analogous of vacuum polarization graphs in QED. They contain loops of quarks, instead of loops of electrons. The contribution of such diagrams to the *renormalization of the strong coupling constant* is given by

$$\begin{aligned}\alpha_s(\underset{\sim}{Q}^2) &= \alpha_s(\Lambda^2) \times (1 + N_f \frac{\alpha_s(\Lambda^2)}{6\pi} \ln(\frac{\underset{\sim}{Q}^2}{\Lambda^2}) + \cdots) \\ &\approx \frac{\alpha_s(\Lambda^2)}{[1 - N_f \frac{\alpha_s(\Lambda^2)}{6\pi} \ln(\frac{\underset{\sim}{Q}^2}{\Lambda^2})]}\end{aligned} \tag{2.30}$$

for $\underset{\sim}{Q}^2 \gg \Lambda^2$, where N_f is the number of flavours of quarks with effective mass smaller than $\sqrt{\underset{\sim}{Q}^2}$ (for $\underset{\sim}{Q}^2 \to \infty$, N_f is the *total* numbers of existing quark flavours), and the parameter Λ sets the basic mass scale for QCD. The running coupling constant $\alpha_s(\underset{\sim}{Q}^2)$ increases with increasing $\underset{\sim}{Q}^2$, and the interpretation is the same as in QED: at short distances, the strong *colour charge* of a quark is probed without any effect due to the quark-antiquark pairs (the quark loops in the Feynman diagrams of the first row in Fig. 2).

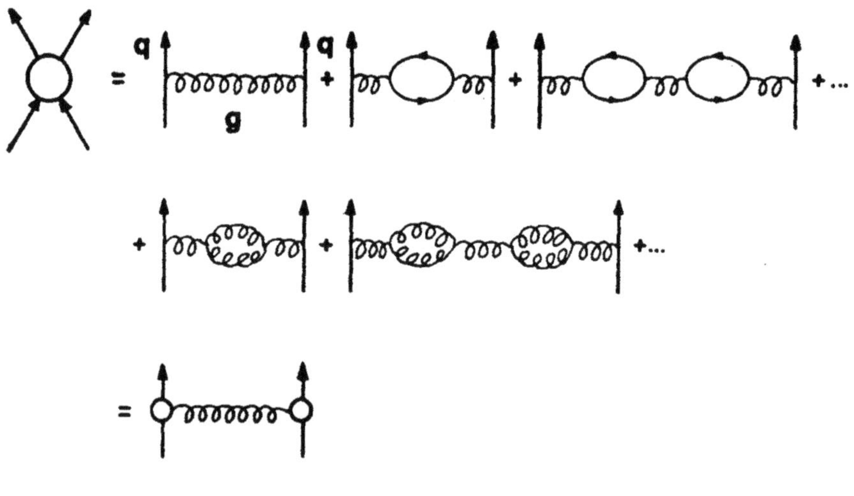

Fig. 2 – Feynman diagrams defining the *renormalized strong coupling constant* (or *effective running constant*) $\alpha_s(\underset{\sim}{Q}^2)$.

The graphs of the second row are peculiar to non-abelian gauge theories, and, therefore, to QCD. Due to the Yang-Mills term $\mathcal{L}_{YM}$ in QCD lagrangian (eq. 2.24), gluons couple among themselves, giving rise to gluon self-interaction diagrams (containing gluon loops). The net contribution of these graphs results in an *anti-screening effect*([19]), i.e. in a *decreasing* of $\alpha_s(\underset{\sim}{Q}^2)$ at large momentum transfers.

It can be shown that, in QCD, the combined effect of all quark and gluon loops is to produce a running coupling constant given by

$$\begin{aligned}\alpha_s(\underset{\sim}{Q}^2) &= \alpha_s(\Lambda^2) \times (1 + N_f \frac{\alpha_s(\Lambda^2)}{6\pi} \ln(\frac{\underset{\sim}{Q}^2}{\Lambda^2}) - \frac{11\alpha_s(\Lambda^2)}{4\pi} \ln(\frac{\underset{\sim}{Q}^2}{\Lambda^2}) + \cdots) \\ &\approx \frac{4\pi}{\beta_0 \ln(\frac{\underset{\sim}{Q}^2}{\Lambda_v^2})}\end{aligned} \tag{2.31}$$

for $\underset{\sim}{Q}^2 \gg \Lambda_v^2$, where $\beta_0 = 11 - 2/3N_f$ and Λ_v is a new parameter defined by

$$\ln \Lambda_v^2 = \ln \Lambda^2 - [4\pi/\beta_0 \alpha_s(\Lambda^2)]. \tag{2.32}$$

It is easily seen, therefore, that (for $N_f \leq 16$) gluon coupling graphs are dominant, and cause the effective strength of the strong coupling constant (colour charge) to decrease logaritmically at large momentum transfers. This implies that QCD is an *asymptotically free* ([20]) gauge theory. Otherwise speaking, at very short distances (i.e. when probed at very high $\underset{\sim}{Q}^2$ values), quarks behave nearly as free particles within hadrons. This *asymptotic freedom* of QCD allows a rigorous perturbative calculation of strong interactions at short distances.

Recently, $\alpha_s(\underset{\sim}{Q}^2)$ has been determined quite unambigously from measurement of the bifurcation ratios for the $\mathcal{Y}$-radiative decay $\mathcal{Y}(b\bar{b}) \rightarrow \gamma X$([21]):

$$\alpha_s(0.157\, M_{\mathcal{Y}}) = \alpha_s(1.5\, GeV) = 0.23 \pm 0.13. \tag{2.33}$$

This implies that — in the non-relativistic limit of infinitely massive quarks — the interaction between the quarks i and j is given by

$$V_{ij}(\underset{\sim}{Q}^2) = [\hat{T}(i) \cdot \hat{T}(j)] \times \frac{4\pi\alpha_s(\underset{\sim}{Q}^2)}{\underset{\sim}{Q}^2} \tag{2.34}$$

where $\alpha_s(\underset{\sim}{Q}^2)$ is given by eq. (2.31) and

$$\Lambda_v = (270 \pm 100) \quad MeV. \tag{2.35}$$

The above estimate of the cutoff parameter Λ_v sets the lenght scale of QCD at distances of the order of a fermi, that is the lenght scale of nuclear interactions. Consequently, not only perturbation theory can be applied to QCD at short distances, but one espects that perturbative QCD calculations become relevant for nuclear physics whenever the nucleon structure is involved. In the next chapter, we shall see this to be indeed the case, by illustrating in detail some applications of perturbative QCD to nuclear physics.

2.5 – Confinement of quarks and gluons.

We have seen in sect. 2.2 that colour $SU(3)_c$ is an exact internal symmetry of hadrons, which do not carry colour quantum numbers. The only coloured entities are quarks and gluons, that, however, have not yet been observed in nature ([22]). Therefore one has to assume that quarks and gluons are permanently *confined* within hadrons.

This is apparently in contrast with the QCD lagrangian (2.26), that — at least in perturbation theory — seems to describe a world of interacting quarks and gluons where nothing prevents them from being asymptotically free. But we have seen that the QCD effective coupling $\alpha_s(\underset{\sim}{Q}^2)$ *decreases* at large momentum transfers and *increases* at small $\underset{\sim}{Q}^2$. The growth of the coupling at large distances is called *infrared slavery*([23]), since it indicates the existence of a domain in which the strong interactions become formidable. Unfortunately, in such an infrared region standard perturbative calculations do no more apply, and therefore strict confinement in QCD remains an open question. A number of different mechanisms have been proposed in order to account for absolute colour confinement ([24]). One of the most popular ones is the bag model (in its various versions) ([25,8]), in which quarks and gluons are confined to a sphere.

But it must be stressed that all proposed confinement models are essentially phenomenological (and rather *ad hoc*, since they do not originate from QCD, but are added to it from the outside). A more promising approach is provided by gauge theories on a lattice ([26]). Indeed, lattice QCD allows to get results in the strong coupling limit, where perturbative techniques fail. However, till now, even lattice formulations have been unable to provide a definite, rigorous proof of absolute colour confinement in QCD, that is yet an outstanding and unsolved problem.

3 – Applications of QCD to nuclear physics.

In this chapter, we will consider some applications of QCD to nuclear physics. First of all, we shall discuss nuclear form factors, and show how deep inelastic scattering provides information on the existence of quarks in nucleons. The so-called EMC effect, recently found at CERN ([5]), is briefly considered. Then we give a short report of nuclear applications of perturbative QCD in the light-cone approach([6]).

3.1 – Electromagnetic structure functions of the nucleon.

A substantial amount of information about nuclear structure is obtained by studying the scattering of charged leptons (electrons and muons) by nuclei. In such processes, both charge and magnetization densities of the nucleus can be directly measured. The scattering process of a lepton by a nucleus is schematically shown in Fig.3. Let us consider the nucleus rest frame (i.e. the laboratory frame, $\underline{p} = 0$), and let $\underset{\sim}{k} = (E, \underline{k})$, $\underset{\sim}{k}' = (E', \underline{k}')$, $\underset{\sim}{p}$, represent the 4-momenta of the initial lepton, the final lepton and the initial nucleus respectively. The difference $\underset{\sim}{q} = \underset{\sim}{k}' - \underset{\sim}{k}$ is the 4-momentum transfer from the lepton to the nucleus. The most frequently used kinematic invariants are $\underset{\sim}{q}^2$ (see eq. (2.28)), its negative $\underset{\sim}{Q}^2$ (eq.(2.29)) and the energy transfer

$$\nu = \frac{p^\mu q_\mu}{M_t} = E - E' \tag{3.1}$$

where M_t is the target mass. When the nucleus recoils as a whole (coherent scattering), one has $\nu \approx \underset{\sim}{q}^2/2M$ (M being the nucleus mass and $q^2 = |\underline{q}|^2$). If the scattering involves a single nucleon inside the nucleus (quasi-elastic scattering), it is $\nu \approx q^2/2M_n$, with M_n the nucleon mass. The following relation is also useful:

$$\underset{\sim}{Q}^2 = 4EE' \sin^2(\theta/2). \tag{3.2}$$

The simplest case is the elastic scattering of charged leptons on a spin 0 nucleus, in which a mere charge scattering occurs. In the non-relativistic limit, one has $\nu = q^2/2M \approx 0$ and the elastic differential cross section reads

$$\left(\frac{d\sigma}{d\Omega}\right)_{el} = \left(\frac{d\sigma}{d\Omega}\right)_{Mott} \cdot |F_1(q^2)|^2. \tag{3.3}$$

Here, $d\Omega$ is, as usual, the element of solid angle, and $(\frac{d\sigma}{d\Omega})_{Mott}$ is the differential Mott cross section (i.e. the quantum cross section for Coulomb scattering by a point charge).

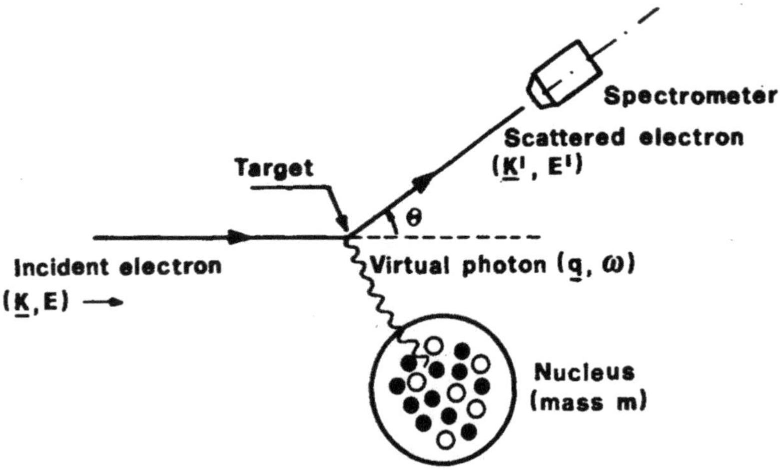

Fig. 3 – Schematic picture of the electron-nucleus scattering.

The function $F_1(q^2)$ is called the *charge form factor* (or *Dirac form factor*) of the nucleus and is given by the Fourier transform of the nuclear charge density $\rho(\underline{r})$:

$$F_1(q^2) = \frac{1}{Ze} \int d^3\underline{r}\rho(\underline{r})e^{-i(\underline{q}\cdot\underline{r})}. \tag{3.4}$$

Clearly, $\rho(\underline{r})$ can be obtained from $F_1(q^2)$ by means of an *inverse Fourier transform.* By the mathematical properties of Fourier transforms, it follows that, if we want to probe the charge structure of the nucleus inside a lenght Δr, we need the form factor $F_1(q^2)$ at momentum transfers of the order

$$q \approx \frac{\hbar}{\Delta r} \approx \frac{200\, MeV \cdot fm}{\Delta r} \tag{3.5}$$

At very high momentum transfers, the form factor is able to provide information not only on the nucleus, but also on the nucleon structure †.

In the more physical case of a target with spin (for instance, in lepton-nucleon scattering), one has to take into account the magnetic interaction between the electron spin and the target spin. The non-relativistic cross section for elastic scattering is now given by the Rosenbluth formula:

$$\left(\frac{d\sigma}{d\Omega}\right)_{el} = \left(\frac{d\sigma}{d\Omega}\right)_{Mott} \times \\ \{F_1(q^2) + \frac{q^2}{4M^2}[2F_1(q^2) + KF_2(q^2)]^2 \tan^2(\theta/2) + K^2F_2(q^2)\}, \tag{3.6}$$

† In the non-relativistic case, analogous informations are provided by inelastic scattering processes when $\underset{\sim}{Q}^2, \nu \to \infty$.

where K is the anomalous part of the magnetic moment, $F_1(q^2)$ is still given by eq.(3.4), and the *magnetic form factor* (or *Pauli form factor*) $F_2(q^2)$ writes

$$F_2(q^2) = \frac{1}{K}\int d^3\underline{r}\,\mu(\underline{r})e^{-i(\underline{q}\cdot\underline{r})}. \tag{3.7}$$

with $\mu(\underline{r})$ being the density of anomalous magnetic moment.

It can be shown that the most general cross section for elastic scattering can be written in the form

$$\left(\frac{d\sigma}{d\Omega}\right)_{el} = \left(\frac{d\sigma}{d\Omega}\right)_{Mott} \cdot [A(\underset{\sim}{Q}^2) + B(\underset{\sim}{Q}^2)\tan^2(\theta/2)] \tag{3.8}$$

where the relativistic invariant functions $A(\underset{\sim}{Q}^2)$ and $B(\underset{\sim}{Q}^2)$ are the (elastic) *structure functions* of the target. For spin 1/2- particles and in the non-relativistic limit they are linear combinations (Sachs combinations) of the functions $F_1(q^2)$ and $F_2(q^2)$ given by eq.(3.6).

Let us now consider the inelastic case. The (relativistic) differential cross section for inelastic scattering can be put in the form (α = fine structure constant)

$$\frac{d\sigma}{d\Omega d\nu} = \frac{4\alpha^2}{\underset{\sim}{Q}^2}E'^2\left[\frac{F_2^e(\underset{\sim}{Q}^2,\nu)}{\nu}\cos^2(\theta/2) + 2\frac{F_1^e(\underset{\sim}{Q}^2,\nu)}{M}\sin^2(\theta/2)\right] \tag{3.9}$$

where the new invariants $F_i^e(\underset{\sim}{Q}^2,\nu)$ $(i = 1,2)$ are called *inelastic structure functions* and are generalizations of the form factors that appear in elastic scattering. That there are only two structure functions is due to the fact that the exchanged virtual photon can be absorbed by the target in two independent (longitudinal and transverse) polarization states.

Let us notice that there will be different structure functions depending on whether a proton or a neutron is the target nucleon. They will be denoted by F_i^{ep}, F_i^{en}, respectively. The averages of the structure functions for p and n are called the *nucleon electromagnetic structure functions*

$$F_i^{eN} = (1/2)(F_i^{ep} + F_i^{en})\ . \tag{3.10}$$

For isoscalar targets (i.e. targets with an equal number of protons and neutrons: $N = (1/2)(N_p + N_n)$), the averages (3.10) are the relevant structure functions. Since real targets are generally isoscalar in good approximation, it is usually assumed that F_i^{eN} are the structure functions actually measured experimentally.

It is customary to say that F_1^e and F_2^e determine the *response function* of the target to the excitation connected to the photon absorption. Let us note that, in the absorption of real photons, one has obviously $\underset{\sim}{Q}^2 = 0$. Since lepton-nucleus scattering involves the absorption of virtual photons, $\underset{\sim}{Q}^2$ and ν are independent variables and, therefore, it is possible to measure F_1 and F_2 on the whole plane $(\underset{\sim}{Q}^2,\nu)$. Moreover, real photons are always *transverse*, while virtual photons may be either *transverse* or *longitudinal*. By varying the kinematic conditions, it is possible to change the polarization of the exchanged photons and measure separately the structure functions F_1 and F_2. Typical response functions of a nucleus and a single nucleon to the absorption of real photons and to the scattering of leptons are shown in Fig.4 as a function of ν at constant $\underset{\sim}{Q}^2$. The absorption of real photons occurs at $\underset{\sim}{Q}^2 = 0$ and is dominated by the giant dipole resonance (GDR) at low energies and by the Δ resonance (excited state of nucleons with $T = J = 3/2$) above the treshold for pion creation. For lepton scattering, the relative importance of the different processes changes with $\underset{\sim}{Q}^2$[27]. A special role is played, to our aims, by the region $\underset{\sim}{Q}^2, \nu \to \infty$ (deep inelastic scattering)[28]. Indeed, as we shall see in the next section, deep inelastic processes directly probe the nucleon structure and thereby provide evidence for the existence of quarks inside nucleons.

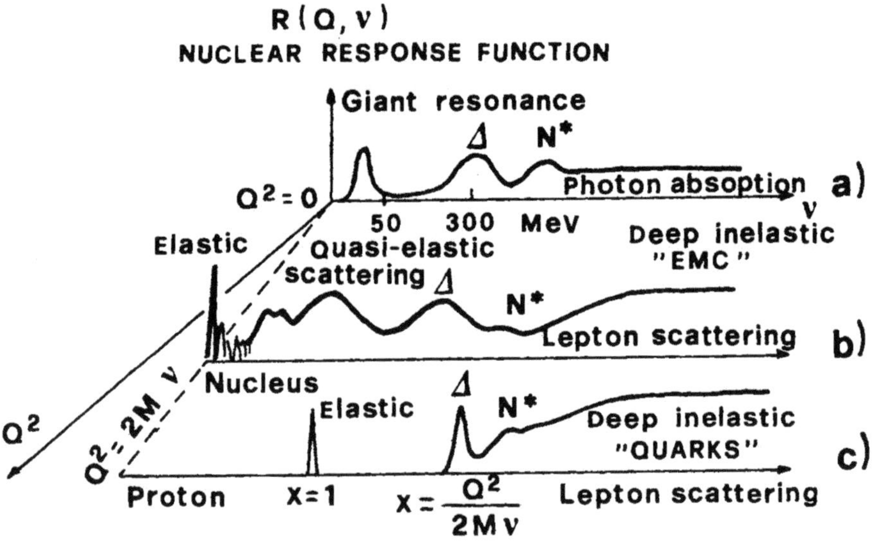

Fig. 4 – Response function of a nucleus to the excitation due to electromagnetic probes: a) photons ($\underset{\sim}{Q}^2 = 0, \underset{\sim}{q}^2 = \omega^2$) b),c) leptons ($\underset{\sim}{Q}^2 = \underset{\sim}{q}^2 - \omega^2 > 0$). The comparison between the lepton-proton and lepton-nucleus scattering processes clearly shows the effect of the Fermi motion of the nucleons in the nucleus (Fermi smearing).

3.2 – Scale invariance and the parton model.

Let us consider eq. (3.4) for the nuclear form factor $F_1(q^2)$. It is easy to see that, for increasing q^2, $F_1(q^2)$ decreases and is vanishing for $q^2 \to \infty$ for a nuclear charge distribution extended in space. The only way to get a nonvanishing constant value for $F_1(q^2)$ as $q^2 \to \infty$ is to have a pointlike structure. Analogous results hold true for the structure functions $F_i^e(\underset{\sim}{Q}^2,\nu)$. One expects a decrease of $F_i^e(\underset{\sim}{Q}^2,\nu)$ as $\underset{\sim}{Q}^2$ increases, and $\lim_{Q^2\to\infty} F_i^e(\underset{\sim}{Q}^2,\nu) = 0$ for an extended target charge. Since, in the deep inelastic region, lepton scattering directly probes the structure of nucleons inside the nucleus, such a behaviour of the structure functions corresponds to nucleon constituents with finite size. On the contrary, a non zero limiting value of $F_i^e(\underset{\sim}{Q}^2,\nu)$ for $\underset{\sim}{Q}^2 \to \infty$ is a clear signal for pointlike constituents inside nucleons. In this case, there is no distance scale (like the size of nucleon constituents) that can enter the dependence of the structure functions on ν and $\underset{\sim}{Q}^2$. Then, one expects the F_i^e's to depend asymptotically only on the dimensionless ratio (*Bjorken variable*):

$$x = \frac{Q^2}{2M\nu} \, . \tag{3.11}$$

This property of the structure functions, known as *Bjorken scaling* ([29,30]), was indeed experimentally observed at SLAC ([31]), and it supported the picture that deep inelastic electron scattering is characterized by scattering from pointlike constituents (called *partons* by Feynman([30])). Moreover, one assumes that partons are quasi-free, i.e. their mutual interaction can be neglected during and after the scattering process. The parton model provides also the meaning of the Bjorken variable x. Let us introduce the definition $\underset{\sim}{p}_{part} = \xi \, \underset{\sim}{p}$, where ξ is the momentum fraction carried by a parton before being struck by a photon whit momentum $\underset{\sim}{q}$. If m_p is the parton mass, conservation of energy and momentum at parton vertex gives

$$m_p^2 = (\xi \, \underset{\sim}{p} + \underset{\sim}{q})^2 \tag{3.12}$$

from which it is possible to calculate ξ in terms of $\underset{\sim}{Q}^2$ and ν:

$$\xi = \frac{\underset{\sim}{Q}^2 + m_p^2}{M(\nu + \sqrt{\nu^2 + \underset{\sim}{Q}^2 + m_p^2})} \approx \frac{2x}{1 + \sqrt{1 + \frac{4M^2x^2}{\underset{\sim}{Q}^2}}} \quad (\underset{\sim}{Q}^2 \gg m_p^2) \; ; \tag{3.13}$$

ξ is called the Nachtmann variable. It is easily seen that, for small m_p, in the Bjorken limit $\underset{\sim}{Q}^2, \nu \to \infty$, $\xi \to x$. In other words, for large values of $\underset{\sim}{Q}^2$ and ν the Bjorken variable represents the momentum fraction carried by the struck parton. Let us introduce the new kinematic variables

$$y = \frac{\nu}{E} = \frac{p^\mu q_\mu}{p^\mu k_\mu} \tag{3.14}$$

$$s = (\underset{\sim}{k} + \underset{\sim}{p})^2 = 2ME + M^2 \tag{3.15}$$

that represent, respectively, the fractional energy transfer to the target and the energy squared in the CM system. In terms of x, y and s the doubly-differential inelastic cross section (3.9) reads

$$\frac{d^2\sigma}{dxdy} = \frac{2\pi\alpha^2 s}{\underset{\sim}{Q}^4} F_2^e(x, \underset{\sim}{Q}^2)[1 + (1-y)^2 - \frac{y^2(F_2^e - 2xF_1^e)}{F_2^e}] \tag{3.16}$$

3.3 – Weak structure functions of the nucleon.

Results analogous to those derived in sect. 3.1 for the inelastic scattering of charged leptons and nucleons can be obtained for the scattering of neutrinos and antineutrinos. In this case, scattering takes place through the exchange of the charged weak bosons $W^{\pm}$.

The inelastic differential cross section for such processes can be written as (by neglecting the term $Mxy/2E$)

$$\begin{aligned}\frac{d^2\sigma^{\nu,\bar{\nu}}}{dxdy} = \frac{G^2 ME}{\pi} \times \{&(2xF_1^{\nu,\bar{\nu}} \pm xF_3^{\nu,\bar{\nu}}) \\ &+ (1-y)^2(2xF_1^{\nu,\bar{\nu}} \mp xF_3^{\nu,\bar{\nu}}) + 2(1-y)(F_2^{\nu,\bar{\nu}} - 2xF_1^{\nu,\bar{\nu}})\}\end{aligned} \tag{3.17}$$

where the upper (lower) sign refers to neutrino (antineutrino) scattering. In this case there are three *weak structure functions* $F_i^{\nu,\bar{\nu}}(\underset{\sim}{Q}^2, \nu)$ since there are three possible helicity states for the intermediate vector bosons $W^{\pm}$ and, unlike the e.m. case, weak interactions do not conserve parity (otherwise speaking, the third structure function is connected to parity violation).

As in the electromagnetic case, there will be different structure functions for proton and neutron targets, and we shall denote them by $F_i^{\nu p}, F_i^{\nu n}$ and $F_i^{\bar{\nu}p}, F_i^{\bar{\nu}n}$ $(i = 1, 2, 3)$. For isoscalar targets, one has the nucleon structure functions for neutrino and antineutrino scattering

$$\begin{aligned}F_i^{\nu N} &= 1/2(F_i^{\nu p} + F_i^{\nu n}) \\ F_i^{\bar{\nu} N} &= 1/2(F_i^{\bar{\nu} p} + F_i^{\bar{\nu} n}).\end{aligned} \tag{3.18}$$

As for the average electromagnetic structure functions, it is often assumed that are the $F_i^{\nu,\bar{\nu}N}$ which are measured in experiments.

3.4 – Quarks inside nucleons: the quark-parton model.

We have seen in sect.3.2 that Bjorken scaling indicates that nucleon constituents are pointlike (partons). However, it remains to prove that partons do indeed coincide with quarks,i.e. they are particles with spin 1/2 and fractional charge −1/3,+2/3. This can be experimentally checked by noting that the electromagnetic structure functions $F_i^e(\underset{\sim}{Q}^2, \nu)$ are deeply influenced by the spin of partons. In particular, one has

$$\frac{2xF_1^e(x, \underset{\sim}{Q}^2)}{F_2^e(x, \underset{\sim}{Q}^2)} = \begin{cases} 1 & \text{for spin } 1/2; \\ 0 & \text{for spin } 0. \end{cases} \tag{3.19}$$

The first relation is called the *Callan-Gross relation* [32]. Moreover, it can be shown that the ratio of the two photon absorption cross sections for the two helicity states of the absorbed photon can be expressed in terms of F_1^e, F_2^e as follows:

$$R = \frac{\sigma_L}{\sigma_T} = \frac{F_2^e(x, \underset{\sim}{Q}^2) - 2xF_1^e(x, \underset{\sim}{Q}^2)}{2xF_1^e(x, \underset{\sim}{Q}^2)} \tag{3.20}$$

Then

$$R = \begin{cases} 0 & \text{for spin } 1/2; \\ 1 & \text{for spin } 0. \end{cases} \tag{3.21}$$

It turns out experimentally that

$$R = \begin{cases} 0.10 \pm 0.07 : & \nu \text{ scattering at CERN}(^{33}), \\ 0.21 \pm 0.10 : & e \text{ scattering at SLAC}(^{34}). \end{cases} \tag{3.22}$$

Therefore, spin 0 for partons is clearly ruled out.

Further information on partons cannot be obtained by electromagnetic probing, and neutrino and antineutrino scattering must be used. In particular, weak scattering allows to test the parton distribution inside nucleons, as we shall see. Indeed, one expects that, besides the constituent partons (*valence partons*), there is in addition some background *cloud* of parton antiparton pairs in nucleons, which are excited from the Dirac sea by the interaction of valence partons (from the standpoint of perturbative QCD, this corresponds to the vacuum polarization diagrams of Fig.2). Since neutrino and antineutrino scatter off particles and antiparticles in a different manner, neutrino and antineutrino scattering can reveal the existence of antipartons inside the nucleons. Let us denote by $q(x)$, $\overline{q}(x)$ the parton and antiparton x-distribution functions, respectively, (i.e. the number of partons and antipartons with momentum fraction x in a unit interval of x), and by the suffixes v, s the valence and sea partons, respectively. We have the obvious relations:

$$\begin{aligned} q(x) &= q_v(x) + q_s(x) \\ \overline{q}(x) &= \overline{q}_s(x) = q_s(x) \end{aligned} \tag{3.23}$$

Then, it is possible to relate the structure functions to the parton distributions as follows:

$$F_2^\nu(x) = x[q(x) + \overline{q}(x)] = x[q_v(x) + 2q_s(x)]; \tag{3.24}$$

$$\frac{1}{2}[xF_3^\nu(x) + xF_3^{\overline{\nu}}(x)] = \overline{xF_3^\nu(x)} = x[q(x) - \overline{q}(x)] = xq_v(x); \tag{3.25}$$

$$F_2^\nu(x) - \overline{xF_3^\nu(x)} = 2xq_s(x). \tag{3.26}$$

In other words, $F_2^\nu(x)$ measures the sum of all parton distributions; $F_3^{\nu,\overline{\nu}}$ are related to the difference of parton and antiparton distributions (i.e. to valence parton distribution); finally, the difference between F_2^ν and $\overline{xF_3^\nu}$ is twice the momentum distribution of the sea partons. Therefore, by experimental measurements of the $\nu, \overline{\nu}$ scattering cross sections, it is possible to deduce the structure functions $F_2^\nu, F_3^{\nu,\overline{\nu}}$, and then the parton and antiparton distributions inside nucleons. The results so obtained from various experimental groups are shown in Fig.5.

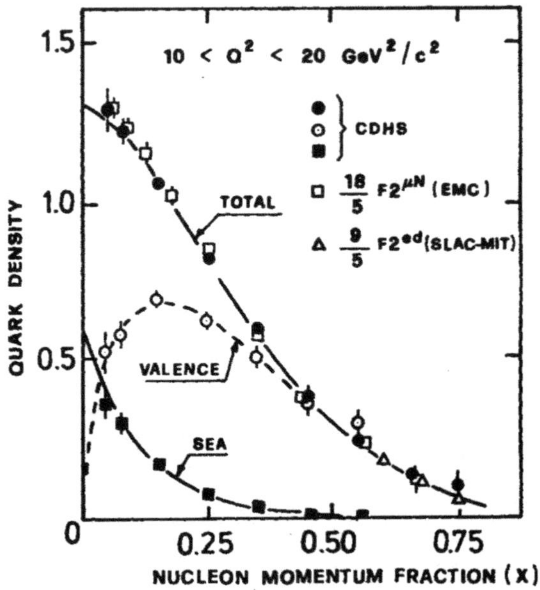

Fig. 5 – Structure functions for partons and antipartons. The function F_2^ν measures the sum of all parton distributions and is given by $F_2^\nu = x[q(x)+\bar{q}(x)] = x[q_v(x)+2q_s(x)]$. Further, one has $F_2^{\nu N}(x) = \frac{18}{5}F_2^{eN} = \frac{18}{5}F_2^{\mu N} = \frac{9}{5}F_2^{ed}$.

From the structure functions F_2^ν and F_3^ν, one can obtain information on the number of valence partons, and on the fraction of momentum carried by valence and sea partons. The number of valence partons is obviously given by

$$N_v = \int_0^1 (q - \bar{q})dx = \int_0^1 \frac{xF_3^\nu}{x}dx. \tag{3.27}$$

The experimental results obtained at CERN([35]) give the estimate $N_v = 3.2 \pm 0.5$, in astonishing agreement with the quark model. The fraction of momentum carried by all partons is related to F_2^ν. One has, from the experimental data([35])

$$\int_0^1 x(q + \bar{q})dx = \int_0^1 F_2^\nu(x)dx = 0.45 \pm 0.01. \tag{3.28}$$

This is a small fraction of the total momentum. It follows that the missing momentum is carried by additional particles that do not feel weak or electromagnetic interactions. Such a result is a clear indication for the existence of gluons inside nucleons.

The last property to be checked in order to fully identify partons with quarks is the fractional charge. This last information can be obtained by a comparison of the weak structure function $F_2^{\nu N}$ to the analogous electromagnetic one F_2^{eN} (see (3.10),(3.18)). To this aim, one has to take into account the different nature of electromagnetic and weak probes. In the case of charged lepton scattering, the exchanged photon sees only electrically charged partons. Therefore, parton charges must explicitly enter the generalization to F_2^{ν} of eq. (3.24) for $F_2^e(x)$. Then, if $q_i(x)$ is the distribution of partons of kind i and charge e_i (in units of the electron charge) † one gets

$$F_2^{\nu}(x) = F_2^{\bar{\nu}}(x) = x\sum_i q_i(x); \tag{3.29}$$

$$F_2^e(x) = x\sum_i e_i^2 q_i(x). \tag{3.30}$$

In the framework of the quark model, index i labels the different flavours. Let us assume that only the light quarks u, d play an essential role in the production of quark–antiquark pairs. By taking into account isospin invariance, one gets that the distribution for *up* quarks in the proton is the same as for *down* quarks in the neutron and viceversa. Then, by eqs. (3.29), (3.30) and the definitions of the average structure functions (3.10), (3.18), we find

$$\frac{F_2^{\nu N}}{F_2^{eN}} = \frac{F_2^{\nu p} + F_2^{\nu n}}{F_2^{ep} + F_2^{en}} = \frac{18}{5}\frac{q+\bar{q}}{q+\bar{q}} = \frac{18}{5}, \tag{3.31}$$

in good agreement with experimental observation([35]).

In conclusion, deep inelastic scattering of both charged and neutral leptons is a precious source of information about nucleon structure. It strongly supports the validity of the quark model, by providing definite evidence — within the experimental uncertainties — for a picture of quarks as spin 1/2, fractionally charged particles, that behave as quasi-free particles at high momentum transfers.

3.5 – The EMC effect.

Recent experimental results ([5,36,37]) on deep inelastic scattering of leptons off nuclear targets exhibit a rather unexpected behaviour of the structure functions $F_2^{eN}(x, \underset{\sim}{Q}^2)$ for scattering off heavy nuclei. Precisely, F_2^{eN} shows an enhancement at small values of x and a depletion for larger values as compared to the corresponding structure function for scattering from deuterium (see Fig.6). Such an effect (called EMC effect, from the European Muon collaboration, that was the first to observe this result at CERN ([5])), is contrary to theoretical estimates of the corrections due to Fermi motion of the nucleons inside the nucleus([38]). The EMC effect is a typical nuclear phenomenon not easy to explain from a theoretical viewpoint. The basic point is to understand how and why quarks, though

† Let us note that — as previously mentioned — charged lepton scattering is unable to distinguish between parton and antipartons, due to the square of charge in eq. (3.28). Let us also stress that eq. (3.29) is strictly valid only for vanishing Cabibbo angle (see ref. (28)).

confined inside nucleons, are sensible to the structure of the surrounding medium. Many mechanisms have been suggested to explain this effect, including modifications of the pion cloud $q\bar{q}$ ([39]), formation of Δ's ([40]) or multiquark bags inside the nucleus ([41]). The most likely explanation is that the confinement radius in heavy nuclei is different from what it is in free nucleons or in deuterium ([42]). However, no clear understanding of the EMC effect has yet emerged. The only firm point is that it indicates a modification of nuclear properties due to nuclear medium. Therefore, the EMC effect does support the need of an approach to nuclear physics in terms of QCD (even if it is presently unclear the possible role on standard nuclear physics of an effect observed in the Bjorken limit $Q^2, \nu \to \infty$).

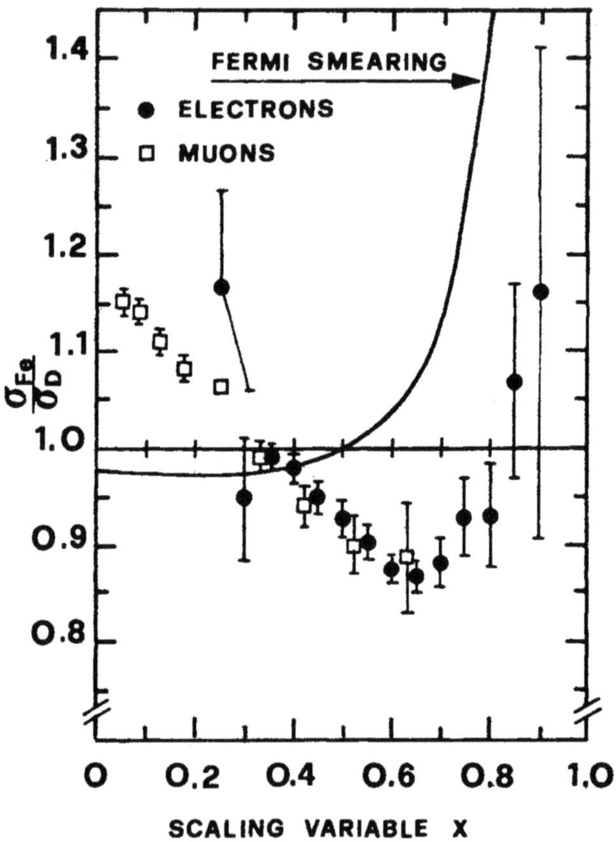

Fig. 6 – The EMC effect. The ratio $\sigma_{Fe}/\sigma_{deut}$ measured for the deep inelastic scattering of electrons (CERN) and muons (SLAC) sensibly deviates from one. This result cannot be explained only in terms of the Fermi nucleon motion.

3.6 – Perturbative QCD on the light-cone: nuclear applications.

We have seen in sect. 2.4 that asymptotic freedom allows to use perturbative techniques for QCD at short distance (lower than the nuclear size), thus permitting the direct application of QCD to a number of nuclear phenomena where the quark and gluon substructure plays an essential role. They include, e.g., the study of amplitudes for meson photo - production ([43]) and for electro - and photo - disintegration of deuterium ([44]), factorization theorems for nucleon ([45]) and deuterium ([46]) form factors, rigorous boundary conditions for nuclear amplitudes ([47]). In this section, we want only to discuss the very basic subject of nucleon and nuclear form factors at large momentum transfer ([46,47]). The basic tool for these asymptotic calculations is the QCD perturbation theory on the light-cone ([6]). Let us briefly review its main features. The idea of specifying dynamics on a light-like plane is an old one (it can be traced back to Dirac) ([48]), but it has acquired new interest in the framework of deep inelastic processes, since — as we shall see very soon — there is a deep connection between light-cone formalism and infinite momentum limit (in other words, light-cone formulations are suitable when energies and longitudinal momenta are large and comparable). Consider a hadron moving in the z-direction, and denote by $\underset{\sim}{p} = (p_0, 0, 0, p)$ its four-momentum. The infinite-momentum limit (characteristic of the deep inelastic region) $p \to \infty$ can be formally attained by a Lorentz boost in the z-direction with speed $v \to 1$. Under such a boost, one gets, e.g. for the time-component x^0:

$$(x')^0 = x^0 \cosh\theta + x^3 \sinh\theta \tag{3.32}$$

where, as usual, $\theta = \tanh^{-1} v$. In the limit $v \to 1$, it is $\cosh\theta \simeq \sinh\theta$, whence

$$(x')^0 \;\propto\; (x^0 + x^3)\,. \tag{3.32a}$$

It can be shown ([49]) that the *infinite-momentum frame* is obtained by choosing new coordinates (*light-cone variables*) related to the normal ones by

$$\tau = \frac{x^0 + z}{\sqrt{2}}; \qquad \varsigma = \frac{x^0 - z}{\sqrt{2}}; \qquad x'_\perp = \underset{\sim}{x}_\perp; \tag{3.33}$$

where $\underset{\sim}{x}_\perp = (x^1, x^2)$. It is easily checked that the new axes lie on the light cone. For a generic four-vector $\underset{\sim}{v}$, the following notation is usually adopted:

$$v^\pm = \frac{v^0 \pm v^3}{\sqrt{2}}\,. \tag{3.34}$$

In this basis, the scalar product of two four-vectors reads

$$\underset{\sim}{a} \cdot \underset{\sim}{b} = a^+ b^- + a^- b^+ - \underset{\sim}{a}_\perp b_\perp; \quad \underset{\sim}{a}^2 = 2a^+a^- - a_\perp^2\,. \tag{3.35}$$

As it is well known, hyperplanes at t = *const* in Minkowski space play a special role in canonical quantization. They are spacelike surfaces with normal four-vectors $\underset{\sim}{n}$ = $(1, 0, 0, 0)$ defined by the equation

$$n^\mu x_\mu = const\,. \tag{3.36}$$

In the infinite momentum frame, the corresponding surfaces are planes $\tau = const$. It is easily seen that they are lightlike hyperplanes; indeed

$$\tau = \frac{x^0 + z}{\sqrt{2}} = \frac{n^\mu x_\mu}{\sqrt{2}} = x^+ = const \tag{3.37}$$

where

$$\underset{\sim}{n} = (1,0,0,-1), \quad n^+ = 0, \quad n^- = \sqrt{2}, \quad \underset{\sim}{n}^2 = 0\,. \tag{3.38}$$

To be concrete, consider $x^+ = 0$. Such an hyperplane is tangent to the light-cone †(see Fig.7) and represents the limit of the surface t=0 for $v \to 1$. Let us now apply the above results to the quark-parton model. The four-momentum of the quark i will be denoted by $\underset{\sim}{k}_i = (k_i^0, \underline{k}_{\perp i}, k_i)$, (with k_i being the z-component, and $\underline{k}_{\perp i}$ the transverse momentum vector). Let x_i be the momentum fraction of the quark i: then, the momentum of the i-th quark in the hadron considered can be written as

$$\underline{k}_i = x_i \underline{p} + \underline{k}_{\perp i} \tag{3.39}$$

where, obviously:

$$\underline{k}_{\perp i} \cdot \underline{p} = 0\,; \sum_i \underline{k}_{\perp i} = 0\,; \sum_i x_i = 1\,. \tag{3.40}$$

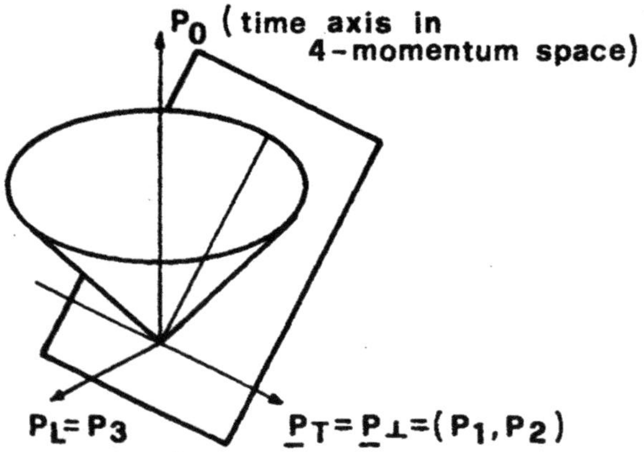

Fig. 7 – Infinite-momentum frame.

In the infinite-momentum frame, we have, explicitly:

$$\begin{aligned} k_i &= k_i^+ = \frac{k_i^0 + k_i^3}{\sqrt{2}}\,; \\ x_i &= \frac{k_i^+}{p^+} = \frac{\underset{\sim}{k} \cdot \underset{\sim}{n}}{\underset{\sim}{p} \cdot \underset{\sim}{n}}\,. \end{aligned} \tag{3.41}$$

† Let us recall that on a lightlike plane all distances are spacelike, except for points on the straight line common with the light-cone.

with $\underset{\sim}{n}$ given by eq. (3.38). The technical usefulness of the infinite-momentum (or light-cone) approach is that it is very similar to the standard non-relativistic many-particle theory. Essentially, all theorems proved in non-relativistic quantum theory hold for the relativistic theory at equal time on the light-cone. In particular, the Feynman relativistic perturbation theory reduces to the old fashioned, non relativistic perturbation theory (as firstly proved by Weinberg ([50])). Then, by following ([6]), the momentum-space n-particle wavefunction of a composite system h (hadron or nucleus) with total momentum $\underset{\sim}{p}$ will be labelled as follows:

$$\psi_n^h = \psi_n^h(x_i, \underline{k}_{\perp i}, \lambda_i) \tag{3.42}$$

where λ_i is the helicity of the quark i and the states are specified on the light-cone at equal time $\tau = t + z$. Any wavefunction (3.42) is the n-particle component of the expansion of the hadron state vector in the Fock state basis. For instance, one has

$$\begin{aligned} |\pi^+\rangle &= \psi_2^\pi |u\bar{d}\rangle + \psi_3^\pi |u\bar{d}g\rangle + \psi_6^\pi |u\bar{d}[q\bar{q}]\rangle + \cdots \\ |p\rangle &= \psi_3^p |uud\rangle + \psi_4^p |uudg\rangle + \cdots \\ |D\rangle &= \psi_6^D |uud\,ddu\rangle + \cdots \end{aligned} \tag{3.43}$$

This is in exact analogy to the Fock state expansion for positronium in QED:

$$|positronium\rangle = \psi_2 |e^+e^-\rangle + \psi_3 |e^+e^-\gamma\rangle + \cdots$$

The functions ψ must satisfy the following *Schrödinger-like light-cone equation*:

$$\left[M^2 - \sum_{i=1}^{n} \left(\frac{\underline{k}_{\perp i}^2 + m_i^2}{x_i} \right) \right] \psi - V_{QCD}\psi \tag{3.44}$$

(see refs. ([6,43]) for the definition of the *QCD potential* V_{QCD}). For the electromagnetic field, it is convenient to choose the light-cone gauge $A^+ = A^0 + A^3 = 0$ (that allows to eliminate unphysical degrees of freedom). Straightforward application of perturbative QCD on the light-cone allows to find general expressions for hadronic amplitudes ([43,44,45]). For instance, the form factor of a composite system of quarks (hadron, nucleon or nucleus) can be written in the following factorized form ([44,45]):

$$F_N(Q^2) = \int_0^1 [dy] \int_0^1 [dx] \Phi_N^+(y_i, Q) T(x, y, Q) \Phi_N(x_i, Q) \tag{3.45}$$

where $[dx] = \delta(1 - \sum_i x_i) \prod_i dx_i$, $\Phi_N(x_i, Q)$ is the probability amplitude for finding inside the system in the initial state a distribution $\{x_i\}$ of light-cone momentum fractions; $\Phi_N(y_i, Q)$ is the analogous distribution for the final state, and $T(x, y, Q)$ is the probability amplitude for scattering quarks collinear with $\underset{\sim}{p}$ (with momentum fractions $\{x_i\}$) to the final direction $\underset{\sim}{p} + \underset{\sim}{q}$ (with momentum fractions $\{y_i\}$), after absorbing a photon with momentum $\underset{\sim}{q}$ (see Fig.8).

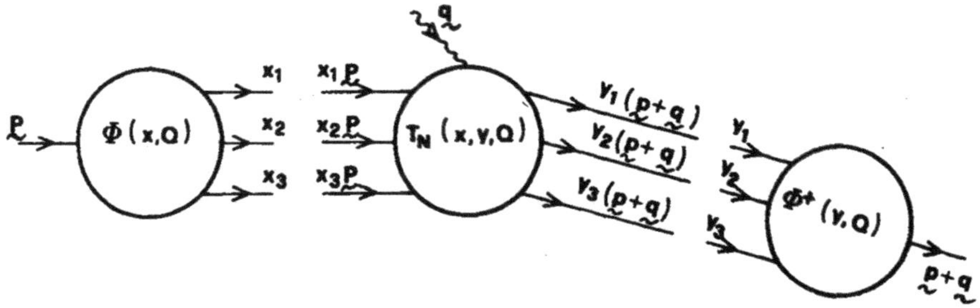

Fig. 8 – Factorization of the nuclear form factors at large Q^2 in QCD.

A few remarks are in order. First, we have put exactly zero the contributions of the quark transverse momenta $\underline{k}_{\perp i}$. Indeed, they are not zero, but, being associated to the Fermi motion of quarks inside the nucleons, their contributions can be neglected in the asymptotic region (at large Q^2 values). Moreover, in the same limit, it is possible to estimate Φ_N by assuming only the valence quarks to contribute, i.e. from the knowledge of the wavefunction ψ_V (e.g., for the nucleon, $\psi_V = \psi_{qqq}$). The other contributions from sea quarks or gluons are suppressed by powers of $(Q^2)^{-1}$ (to leading order in $1/Q$ and in the light-cone gauge $A^+ = 0$). Then, one has [44,45]

$$\Phi_N(x_i, Q) = \int^{k_{\perp i}<Q} [d^2k_\perp] \psi_V(x_i, \underline{k}_{\perp i}). \tag{3.46}$$

The graphs giving the relevant contributions to the scattering amplitude $T(x, y, Q)$ are shown in Fig. 9. To leading order in the coupling constant $\alpha_s(Q^2)$, one gets [44,45]

$$T(x, y, Q) = \left[\frac{\alpha_s(Q^2)}{Q^2}\right]^2 t(x, y) \tag{3.47}$$

where $t(x, y)$ is a rational function of the light-cone momentum fractions $\{x_i\}$ and $\{y_i\}$

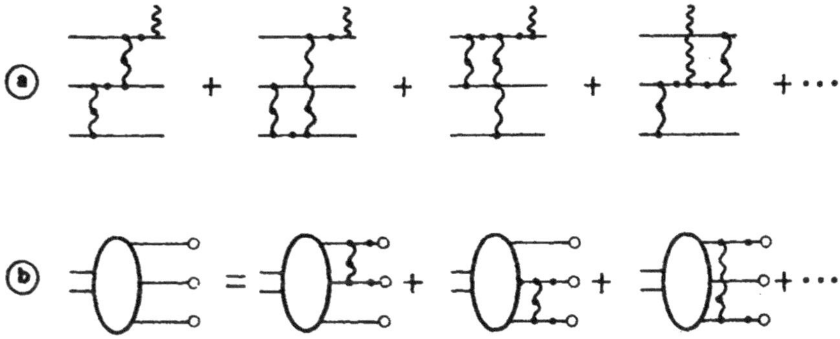

Fig. 9 – a) The leading order diagrams for the hard scattering amplitude T. The dots indicate insertions which enter the renormalization of the coupling constant; b) The leading order diagrams which determine the $\underset{\sim}{Q}^2$ dependence of $\phi_N(x, Q)$.

The explicit calculation of $F_N(Q^2)$ for a bound system of n valence quarks gives, to the leading order in $\alpha_s(Q^2)$, an expression of the type

$$F_N(Q^2) \approx (\frac{1}{Q^2})^{n-1} \times \{logarithmic\ function\ of\ Q^2\} \tag{3.48}$$

namely, the asymptotic behaviour of the form factor is determined by a simple quark counting rule ([51]).

Since the quark helicities are conserved by $T(x, y, Q)$ (again to leading order in $1/Q$) and $\Phi_N(x_i, Q)$ is the $L_z = 0$ angular momentum projection of the wavefunction, overall hadronic helicity is conserved at large momentum transfer for any exclusive reaction described by QCD. This is expressed by the *QCD helicity rule*

$$\sum_{initial} \lambda_N = \sum_{final} \lambda_N \tag{3.49}$$

valid to all orders in perturbation theory †. Then, the leading form factor is actually given by eq. (3.45); the Pauli form factor is suppressed by powers of Q^2. In the deuteron case, the leading form factor corresponds to zero helicity ($\lambda = \lambda' = 0$), and is given by

$$F_D(Q^2) = \sqrt{A_D(Q^2)} \tag{3.50}$$

where $A_D(Q^2)$ is the helicity zero - helicity zero electron - deuteron scattering amplitude. The logarithmic function of (Q^2) that appears in eq. (3.47) depends essentially on the mixing of Fock states entering the wave function Φ_N, and, therefore, on the interaction

† It can be shown ([45]) that the helicity rule (3.49) is a direct consequence of the gluon's spin, and is one of the most characteristic features of QCD.

properties of quarks at large distances and small momentum transfers. Other corrections to the form factors are due to multiple scattering and other higher-order effects. Unfortunately, an exact derivation of Φ_N would require non-perturbative techniques (like lattice QCD or, more directly, the solution of the light-cone evolution equation (3.44)).

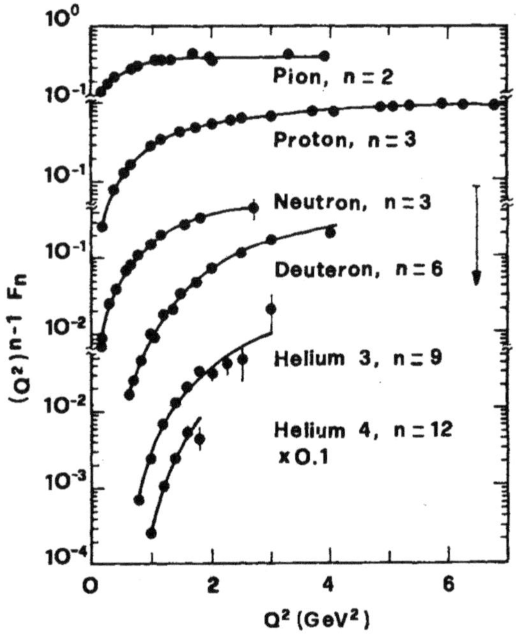

Fig. 10 – Comparison of the experiment with the QCD previsions $(Q^2)^{n-1}F_n(Q^2) \simeq$ *const* for form factors. The proton data extend beyond 30 GeV^2.

As it is shown in Fig. 10, the power-law behaviour of the theoretical QCD predictions for the asymptotic form factors is consistent with the experimental data. For instance, one has indeed for the nucleon $\lim_{Q^2\to\infty} Q^4 F_N(Q) \approx$ *const*. This constitutes a further evidence for the presence of three valence quarks inside the nucleon, and reflects the basic scale invariance of quark and gluon interactions. Such an agreement between theory and experiment for form factors of exclusive processes at large Q^2 clearly shows the effectiveness of ligth-cone perturbative QCD in describing nuclear phenomena at short distances. Other positive tests are provided by the asymptotic behaviour of scattering amplitudes for various hadron processes involving the deuterium or other light nuclei. On the contrary, the standard methods, based on a field theory with an effective nucleon-meson lagrangian, are not satisfactory, since they give wrong asymptotic behaviours for any dynamical variable. Although the situation can be remedied by introducing suitable "ad hoc" form factors at vertices,such an "old-fashioned" approach is clearly unsatisfactory, due to its poor predictive power. We are already familiar with the fact that perturbative QCD is no more applicable at large distances (small Q^2). This can be easily seen by the failure of the validity assumption leading to the equations (3.45), (3.48). Indeed, at small Q^2 the calculation of the form factors would involve a large number of graphs, in order to account for the contributions of the Fock states with sea quarks, radiative corrections and quark and gluon loops.

In the next chapter, we shall discuss approximate *models* of QCD, able to give predictions for nuclear physics also at large distances.

4 – Phenomenological QCD-like models for hadrons and nuclei.

As already noted at the end of the previous chapter, it is a very difficult task to do QCD calculations at intermediate and large distances, even for few-quark systems (baryons and mesons). In order to obviate such a difficulty, and to permit comparing theoretical predictions to the experimental data in many sectors of hadronic and nuclear physics, some phenomenological, non-relativistic quarks models have been constructed, that include some of the peculiar features of QCD. The main aim is to get phenomenological or semi-phenomenological theories able to give predictions similar to those of QCD, at least for the study of few-quark systems at intermediate and large distances ($r \approx 1 - 2$ fm).

These QCD-like teories are aimed, among the others, to reproducing the nuclear and hadronic spectra and to calculating the form factors and the electromagnetic moments of hadrons and nuclei. But the most important goal is to describe the nucleon-nucleon interaction, whose derivation from first principles in QCD is still an open and outstanding problem.

Although — as we shall see — most of the proposed models are partially successful in describing hadron and nuclear properties within restricted frameworks, it is still lacking an exaustive model able to fully account for quark correlations at distances of the order of 1 fermi or more in hadron and nuclear physics. Moreover, it is not an easy task to join together phenomenological quark models and perturbative QCD (effective in describing asymptotic behaviour) in a single, coherent description of nuclear and hadronic phenomena.

4.1 – Phenomenological lagrangians and quark-quark potential.

We are by now acquainted with quarks carrying colour, that is the *strong charge* in QCD, in full analogy with the electric charge in QED. In other words, quarks are the sources of the colour field in QCD, like electric charges are the sources of the e.m. field. By pursuing this analogy, we want to derive phenomenological QCD lagrangians and the quark-quark potential in the non-relativistic limit. As it is well known, the electromagnetic potential four-vector A_ν produced by a pointlike, spin-1/2 particle with charge e satisfies the equation

$$\Box^2 A_\nu = e(\overline{\psi}\gamma_\nu\psi) \tag{4.1}$$

where ψ is the spinor wavefunction of the source and the γ_ν's are the Dirac matrices. If the particle posses also an anomalous magnetic moment μ_a, one has to add a further term at the rigth-hand side of eq.(4.1):

$$\Box^2 A_\nu = e(\overline{\psi}\gamma_\nu\psi) + \mu_a\partial^\lambda(\overline{\psi}\sigma_{\lambda\nu}\psi) \tag{4.2}$$

whith $\sigma_{\lambda\nu} = 1/2(\gamma_\lambda\gamma_\nu - \gamma_\nu\gamma_\lambda)$. In the field-theoretical language, the first term is referred to as *vector coupling* (with coupling constant e) of the potential A_ν to the source, while the second one is called a *tensor derivative coupling* (with coupling constant μ_a). Eqs. (4.1), (4.2) are straightforwardly generalized to the gluon fields, according to the following rule: the source term has to contain the source vawefunction and to posses the same tensor properties of the gluon field. Let A^i_μ be the $SU(3)_c$ octet of vector gluon fields (where i is the colour index: see sect. 2.3 for notation), and ψ^f the ($SU(3)_c$ vector) wavefunction

(2.1) of a quark with flavour f. Then, it is immediate to write down the QCD analougous of eq. (4.1):

$$\Box^2 A^i_\nu = -\alpha_s(\overline{\psi}^f \gamma_\nu \psi^f)\widehat{T}_i \tag{4.3}$$

where $\widehat{T}_i$ is the i-th $SU(3)_c$ generator (see eq. (2.11a)) and α_s is the strong running coupling constant. We have assumed massless gluons.

The above equation can be derived from a Lagrangian density by means of the standard tools of the Lagrange-Hamilton formalism for fields †. The total phenomenological Lagrangian describing the interaction between quarks and gluons has the form

$$\mathcal{L}_{ph} = \mathcal{L}_q + \mathcal{L}_g + \mathcal{L}_{int} \tag{4.4}$$

where $\mathcal{L}_q$, $\mathcal{L}_g$ are the free Lagrangian density for quarks and gluons, respectively, and — as it is easy to verify — the interaction Lagrangian corresponding to the source term in eq.(4.3) reads

$$\mathcal{L}_{int} = -\alpha_s(\overline{\psi}^f \gamma^\nu \widehat{T}_i \psi^f) A^i_\nu. \tag{4.5}$$

The phenomenological Lagrangian (4.4) can be also considered in the second-quantization formalism, but it is obvious that it is not equivalent to the QCD Lagrangian (2.26). For instance, $\mathcal{L}_{ph}$ does not contain the higher-order terms in the gluon fields that, on the contrary, are present in $\mathcal{L}_{QCD}$.

However, this simple phenomenological model provides a good description of the quark-quark interaction ([52]), at least at the lowest order of perturbation theory (Born approximation). By taking into account the electromagnetic analogy, it is easy to see that one gets an expression for q-q interaction in the non-relativistic limit exactly similar to the well-known Breit-Fermi interaction ([53]) for two electrons in the electromagnetic case. The only difference is that the product of charges, $e_1 e_2$, is replaced by the scalar product (in the $SU(3)_c$ space) of two generators $\widehat{T}(i)$ †:

$$e_1 e_2 \rightarrow \alpha_s\{\widehat{T}(1)\cdot\widehat{T}(2)\}. \tag{4.6}$$

---

† Let us remind that the Euler-Lagrange equations for the field component ϕ_k read

$$\frac{\partial\mathcal{L}}{\partial\phi_k} - \frac{\partial}{\partial x_\mu}\left(\frac{\partial\mathcal{L}}{\partial\pi^\mu_k}\right) = 0$$

where $\mathcal{L} = \mathcal{L}(\phi_k, \pi^\mu_k)$ is the Lagrangian density and $\pi^\mu_k = \frac{\partial\phi_k}{\partial x_\mu}$ is the *momentum conjugate* to ϕ_k. The free Lagrangian for a massive, n-component, complex scalar field G reads

$$\mathcal{L}_{free} = \frac{1}{2}\sum_{k=1}^{n}(\partial_\nu\phi_k\partial^\nu\phi_k - m^2|\phi_k|^2).$$

For this Lagrangian, the Lagrange equations become $(\Box^2 + m^2)\phi_k = 0$, i.e. the usual Klein-Gordon equation. The Hamiltonian density can be obtained by the relation $\mathcal{H} = \sum_k \frac{\partial\mathcal{L}}{\partial\pi^0_k}\pi^0_k - \mathcal{L}$.

† This is essentially due to the fact that the U(1)symmetry of QED goes into the $SU(3)_c$ symmetry of QCD.

One has therefore:

$$
\begin{aligned}
V_{ij}^{(1g)} = \alpha_s [& \frac{1}{|\underline{r}_{ij}|} - \frac{1}{2m_i m_j}\left(\frac{\underline{p}_i \cdot \underline{p}_j}{|\underline{r}_{ij}|} + \frac{\underline{r}_{ij} \cdot (\underline{r}_{ij} \cdot \underline{p}_i)\underline{p}_j}{|\underline{r}_{ij}|^3}\right) - \\
& \frac{\pi}{2}\delta^3(\underline{r}_{ij})\left(\frac{1}{m_i^2} + \frac{1}{m_j^2} + \frac{16}{3}\frac{\underline{s}_i \cdot \underline{s}_j}{m_i m_j}\right) - \\
& \frac{1}{2|\underline{r}_{ij}|^3}\{\frac{1}{m_i^2}(\underline{r}_{ij} \times \underline{p}_i) \cdot \underline{s}_i - \frac{1}{m_j^2}(\underline{r}_{ij} \times \underline{p}_j) \cdot \underline{s}_j + \\
& \frac{1}{m_i m_j}[2(\underline{r}_{ij} \times \underline{p}_i) \cdot \underline{s}_j - 2(\underline{r}_{ij} \times \underline{p}_j) \cdot \underline{s}_i - \\
& 2\underline{s}_i \cdot \underline{s}_j + \frac{6(\underline{s}_i \cdot \underline{r}_{ij})(\underline{s}_j \cdot \underline{r}_{ij})}{|\underline{r}_{ij}|^2}]\}] \cdot (\hat{T}(i) \cdot \hat{T}(j))
\end{aligned}
\tag{4.7}
$$

Here, $\underline{r}_{ij} = \underline{r}_i - \underline{r}_j$ and $\underline{s}_i$ is the i-th quark's spin. The symbol $(1g)$ means that this approximation corresponds, from the field-theoretical standpoint, to the so-called *one-gluon exchange*. It is represented by the first diagram at the right-hand side of Fig.11.

In the following, we shall confine ourselves to take into account only the one-gluon quark-quark interaction (4.7), since, as already noted, our phenomenological Lagrangian cannot account for higher-order gluon interactions. However, the following points are worth to be stressed right now:

1) $V_{ij}^{(1g)}$ contains singular pieces, like the spin-spin interaction. This is due to having approximated the sources as pointlike (by neglecting higher-order renormalization effects), and can be eliminated by introducing a suitable cut off in the source term at large momentum tranfers ($\alpha(Q^2) \to 0$ for $Q^2 \to \infty$), so that

$$\alpha_s \to \alpha_s(Q^2) \tag{4.8}$$

where

$$\alpha_s(0) = \alpha_s, \qquad \lim_{Q^2 \to \infty} \alpha_s(Q^2) = 0 \tag{4.9}$$

Then, one *reduces* the interaction for $Q^2 \to \infty$ ($r \to 0$), and gets a finite range for the spin-spin term. The introduction of $\alpha_s(Q^2)$ in the source term modifies the radial dependence of the various terms in (4.7), but the overall structure of $V_{ij}^{(1g)}$ does not change.

2) Obviously, the one-gluon approximation is unable to generate the confinement of quarks inside hadrons. An ad hoc way to obviate this difficulty is adding to $V_{ij}^{(1g)}$ a *confining term* of the kind

$$V_{ij}^{conf} = -kr^{\nu}(\hat{T}(i) \cdot \hat{T}(j)) \tag{4.10}$$

with $k > 0$ and $\nu = 1, 2$ (as we shall see, it is $\langle \hat{T}(i) \cdot \hat{T}(j) \rangle < 0$). In the following, we shall criticize this assumption in detail. For the moment, we will assume as starting point the quark-quark potential

$$V_{qq} = \sum_{i=1}^{n} \sum_{j>i} (V_{ij}^{(1g)} + V_{ij}^{conf}) \tag{4.11}$$

where n is the number of valence quarks inside the hadron considered. Indeed, as it will be seen, it allows to give a very good description of the hadron spectroscopy and seems to constitute also a promising approach for nuclear physics.

4.2 – Constituent quark model for hadrons.

It has been seen in sect.3.6 that the structure of a composite system (hadrons or nucleons), in terms of the QCD Fock states, is very complex (see eq. (3.43)). This is due to the fact that, in general, one cannot neglect the Fock states with sea quarks and gluons †.

A more simplified structure is obtained in the so-called non-relativistic *constituent quark model* (7), in which only valence quarks are taken into account. Moreover, one assumes ab initio that quarks are massive ‡ (typical mass values are, e.g. $m_u \simeq m_d \simeq 330$ MeV; $m_s \simeq 455$ MeV; $m_c \simeq 1500$ MeV, ...).

This section will be concerned with the application of the constituent quark model to hadrons, specially to low-energy baryon states (nucleons and Δ-resonance). First of all, let us briefly review the basic notions about the low-lying hadrons, built up by the light quarks (u,d,s).

4.2.1 – Meson spectroscopy.

The meson wavefunctions can be written in the form

$$\psi_M = \psi^C \sum_{\alpha\beta} (\psi_\alpha^{f,s} \otimes \psi_\beta^L)_J^{M_J} C_{\alpha\beta}^{LS,J} \tag{4.12a}$$

where ψ^C is the (singlet) colour wavefunction, $\psi_\alpha^{f,s}$ is the flavour and spin wavefunction (labelled by the quantum numbers Y, T and T_3) and ψ^L is the usual orbital wavefunction. The symbol $\otimes$ denotes the direct product of vectors ††.

The peculiar features of the constituent quark model are embodied in the ψ^f structure. As we have already mentioned in chapt. 2, low-lying hadrons are classified according to the different representations of $SU(3)_f$ ‡‡.

Henceforth, we shall denote by $\underline{N}$ a multiplet of states belonging to the same N-dimensional *irreducible representation* of $SU(3)_f$. Essentially, irreducible representations correspond, from a physical point of view, to set of states that are degenerate in energy

---

† However, we have seen that, in calculations of the form factors, one can consider only state with valence quarks in the asymptotic region.

‡ Let us recall that, in QCD, quarks are assumed to be massless (at the fundamental level) (cf. sect. 2.3). Quark masses are generated by spontaneous symmetry breaking and renormalization mechanisms, that are characteristic of a non-abelian gauge theory like QCD.

†† The direct product of two vectors $\underline{u}, \underline{v}$ is obtained by taking all possible products of their components: $\underline{u} \otimes \underline{v} = \{u_i v_j\}$.

‡‡ Remember that, as already stressed in chapt. 2, the $SU(3)_f$ symmetry (i.e. the symmetry under rotations in $SU(3)_f$ space) is only approximate (unlike $SU(3)_c$, that is an exact symmetry), as it is shown by the mass difference among quarks ($m_s \neq m_u, m_d$).

(but see ref.(16) for a more rigorous mathematical definition). Moreover, it can be shown that all the state vectors of the same multiplet have the same permutation symmetry under exchange of the quark indices.

Quarks belong to the 3-dimensional representation, $\underline{3}$, of $SU(3)_f$, while antiquarks belong to the coniugate representation, $\underline{\bar{3}}$ (this is essentially due to the fact that antiquarks have opposite value of the hypercharge). The $q\bar{q}$ structure of a meson corresponds, from a mathematical point of view, to the Kronecker product of the corresponding irreducible representations ([3,16]). The representation thus obtained is, in general, reducible, i.e. it can be written as the direct sum of irreducible representations †. In this case, one has ([3,16])

$$q\bar{q} \rightarrow \underline{3} \otimes \underline{\bar{3}} = \underline{1} \oplus \underline{8} \tag{4.13}$$

Then, mesons are classified in $SU(3)_f$ *nonets* (sum of a singlet and an octet), with different configurations ${}^{(2S+1)}L_J$. The best known meson nonet is the *vector meson* one, that

† In order to clarify the concepts of product of representations and reducible representation, let us consider the familiar example of matrices. Given a $N_1 \times N_1$ matrix $\hat{A}$ and a $N_2 \times N_2$ matrix $\hat{B}$, their *Kronecker* (or *direct*) *product* is the $(N_1N_1) \times (N_2N_2)$ matrix with elements $(\hat{A} \otimes \hat{B})_{pq,rs} = A_{pr}B_{rs}$, i.e. a matrix whose rows and columns are labeled by a pair of indices. The *direct sum* of $\hat{A}$ and $\hat{B}$ is the (block-diagonal) $(N_1 + N_1) \times (N_2 + N_2)$ matrix

$$\hat{A} \oplus \hat{B} = \begin{pmatrix} \hat{A} & 0 \\ 0 & \hat{B} \end{pmatrix}$$

In other words, direct summation may be loked upon as a manipulation that strings together operators along a main diagonal . Given an N-dimensional matrix representation $\hat{D}(\alpha)$ of a group $\mathcal{G}$ ($\alpha \in \mathcal{G}$), we shall say that $\hat{D}(\alpha)$ is reducible if, $\forall\alpha \in \mathcal{G}$, $\hat{D}(\alpha)$ can be put in the block-diagonal form

$$\hat{D}(\alpha) = \begin{pmatrix} \hat{D}_1(\alpha) & \cdots & \cdots & \cdots \\ \cdots & \hat{D}_2(\alpha) & \cdots & \cdots \\ \cdots & \cdots & \cdots & \cdots \\ \cdots & \cdots & \cdots & \hat{D}_m(\alpha) \end{pmatrix}$$

or

$$\hat{D}(\alpha) = \bigoplus_{i=1}^{m} \hat{D}_i(\alpha)$$

where $\hat{D}_i(\alpha)$ is the n_i-dimensional irreducible matrix representation of $\mathcal{G}$ and $\sum_{i=1}^{m} n_i = N$. In the abstract notation $\underline{N}$ for the N-dimensional representation, one has

$$\underline{N} = \bigoplus_{i=1}^{m} \underline{n}_i$$

In general, the direct product of irreducible representations of a given group is reducible and therefore it will admit a decomposition of the above kind . See ref. ([3,16]).

includes $K^*(892)$, $\rho(770)$, $\omega(783)$ and $\phi(1020)$, all essentially in configuration ${}^3S_1$ (with components ${}^3D_1$, as in deuterium).

The *weight diagram* of the vector meson nonet is shown in Fig. 11. Their mass differences are dominated by the difference between strange and up (down) quark mass, i.e. $m_s - m_u(m_d) \simeq (m_\phi - m_\rho) \simeq 125$ MeV. The orbital excitations have configurations ${}^3L_J$ with $J = L \pm 1, L$ and the same flavour content. Two other tensor nonets have been observed, respectively with $L = 1$ and $J = 2$ (containing the resonances $A_2(1310)$, $f(1270)$, $K^*(1430)$ and $f'(1515)$) and $L = 2$, $J = 3$ (including $g(1680)$, $\omega(1670)$ and $K^*(1780)$. Let us notice that, in general, there is a mixing between the states with $Y = I_3 = 0$ (ϕ and ω for the vector mesons, and f, f' for the tensor ones).

The meson multiplet of lowest average mass is the *pseudoscalar meson nonet* (K, π, η, η') with $J^P = 0^-$. In this case, too, η and η' mixing is to be taken into account. However, it is smaller than in the previous case. Excited states with configuration ${}^0L_1$, have been also observed, but we shall not consider them for brevity's sake.

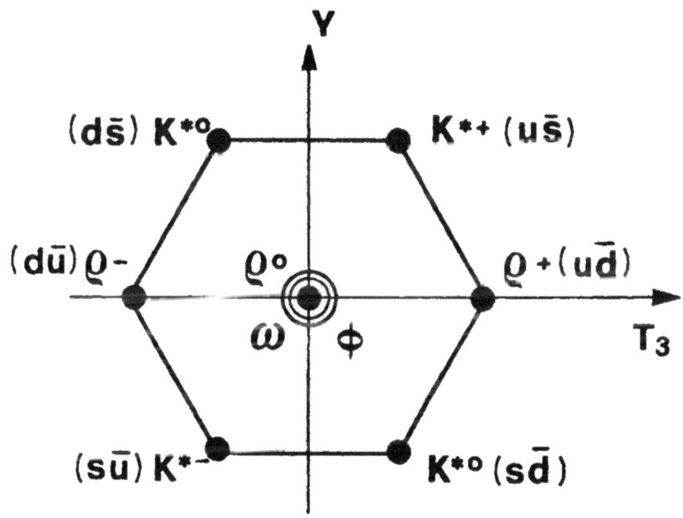

Fig. 11 - $SU(3)_f$ weight diagram of the vector meson nonet (k, ρ, ω, ϕ). Notice that: $\omega = (u\bar{u} + d\bar{d})/\sqrt{2}$; $\rho^0 = (u\bar{u} - d\bar{d})/\sqrt{2}$; $\phi = s\bar{s}$.

4.2.2 — Baryon spectroscopy.

The barion wave function ψ^B is now given by

$$\psi_B = \psi^C \hat{A} \sum_{\alpha\beta} (\psi_\alpha^{f,s} \otimes \psi_\beta^L)_J^{M_J} C_{\alpha\beta}^{LS,J} \tag{4.12b}$$

where $\hat{A}$ is the antisymmetrizing operator. Since the colour wave function ϕ^C is a completely antisymmetric singlet state, then $(\psi_\alpha^{f,s} \otimes \psi_\beta^L)$ must be completely symmetric. Hence, only functions $\psi^{f,s}$ and ψ^{L} with suitable symmetries can be used to build up baryon wave functions.

By analogy to mesons, to the baryon structure q^3 it corresponds the Kronecker product ([3,16])

$$q^3 \rightarrow \underline{3} \otimes \underline{3} \otimes \underline{3} = \underline{1} \oplus \underline{8} \oplus \underline{8}' \oplus \underline{10}. \tag{4.14}$$

Therefore, baryons are classified in singlets, octets and decuplets. The octet wave functions have mixed symmetry for permutation of the quark indices, while the decuplet states are symmetric.

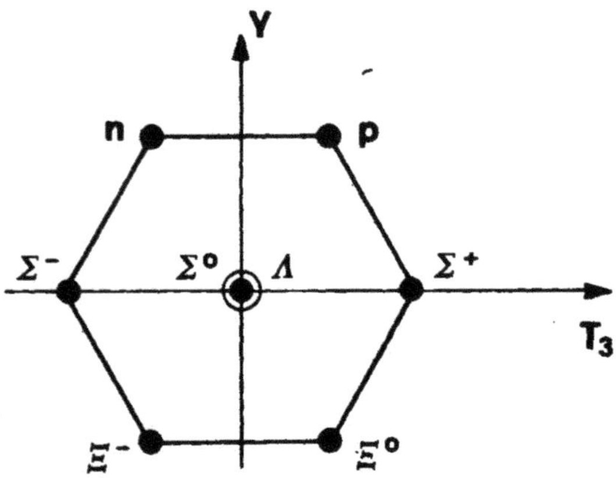

Fig. 12 – SU(3)$_f$ weight diagram of the 1/2$^+$ barion octect.

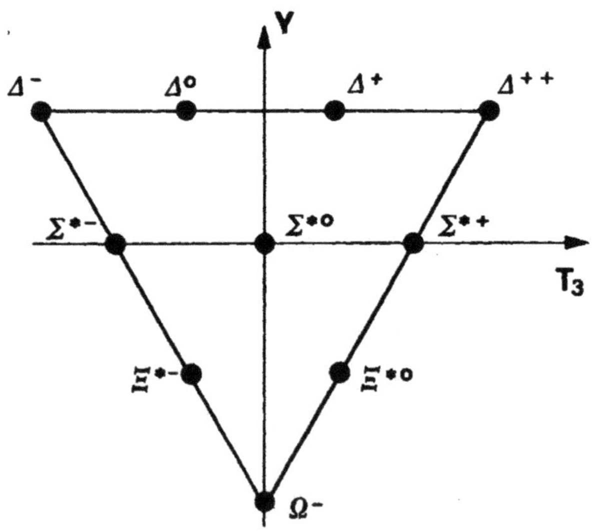

Fig. 13 – SU(3)$_f$ weight diagram of the 3/2$^+$ barion decuplet.

The best known baryon multiplets are the octet with $J^P = 1/2^+$, and $L = 0$, $S = 1/2$ (to which it belongs the nucleon, $N(939)$), and the decuplet $J^P = 3/2^+$, and $L = 0$, $S = 3/2$ (that contains the well-known resonance $\Delta(1232)$). Their weight diagrams are shown in Figs. 12,13. A number of baryon states with orbital excitations have been experimentally observed.

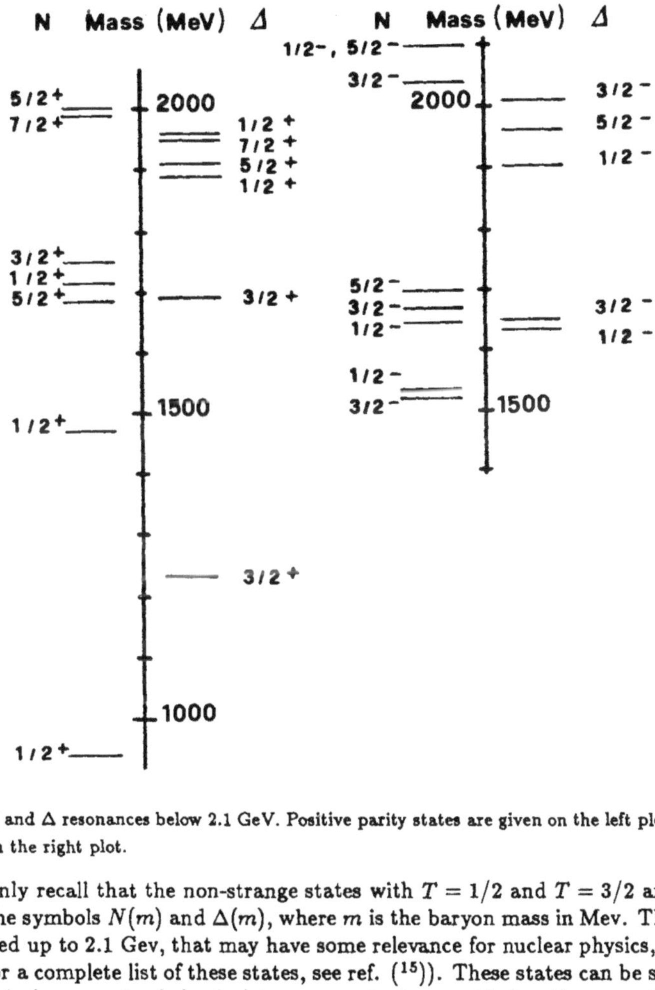

Fig. 14 – All N and Δ resonances below 2.1 GeV. Positive parity states are given on the left plot, negative parity states on the right plot.

Let us only recall that the non-strange states with $T = 1/2$ and $T = 3/2$ are usually denoted by the symbols $N(m)$ and $\Delta(m)$, where m is the baryon mass in Mev. The baryon states observed up to 2.1 Gev, that may have some relevance for nuclear physics, are given in Fig. 14 (for a complete list of these states, see ref. ([15])). These states can be studied by a spectroscopical approach of classical type, namely, by considering them as *atoms* built up by quarks (precisely, as ^{3}H and ^{3}He). The first step is obviously the construction of a complete set of basis vectors, in terms of which one can expand the wavefunction (4.12b). In the following, we shall explicitly discuss the cases of the nucleon and of the resonance $\Delta(1232)$, in which only u and d quarks are involved. Moreover, in this case it is enough to consider — instead of $SU(3)_f$ — the isospin subgroup $SU(2)_{is}$. Clearly, the up and down quarks belong to the (2-dimensional) fundamental representation of $SU(2)_{is}$.

To construct the spatial wavefunction for baryons, it is useful to introduce the following coordinates:

$$\begin{cases} \underline{\rho} = \frac{1}{\sqrt{3}}(\underline{r}_1 - \underline{r}_2) & ; \\ \underline{\lambda} = \frac{1}{3}(\underline{r}_1 + \underline{r}_2 - 2\underline{r}_3) & ; \\ \underline{R} = \frac{1}{3}\sum_{i=1}^{3} \underline{r}_i & . \end{cases} \tag{4.15}$$

Here, $\underline{\rho}$ and $\underline{\lambda}$ are the usual Jacobi coordinates (describing the internal motion) and $\underline{R}$ is the center-of-mass coordinate. Let us note that $\underline{\rho}$ and $\underline{\lambda}$ have mixed symmetry under permutation P_{ij} $(i,j = 1,2,3)$ of the quarks indices. Moreover, under P_{12}, $\underline{\rho}$ and $\underline{\lambda}$ are antisymmetric and symmetric, respectively. For simplicity's sake, in the following we shall denote by the indices ρ and λ (e.g. ψ^ρ and ψ^λ) any function with the same permutation symmetries of the Jacobi coordinates. As usual, the symbol ψ^S (ψ^A) is adopted for a totally symmetric (antisymmetric) function. All possible spin (χ) and isospin (B) wavefunctions for three quarks with $J = 1/2, T = 3/2$ are listed in tables III and IV, respectively. The last column of these tables reports, for completeness' sake, the various sets of integer numbers $[\lambda_i]$ ("Young frames") denoting the *Young tableaux* needed to classify the baryon states according to the irreducible representations of the symmetric group S_3([3,16]) †.

---

† Henceforth, we shall denote by S_n the symmetric group, i.e. the group of permutations of n objects. It is outside the scope of the present lectures to develop the details of the representation theory of the symmetric group, and we refer the reader to refs. ([3,16]). However, let us recall the main definitions and properties. A *Young diagram* $Y_{[\lambda]}$ of *shape* (or frame) $[\lambda] = [\lambda_1\lambda_2\ldots\lambda_n]$, where the λ_i are non-negative integers satisfying $\lambda_1 \geq \lambda_2 \geq \ldots \lambda_n$, is a diagram consisting of λ_1 boxes in row 1, λ_2 boxes in row 2, ... λ_n boxes in row n, arranged as illustrated below:

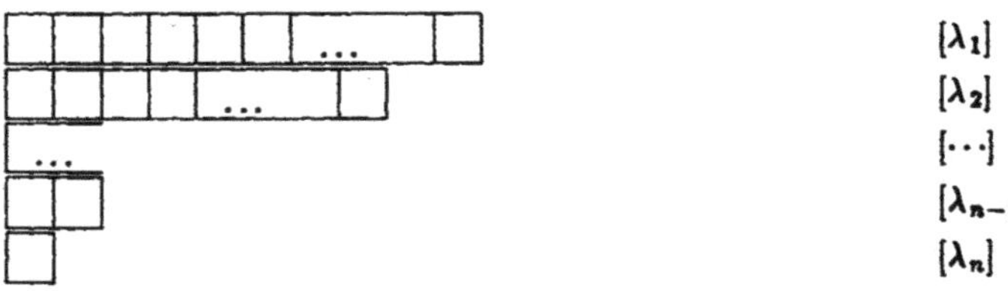

A *standard Young tableaux* is a Young diagram in which the boxes have been filled with positive integers $1, 2, \ldots, n$ such that the numbers do not decrease in going from left to right in a row, and increase in going from top to bottom in a column. If $\sum_{i=1}^{n} \lambda_i = N$, we shall call $[\lambda]$ a *partition* of N into n parts. One generally counts the zeros in determining the parts of a partition. For instance, the partitions of 4 into 3 parts are [400], [310], and [211]. When the number of parts is understood, we shall omit the zeros and write, for instance, [4], [3,1] and [2,2]. The important results for the symmetric group S_n are as follows: (i) the set of irreducible representations of S_n is in one-to-one correspondence with the set of partitions $\{[\lambda]\}$ of n into n parts; (ii) the set of basis vectors of the carrier space of the irreducible representation $[\lambda]$ of S_n is in one-to-one correspondence with the

TABLE III – Spin states for baryons				
S	S_{12}	name	structure	S_3 symmetry $[\lambda]$
1/2	1	χ^λ	$\frac{1}{\sqrt{6}}[-(\uparrow\downarrow+\downarrow\uparrow)+2\uparrow\uparrow\downarrow]$	[21]
1/2	0	χ^ρ	$\frac{1}{\sqrt{2}}(\uparrow\downarrow-\downarrow\uparrow)$	
3/2	1	χ^s	$(\uparrow\uparrow\uparrow)$	[3]

TABLE IV – Isospin states for baryons(T=T_3)				
T	T_{12}	name	structure	S_3 symmetry $[\lambda]$
1/2	1	B^λ	$\frac{1}{\sqrt{6}}[-(ud+du)+2uud]$	[21]
1/2	0	B^ρ	$\frac{1}{\sqrt{2}}(ud-du)$	
3/2	1	B^s	(uuu)	[3]

Finally, table V lists all possible spin-isospin state vectors ψ for baryons (obviously, they are orthonormal and have definite symmetry).

TABLE V – Spin-Isospin baryonic states with definite symmetry.			
(S,T)	name	structure	S_3 symmetry $[\lambda]$
1/2,1/2	Φ^s	$1/2[B^\lambda\chi^\lambda+B^\rho\chi^\rho]$	[3]
1/2,1/2	Φ^A	$1/2[B^\lambda\chi^\rho-B^\rho\chi^\lambda]$	[111]
1/2,1/2	Φ^λ	$1/2[B^\lambda\chi^\lambda-B^\rho\chi^\rho]$	[21]
1/2,1/2	Φ^ρ	$1/2[B^\lambda\chi^\rho+B^\rho\chi^\lambda]$	[21]
1/2,3/2	$\Phi^{\lambda'}$	$B^s\chi^\lambda$	[21]
1/2,3/2	$\Phi^{\rho'}$	$B^s\chi^\rho$	[21]
3/2,1/2	$\Phi^{\lambda''}$	$B^\lambda\chi^s$	[21]
3/2,1/2	$\Phi^{\rho''}$	$B^\rho\chi^s$	[21]
3/2,3/2	$\Phi^{s'}$	$B^s\chi^s$	[3]

set of standard Young tableaux of shape $[\lambda]$ (the number of standard tableaux is then the dimension of such representation).

In order to write down the orbital wavefunctions, one has to choose a suitable orthonormal basis, whose vectors depend on the Jacobi variables $\underline{\rho}$ and $\underline{\lambda}$. To simplify calculations of the radial matrix elements, one often uses harmonic oscillator wavefunctions. In tables VI and VII we give, e.g., all possible states R with $L = 0, 2$ and a number of harmonic oscillator quanta $N = N_\rho + N_\lambda \leq 4$.

TABLE VI – Harmonic oscillator radial functions with $L^\pi = 0^+$ for baryons ($N = N_\lambda + N_\rho \leq 4$).

name	structure*	S_3 [λ]
R_1^s	$K \cdot Exp$	[3]
R_2^s	$(K/\sqrt{3})(\alpha_h^2[\lambda^2+\rho^2]-3)\cdot Exp$	
R_3^s	$(K/2\sqrt{6})(12-8\alpha_h^2[\lambda^2+\rho^2]+\alpha_h^4[\lambda^2+\rho^2]^2)\cdot Exp$	
R_4^s	$\alpha_h^4(K/2\sqrt{30})\cdot$ $([\lambda^2+\rho^2]^2-2[\lambda^2-\rho^2]^2+4(\underline{\lambda}\cdot\underline{\rho})^2)\cdot Exp$	
R_1^λ	$\alpha_h^2(K/\sqrt{3})(\lambda^2-\rho^2)\cdot Exp$	[21]
R_2^λ	$\alpha_h^2(K/\sqrt{15})(\lambda^2-\rho^2)(\alpha_h^2[\lambda^2-\rho^2]-5)\cdot Exp$	
R_1^ρ	$\alpha_h^2(2K/\sqrt{3})(\underline{\lambda}\cdot\underline{\rho})\cdot Exp$	[21]
R_2^ρ	$\alpha_h^2(2K/\sqrt{15})(\underline{\lambda}\cdot\underline{\rho})(\alpha_h^2[\lambda^2+\rho^2]-5)\cdot Exp$	[21]

* Adopted symbols: $K = \alpha_h^3/\sqrt{\pi^3}$; $Exp = e^{-\frac{1}{2}\alpha_h^2(\lambda^2+\rho^2)}$

TABLE VII – Harmonic oscillator radial functions with $L^\pi = 2^+$ for baryons ($N \leq 4$).

name	structure*	S_3 [λ]
D_1^s	$(\overline{K}\sqrt{2/15})(\lambda^2 Y_2^m(\lambda)+\rho^2 Ym_2(\rho))\cdot Exp$	[3]
D_2^s	$(\overline{K}\sqrt{2/75})(\lambda^2 Y_2^m(\lambda)+\rho^2 Ym_2(\rho))(\alpha_h^2[\lambda^2+\rho^2]-5)\cdot Exp$	[3]
D_1^λ	$(\overline{K}\sqrt{2/15})(\lambda^2 Y_2^m(\lambda)-\rho^2 Ym_2(\rho))\cdot Exp$	[21]
D_2^λ	$(\overline{K}\sqrt{2/75})(\lambda^2 Y_2^m(\lambda)-\rho^2 Ym_2(\rho))(\alpha_h^2[\lambda^2+\rho^2]-5)\cdot Exp$	[21]
D_1^ρ	$\alpha_h^2(2\overline{K})(\rho\otimes\lambda)_2^m\cdot Exp$	[21]
D_2^ρ	$\alpha_h^2(2\overline{K}/\sqrt{5})(\rho\otimes\lambda)_2^m(\alpha_h^2[\lambda^2+\rho^2]-5)\cdot Exp$	[21]

* Adopted symbols: $K = \alpha_h^3/\sqrt{\pi^3}$; $\overline{K} = 2\alpha_h^5/\pi$; $Exp = e^{-\frac{1}{2}\alpha_h^2(\lambda^2+\rho^2)}$

Then, one has to combine spin-isospin states and orbital states to get a totally symmetric wavefunction. This can be done, in general, by standard techniques of the permutation group, by putting together the ψ's belonging to a given S_3 representation and the R's of its conjugate one. In the present case, however, this construction can be accomplished in a very simple way by elementary tools. The wavefunctions so obtained are given in tables VIII and IX, for the cases $J^P = 1/2^+, T = 1/2$ and $J^P = 3/2^+, T = 3/2 (L = 0, 2)$, respectively.

TABLE VIII – Totally simmetric states for baryons with $(J,T) = (1/2, 1/2)$ and $\pi = +1$ $(J = J_3)$.

name	L	structure	conventional name
$\Psi_i(L=0)$	0	$R_i^s \Phi^s$	fully symmetric component
$\Psi'_i(L=0)$	0	$\frac{1}{\sqrt{2}}(R_i^\lambda \Phi^\lambda + R_i^\rho \Phi^\rho)$	mixed symmetry component
$\Psi_i(L=2)$	2	$\frac{1}{\sqrt{2}}[(D_i^\lambda \otimes \Phi^\lambda)^{1/2}_{1/2} + (D_i^\rho \otimes \Phi^\rho)^{1/2}_{1/2}]$	D-wave
$\Psi'_i(L=2)$	2	$\frac{1}{\sqrt{2}}[(D_i^\lambda \otimes \chi^s)^{1/2}_{1/2} B^\lambda + (D_i^\rho \otimes \chi^s)^{1/2}_{1/2} B^\rho]$	components

TABLE IX – Totally simmetric states for baryons with $(J,T) = (3/2, 3/2)$ and $\pi = +1 (J = J_3)$.

name	L	structure	conventional name
$\Phi_i(L=0)$	0	$R_i^s(B^s \chi^s)$	S-wave component of Δ
$\Phi_i(L=2)$	2	$(D_i^s \otimes \chi^s)^{3/2}_{3/2} B^s$	D-wave symmetric component
$\Phi'_i(L=2)$	2	$\frac{1}{\sqrt{2}}[(D_i^\lambda \otimes \chi^\lambda)^{3/2}_{3/2} - (D_i^\rho \otimes \chi^\rho)^{3/2}_{3/2}] B^s$	mixed-symmetry D-wave component

Then, let us discuss the dynamical aspect of the problem. In non-relativistic quark models, the internal motion Hamiltonian is taken in the form

$$\hat{\mathcal{H}} = \hat{\mathcal{M}} + (\hat{\mathcal{T}} - \hat{\mathcal{T}}_{CM}) + V_{qq} \tag{4.16}$$

with (n = number of constituent quarks in the hadron considered)

$$\hat{\mathcal{M}} = \sum_{i=1}^{n} \hat{m}_i; \tag{4.16a}$$

$$\hat{\mathcal{T}} = \sum_{i=1}^{n} \hat{\mathcal{T}}_i. \tag{4.16b}$$

Here, $\hat{m}_i$ and $\hat{T}_i$ are the mass operator and the kinetic energy of the i-th quark (usually, one takes $\hat{T}_i$ in the non-relativistic form: $\hat{T}_i = -\hbar^2\nabla_i^2/2m_i$), $\hat{T}_{CM}$ is the center-of-mass energy ($\hat{T}_{CM} = |\underline{P}|^2/2M$ with $\underline{P} = \sum_{i=1}^n \underline{p}_i$, $M = \sum_{i=1}^n m_i$) and the quark-quark potential V_{qq} is given by eqs. (4.11), (4.7). One easily sees, by the explicit form of the $SU(3)_c$ generators (eqs. (2.6), (2.11a)), that it is

$$\langle \hat{T}(i) \cdot \hat{T}(j) \rangle = \begin{cases} -4/3 & \text{in a baryon;} \\ -2/3 & \text{in a meson.} \end{cases} \tag{4.17}$$

The solution of the Schrödinger equation

$$\hat{\mathcal{H}}\psi = E\psi \tag{4.18}$$

can be found either by suitable variational methods or by diagonalizing the hamiltoniah $\hat{\mathcal{H}}$ in the basis of the antisymmetric states (4.12a). In this second case, solving (4.18) amounts to determine the coefficients C in (4.12a), while the dimensional parameter $\alpha_h(\mathrm{fm}^{-1})$ of the harmonic-oscillator wavefunctions is kept fixed (or changed by attempts).

Many authors(54,55) have studied hadron properties by means of the methods dicussed above. For instance, Isgur and coworkers(55) have extensively used an Hamitonian of the Fermi-Breit type (see eq. (4.7)). They neglect the terms dependent on momentum (like the spin-orbit interaction), as being not essential in the description of the low-lying states, and assume a quadratic confining potential (ν=2 in eq. (4.10)). Some results of these papers(55) are given in fig. 15. In particular, Isgur et al. have studied the fundamental states of hadrons (i.e. the minimal energy states for any J^P,T). They get, for the nucleons and the Δ, wavefunctions of the kind(55)

$$|N\rangle \simeq 0.93|\phi_1(L=0)\rangle - 0.29|\phi_2(L=0)\rangle - 0.23|\phi_1'(L=0)\rangle - 0.04|\phi_1(L=2)\rangle; \tag{4.19a}$$

$$|\Delta\rangle \simeq 0.97|\phi_1(L=0)\rangle + 0.20|\phi_2(L=0)\rangle - 0.01|\phi_1(L=2)\rangle + 0.07|\phi_1'(L=2)\rangle; \tag{4.19b}$$

in both cases, the dominant component is the totally simmetric orbital wavefunction with $L = 0$ ($\phi_1(L = 0)$), with small contributions from the first radial excitation ($\phi_2(L = 0)$), mixed simmetry states ($\phi_1'(L = 0)$) and states with $L = 2$ ($\phi_1(L = 2)$). By the wavefunctions (4.19) it is possible to reproduce in detail many hadron properties, like the mean squared radii of nucleons, the amplitude for weak transitions, and so on.

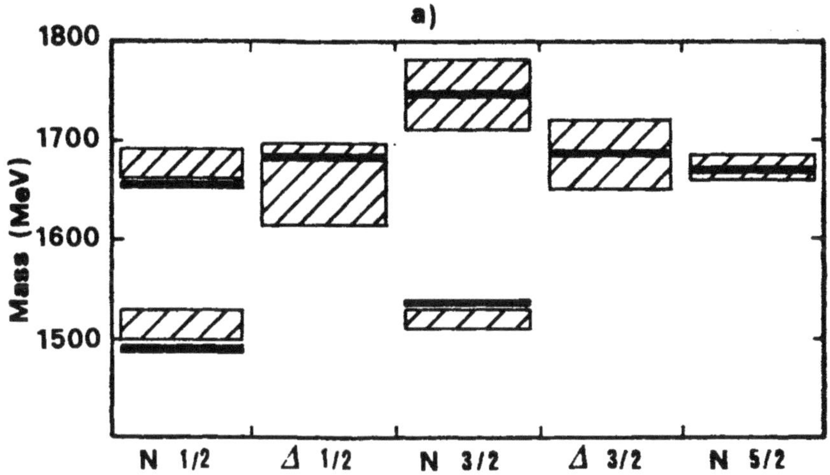

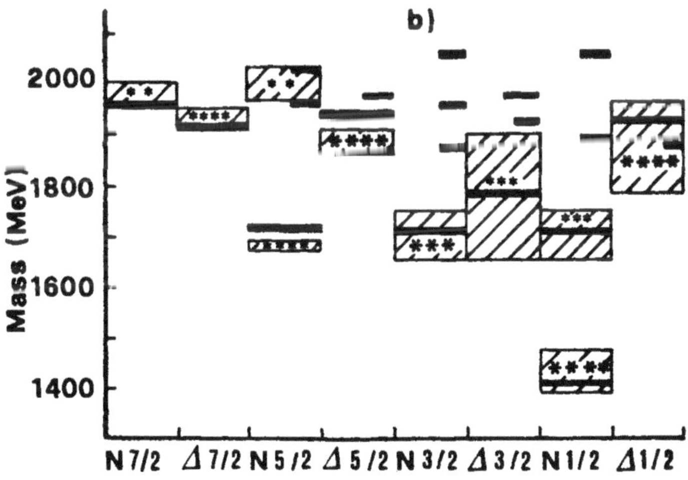

Fig. 15 – a) The experimental spectrum of $S = 0$ negative parity baryons compared to the constituent-quark model predictions[13]. The regions in which the masses of the resonances probably lie are denoted by shaded boxes. b) The spectrum of $S = 0$ positive parity excited baryons compared to the model predictions[13]. States that are predicted to decouple from πN are shown as stubs.

In conclusion, the non-relativistic constituent quark model is remarkably successful in describing many phenomena in hadron physics. However, the model suffers from some drawbacks, concerning both its foundations (it has no sound theoretical basis) and its phenomenological applications (it is spectacularly unsuccessful in many areas, like the description of multiquark states). Since we are mainly interested to nuclear applications, let us only quote that when the model is used to deriving the nucleon-nucleon interaction, V_{NN}, it generates terms diverging as powers of the distance r between nucleons ($\lim_{r\to\infty} V_{NN}(r) \approx r^{-n}$), analogous to the well-known *Van der Waals forces*[56].

4.3 – Application to deuterium.

We have already noted at the beginning of chapt. 2 that hadrons and nuclei can be looked upon, respectively, as *atoms* and *molecules* of quarks. In particular, deuterium can be considered as a quark *biatomic molecule*, i.e. as the set of two three-quark cluster I and II, with $S = J = 1$ and $\pi = +1$ (see fig. 16).

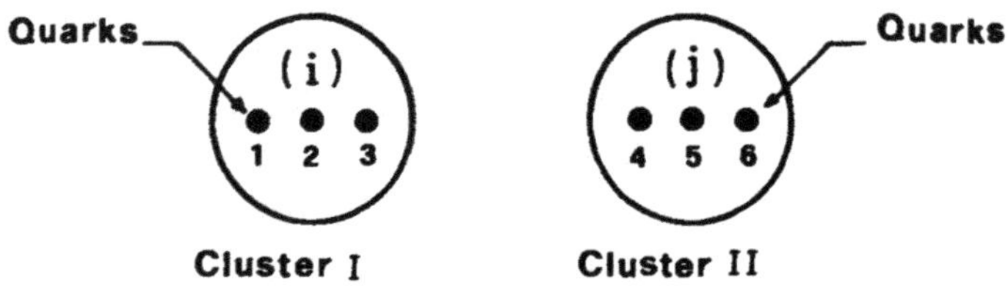

Fig. 16 – The six-quark description of two-nucleon systems.

Let us denote by $q_i (i = 1,2,3)$ and $q_j (j = 4,5,6)$ the quarks of the cluster I and II respectively. As usual, it is convenient to introduce the Jacobi variables for the two clusters:

$$\underline{\rho}_1 = \frac{1}{\sqrt{3}}(\underline{r}_1 - \underline{r}_2); \quad \underline{\lambda}_1 = \frac{1}{3}(\underline{r}_1 + \underline{r}_2 - 2\underline{r}_3); \quad \underline{R}_1 = \frac{1}{3}\sum_{i=1}^{3} \underline{r}_i \quad (cluster \quad I); \tag{4.20a}$$

$$\underline{\rho}_2 = \frac{1}{\sqrt{3}}(\underline{r}_4 - \underline{r}_5); \quad \underline{\lambda}_2 = \frac{1}{3}(\underline{r}_4 + \underline{r}_5 - 2\underline{r}_6); \quad \underline{R}_2 = \frac{1}{3}\sum_{j=4}^{6} \underline{r}_j \quad (cluster \quad II). \tag{4.20b}$$

Moreover, let us define

$$\underline{r} = \underline{R}_1 - \underline{R}_2; \qquad \underline{R} = (1/2)(\underline{R}_1 + \underline{R}_2), \tag{4.21}$$

that represents, respectively, the distance between the center-of-mass of the two clusters, and the total CM coordinate. The internal motion of the deuterium is therefore described by the five variables $\underline{r}$, $\underline{\rho}_k$, $\underline{\lambda}_k$, $(k = 1,2)$.

In this scheme, the deuterium wavefunction can be put in the form

$$\psi = \hat{\mathcal{A}} \sum_{\alpha\beta\gamma} \{\psi_\alpha^C (\psi_\beta^{S,T} \otimes \psi_\gamma^L)_J^M\} C_{\alpha\beta}^{SL(T)J} , \tag{4.22}$$

with self-clear meaning of the symbols.

As in the hadron case, let us firstly construct all possible singlet colour states ψ_α^C. The six-quark structure corresponds, from a mathematical viewpoint, to the following Kronecker product of $SU(3)_c$ irreducible representations:

$$(\underline{3} \otimes \underline{3} \otimes \underline{3}) \otimes (\underline{3} \otimes \underline{3} \otimes \underline{3}) = (\underline{1} \oplus \underline{8} \oplus \underline{8}' \oplus \underline{10}) \otimes (\underline{1} \oplus \underline{8} \oplus \underline{8}' \oplus \underline{10}) \ . \tag{4.23}$$

It can be shown that the above product contains only five singlets, given by the Kronecker products

$$(\underline{1} \otimes \underline{1}), (\underline{8} \otimes \underline{8}), (\underline{8} \otimes \underline{8}'), (\underline{8}' \otimes \underline{8}), (\underline{8}' \otimes \underline{8}'). \tag{4.24}$$

There are therefore five ways to build deuterium singlet states out of the colour multiplets in eq. (4.23). The first one, $(\underline{1} \otimes \underline{1})$, corresponds to the usual picture (in ordinary nuclear physics) of the deuterium as a two-nucleon state, whereas the other four (represented by products of two octets) are peculiar of the quark model and are called *hidden colour states.*

TABLE X – Colour octet states for 3-quark systems.

state	quantum numbers $\nu = (Y, T, T_3)$	vectors $\vert\underline{8}^\rho, \nu\rangle$	vectors $\vert\underline{8}^\lambda, \nu\rangle$
α^+	$1, 1/2, +1/2$	$\frac{1}{\sqrt{2}}(vg - gv)v$	$\frac{1}{\sqrt{6}}[(vg + gv)v - 2vvg]$
α^0	$1, 1/2, -1/2$	$\frac{1}{\sqrt{2}}(vg - gv)g$	$\frac{-1}{\sqrt{6}}[(vg + gv)g - 2ggv]$
β^+	$0, 1, +1$	$\frac{1}{\sqrt{2}}(vb - bv)v$	$\frac{1}{\sqrt{6}}[(vb + bv)v - 2vvb]$
β^0	$0, 1, 0$	$\frac{1}{2}[(vb - bv)g+$ $(gb - bg)v]$	$\frac{1}{2\sqrt{3}}[(vb + bv)g+$ $(gb + bg)v - 2(vg + gv)b]$
β^-	$0, 1, -1$	$\frac{1}{\sqrt{2}}(gb - bg)g$	$\frac{1}{\sqrt{6}}[(gb + bg)g - 2ggb]$
γ^0	$-1, 1/2, +1/2$	$\frac{1}{\sqrt{2}}(bv - vb)b$	$\frac{1}{\sqrt{6}}[(bv + vb)b - 2bbv]$
γ^-	$-1, 1/2, -1/2$	$\frac{1}{\sqrt{2}}(bg - gb)b$	$\frac{1}{\sqrt{6}}[(bg + gb)b - 2bbg]$
δ^0	$0,0,0$	$\frac{1}{2\sqrt{3}}[(bv - vb)g-$ $(bg - gb)v - 2(vg - gv)b]$	$\frac{1}{2}[(bv + vb)g-$ $-(bg + gb)v]$

The singlet $\nu = (Y, T, T_3) = (0,0,0)$ is: $\vert\underline{1}\rangle = \frac{1}{\sqrt{6}}[v(bg - gb) - g(bv - vb) - b(vg - gv)]$

We shall denote the deuterium singlet corresponding to the product $\underline{n}_1 \otimes \underline{n}_2$ by the symbol $\vert\underline{n}_1 \otimes \underline{n}_2, \underline{1}; \nu_1, \nu_2, \nu\rangle$ where $\nu_k = (Y_k^c, t_k^c, t_{k3}^c)$ $(k = 1,2)$ is the set of $SU(3)_c$ quantum numbers needed to specify the single cluster states, and $\nu = (Y, t, t_3) = (0,0,0)$ for a deuterium singlet state. Then, the deuterium singlet wavefunctions can be obtained from the colour states of the individual clusters (listed in table X), by means of a $SU(3)_c$ *Clebsch-Gordan series* ([57]) as follows:

$$\vert\underline{1} \otimes \underline{1}; \underline{1}\rangle = \vert\underline{1}\rangle_I \vert\underline{1}\rangle_{II}; \tag{4.25a}$$

$$
\begin{aligned}
|\underline{8} \otimes \underline{8}; \underline{1}\rangle &= \sum_{\nu_1,\nu_2} C^{\nu_1,\nu_2}_{\underline{8}\otimes\underline{8};\underline{1}} |\underline{8}, \nu_1\rangle_I |\underline{8}, \nu_2\rangle_{II} = \\
&= \frac{1}{2\sqrt{2}} \{ \quad |\alpha^+\rangle_I |\gamma^-\rangle_{II} + |\gamma^-\rangle_I |\alpha^+\rangle_{II} - |\alpha^0\rangle_I |\gamma^0\rangle_{II} - |\gamma^0\rangle_I |\alpha^0\rangle_{II} + \\
&\qquad |\beta^+\rangle_I |\beta^-\rangle_{II} - |\beta_0\rangle_I |\beta_0\rangle_{II} + |\beta^-\rangle_I |\beta^+\rangle_{II} - |\delta^0\rangle_I |\delta^0\rangle_{II} \}
\end{aligned}
\tag{4.25b}
$$

(see table X for the above symbols). All of these states belong to the representation [2,2,2] of the permutation group S_6.

As to the spin-isospin wavefunctions $\psi^{S,T}_{\beta}$, they are classified according to the irreducible representations of the spin-isospin group $SU(4)_{S,T}$, or, alternatively, of the symmetric group S_6. To get them, one can, for instance, take all possible products of the spin-isospin states for the two clusters: $(\psi_{\alpha I} \otimes \psi_{\alpha II})^{S,T}$. In the case of interest to us, (S,T) = (1,0) (and analogously for (S,T) = (0,1)), we obtain 45 states, of which the following

$$
[\Phi^s(I) \otimes \Phi^s(II)]^{S=1,T=0}, \qquad (S_1T_1) = (S_2T_2) = (\frac{1}{2}, \frac{1}{2}), \tag{4.26a}
$$

$$
[\Phi^{s\prime}(I) \otimes \Phi^{s\prime}(II)]^{S=1,T=0}, \qquad (S_1T_1) = (S_2T_2) = (\frac{3}{2}, \frac{3}{2}). \tag{4.26b}
$$

have a special meaning. Indeed, these states represent a two-nucleon (NN) and a two-delta system ($\Delta\Delta$), respectively. Let us notice that, from the symmetric group standpoint, both of them are associated to the Young tableaux [3,3], whereas the remaining wavefunctions are linear combinations of mixed-symmetry states. †.

Lastly, let us consider the orbital wavefunctions ψ^L_γ. Possible sets of harmonic oscillator functions with $L = 0, 2$ are given in tables XI and XII. To specify the symmetry properties, we have adopted symbols of the type $R^{s,s}$, $R^{s,\lambda}$, ..., in which the two indices give, respectively, the behavior of the states under separate permutations of the quarks in clusters I and II. For instance, for $L = 0$ one gets 16 states with harmonic-oscillator quantum number $N \leq 2$, among which 12 have $L_r = 0$ and 4 $L_r = 1$ (due to parity conservation, the last ones do not play any role in standard nuclear physics). The fully symmetric states are associated to the Young tableaux [6], whereas the remaining ones are linear combinations of states of [5,1],[4,2],[3,3].

---

† All these states vectors belong to the products of the completely symmetric S_3 representations $[3]_k$ ($k = 1, 2$) of the two clusters, i.e. $[3]_1 \otimes [3]_2 = [6] \oplus [5,1] \oplus [4,2] \oplus [3,3]$.

TABLE XI – Harmonic oscillator orbital functions with L=0 and $\pi = +1$ ($N = \sum_{i=1}^{2}[N_{\lambda_i} + N_{\rho_i}] + N_r \leq 2$) for deuteron.

name	structure	simmetry for $I \leftrightarrow II$
R_1^s	$K \cdot Exp$	S
R_2^s	$(\frac{K}{\sqrt{30}})(15 - 2\alpha_h^2 X^2) \cdot Exp$	S
$R_1^{s,s}$	$\alpha_h^2(\sqrt{3}K/\sqrt{32})(X_1^2 - X_2^2) \cdot Exp$	A
$R_2^{s,s}$	$\alpha_h^2(\sqrt{3}K/4\sqrt{10})(X_1^2 + X_2^2 - 4r^2) \cdot Exp$	S
$R_1^{s,\lambda}$	$\alpha_h^2(\sqrt{3}K/4)(\lambda_2^2 - \rho_2^2) \cdot Exp$	$\mapsto R_1^{\lambda,s}$
$R_2^{s,\lambda}$	$\alpha_h^2(\sqrt{3}K/2)(\underline{r} \cdot \underline{\lambda}_2) \cdot Exp$	$\mapsto -R_2^{\lambda,s}$
$R_1^{\lambda,s}$	$\alpha_h^2(\sqrt{3}K/4)(\lambda_1^2 - \rho_1^2) \cdot Exp$	$\mapsto R_1^{s,\lambda}$
$R_2^{\lambda,s}$	$\alpha_h^2(\sqrt{3}K/2)(\underline{r} \cdot \underline{\lambda}_1) \cdot Exp$	$\mapsto -R_2^{s,\lambda}$
$R_1^{s,\rho}$	$\alpha_h^2(\sqrt{3}K/2)(\underline{\rho}_2 \cdot \underline{\lambda}_2) \cdot Exp$	$\mapsto R_1^{\rho,s}$
$R_2^{s,\rho}$	$\alpha_h^2(\sqrt{3}K/2)(\underline{\rho}_2 \cdot \underline{r}) \cdot Exp$	$\mapsto -R_2^{\rho,s}$
$R_1^{\rho,s}$	$\alpha_h^2(\sqrt{3}K/2)(\underline{\rho}_1 \cdot \underline{\lambda}_1) \cdot Exp$	$\mapsto R_1^{s,\rho}$
$R_2^{\rho,s}$	$\alpha_h^2(\sqrt{3}K/2)(\underline{\rho}_1 \cdot \underline{r}) \cdot Exp$	$\mapsto -R_2^{s,\rho}$
$R^{\rho,\rho}$	$\alpha_h^2(\sqrt{3}K/2)(\underline{\rho}_1 \cdot \underline{\rho}_2) \cdot Exp$	S
$R^{\lambda,\lambda}$	$\alpha_h^2(\sqrt{3}K/2)(\underline{\lambda}_1 \cdot \underline{\lambda}_2) \cdot Exp$	S
$R^{\rho,\lambda}$	$\alpha_h^2(\sqrt{3}K/2)(\underline{\rho}_1 \cdot \underline{\lambda}_2) \cdot Exp$	$\mapsto R^{\lambda,\rho}$
$R^{\lambda,\rho}$	$\alpha_h^2(\sqrt{3}K/2)(\underline{\rho}_2 \cdot \underline{\lambda}_1) \cdot Exp$	$\mapsto R^{\rho,\lambda}$

Adopted symbols: $K = (\sqrt{3}\alpha_h/2\sqrt{\pi})^{15/2}$; $X_i^2 = \rho_i^2 + \lambda_i^2$;
$X^2 = X_1^2 + X_2^2 + r^2$; $Exp = e^{-\frac{1}{2}\alpha_h^2 X^2}$

To build up antisymmetric states of the kind (4.22) from the various pieces it is often convenient to use general expressions for the Kronecker products of S_6 representations. Because the colour wavefunctions have permutation symmetry $[2,2,2] = \widetilde{[3,3]}$ (where the tilde denotes, as usual, the conjugate representation), the spin-isospin-orbital wavefunction must have the symmetry $[3,3]$. In turn, this last piece of the deuterium wavefunction with the right symmetry is built up by suitable combinations of the $SU(4)_{S,T}$ states and the orbital states belonging to the various S_6 multiplets.

Incidentally, let us notice that some authors ([58]), by neglecting the orbital components of low symmetry (i.e. the $[3,3]$), have considered only three orthogonal states, namely $\psi([6]\{3,3\}), \psi([4,2]\{3,3\})$ and $\psi([4,2]\{5,1\})$ (where each state is an antisymmetric, normalized colour singlet, with orbital symmetry $[\lambda]$ and spin-isospin symmetry $\{\lambda'\}$). Then, it is possible to write the following orthogonal superpositions

$$\psi_{NN} = \frac{1}{3}[\psi([6]\{3,3\}) + 2\psi([4,2]\{3,3\}) - 2\psi([4,2]\{5,1\})] \qquad (4.27)$$

$$\psi_{\Delta,\Delta} = \sqrt{\frac{1}{45}}[2\psi([6]\{3,3\}) + 4\psi([4,2]\{3,3\}) + 5\psi([4,2]\{5,1\})]; \tag{4.28}$$

$$\psi_{CC} = \sqrt{\frac{1}{5}}[2\psi([6]\{3,3\}) - \psi([4,2]\{3,3\}). \tag{4.29}$$

These combinations can be considered as representatives of the channels NN, $\Delta\Delta$ and CC (hidden colour channel). Eq.s (4.27) – (4.29) will be useful later on to analize the results of various authors.

TABLE XII – Harmonic oscillator orbital functions with $L = 2$ and $\pi = +1$ for deuteron ($N \leq 2$).

name	*structure*
D^s	$(\sqrt{2/5})X_2^m \cdot Exp$
$D_1^{s,s}$	$\alpha_h^2(3K/4\sqrt{2})[(X_1)_2^m - (X_2)_2^m] \cdot Exp$
$D_2^{s,s}$	$\alpha_h^2(3K/4\sqrt{2})[(X_1)_2^m + (X_2)_2^m - 4(r \otimes r)_2^m] \cdot Exp$
$D_1^{s,\lambda}$	$\alpha_h^2(3K/4)[(\lambda_2 \otimes \lambda_2)_2^m - (\rho_2 \otimes \rho_2)_2^m] \cdot Exp$
$D_2^{s,\lambda}$	$\alpha_h^2(3K/2)(r \otimes \lambda_2)_2^m \cdot Exp$
$D_1^{\lambda,s}$	$\alpha_h^2(3K/4)[(\lambda_1 \otimes \lambda_1)_2^m - (\rho_1 \otimes \rho_1)_2^m] \cdot Exp$
$D_2^{\lambda,s}$	$\alpha_h^2(3K/2)(r \otimes \lambda_1)_2^m \cdot Exp$
$D_1^{s,\rho}$	$\alpha_h^2(3K/2)(\rho_2 \otimes \lambda_2)_2^m \cdot Exp$
$D_2^{s,\rho}$	$\alpha_h^2(3K/2)(\rho_2 \otimes r)_2^m \cdot Exp$
$D_1^{\rho,s}$	$\alpha_h^2(3K/2)(\rho_1 \otimes \lambda_1)_2^m \cdot Exp$
$D_2^{\rho,s}$	$\alpha_h^2(3K/2)(\rho_1 \otimes r)_2^m \cdot Exp$
$D^{\rho,\rho}$	$\alpha_h^2(3K/2)(\rho_1 \otimes \rho_2)_2^m \cdot Exp$
$D^{\lambda,\lambda}$	$\alpha_h^2(3K/2)(\lambda_1 \otimes \lambda_2)_2^m \cdot Exp$
$D^{\rho,\lambda}$	$\alpha_h^2(3K/2)(\rho_1 \otimes \lambda_2)_2^m \cdot Exp$
$D^{\lambda,\rho}$	$\alpha_h^2(3K/2)(\rho_2 \otimes \lambda_1)_2^m \cdot Exp$

Adopted symbols: $(X_i)_2^m = (\lambda_i \otimes \lambda_i)_2^m + (\rho \otimes \rho)_2^m$;
$X_2^m = \sum_{i=1}^2 (X_i)_2^m + (r \otimes r)_2^m = \sqrt{8\pi/15}\{\sum_{i=1}^2[\lambda_i^2 Y_2^m(\lambda_i) + \rho_i^2 Y_2^m(\rho_i)] + r^2 Y_2^m(r)\}$

A more direct way to get the antisymmetric wavefunctions (4.22) is to first build up all possible products of colour-spin-isospin and orbital states, that are then antisymmetrized by exploiting the explicit form of the operator $\widehat{\mathcal{A}}$; finally, the resulting set is orthonormalized by standard methods (Gram-Schmidt procedure). In particular, if one starts from states antisymmetric inside any cluster and antisymmetric under cluster exchange ($I \leftrightarrow II$), $\widehat{\mathcal{A}}$ can be put in the simple form:

$$\widehat{\mathcal{A}} = 1 - \widehat{\mathcal{A}}' = 1 - \sum_{i=1}^{3}\sum_{j=4}^{6} \widehat{\mathcal{P}}_{ij} \tag{4.30}$$

where $\hat{P}_{ij}$ is the usual permutation operator of the quarks i, j and, by definition, $\hat{\mathcal{A}}'$ is the antisymmetrizing operator containing only the contributions from all possible non trivial permutations of indices (i.e. the identical permutation is ruled out). Similar procedures can be obviously employed for other channels with $(S,T)J^P \neq (1,0)1^+$.

Let us notice that other methods can be used to build up the basis states for deuterium. However, the results do not appreciably depend on the particular method employed. For instance, Maltman and Isgur ([59]) have used a basis in which the quarks u and d are considered as *different* particles (due to their different masses), and therefore the Pauli principle applies separately to the two quark sets (uuu) and (ddd) (remember that, in standard nuclear physics, nuclei can be studied both with and without the isospin formalism).

Finally, as in the hadron case, one has to solve a Schrödinger equation for deuterium with an Hamiltonian of the type (4.16). Again, this can be done by variational tecniques (with wavefunctions containing non-linear free parameters), or by directly diagonalizing the Hamiltonian. In the following we shall discuss the main results obtained by various authors, without entering into details of calculations.

The first point worth considering concerns the quark-quark potential V_{qq} (see eqs. (4.11), (4.7)). In the first works on deuterium within the constituent quark model ([60]). many authors drew the conclusion that the quark-quark interaction is unable to give a bound deuterium, since it gives rise to a nucleon-nucleon interaction $V_{NN}(r)$ not enough attractive at large and intermediate distances ($r \gg 1\,fm$). However, such a result has been questioned in more recent papers. For instance, Williams et al. ([61]), by using the Breit-Fermi-like interaction (4.7), with corrective terms to the kinetic energy of the order $(v/c)^2$, have achieved a fair reproduction of deuterium parameters (see Table XIII). The only evident discrepancy is the value of the quadrupole moment Q (that, incidentally, is connected to the D-wave component of deuterium, for which the model provides the too low value $\approx 4\%$).

TABLE XIII – Deuteron properties calculated by S. A. Williams et al.([61]).

quantity	exp. value	prediction
B.E. (MeV)	2.226	2.1 – 2.28
Q (mb)	2.82	1.3 ± 0.5
μ (μ_N)	0.857	0.90 ± 0.01
r.m.s. (fm)	2.1	1.08 ± 0.2

Further clarifications of this problem can be found in the already quoted papers by Maltman and Isgur([59]). They, too, assume a quark-quark potential of the type (4.7), but neglect the momentum-dependent terms (by analogy to the hadron case in ref. ([55]), see subsect. 4.2.2), and replace the Coloumb term $\propto r_{ij}^{-1}$ by a contact interaction $\propto \delta(\underline{r}_{ij})$. Their procedure can be basically divided in three steps:

i) they solve the ordinary Schrödinger equation for deuterium by a standard variational approach, by assuming a test function of the type:

$$\psi(\lambda_i, \underline{\rho}_i; \underline{r}) = \sum_{\alpha,\beta} C_{\alpha,\beta} \hat{\mathcal{A}}\{\psi_{1(\alpha)}\psi_{2(\beta)}\}\psi(r), \tag{4.31}$$

where the $\psi_{i(\alpha)}(\underline{\lambda}_i, \underline{\rho}_i)$ $(i = 1, 2)$ are the wavefunctions of the two free nucleons and explicitly read

$$\psi_{i(\alpha)} = k \exp(-\alpha_h^2[\lambda^2 + \rho^2])\phi_\alpha. \tag{4.32}$$

Here, the ϕ_α's are suitable colour-spin-isospin states (see the previous sections), that can be taken as $N\frac{1}{2}N\frac{1}{2}$, $\Delta\frac{3}{2}\Delta\frac{3}{2}$, $N_c\frac{3}{2}N_c\frac{3}{2}$, $N_c\frac{1}{2}N_c\frac{1}{2}$, $N_c\frac{3}{2}N_c\frac{1}{2}$, $\Delta_c\frac{1}{2}\Delta_c\frac{1}{2}$ (the notation gives the isospin and spin of the clusters as T_1, S_1, T_2, S_2, with the symbols N and Δ for $T = 1/2, 3/2$; the subscript c denotes a cluster in a colour octet state). The intercluster wavefunction $\psi(r)$ is taken to be of the form

$$\psi(r) = \sum_{i=1}^{i_{max}} \xi_i \exp(-\beta_i^2 r^2) \tag{4.33}$$

where ξ_i, β_i are variational parameters (while the parameter α_h in eq. (4.32) is kept fixed at the value suggested by the study of free nucleons).

ii) Once the parameters β_i, ξ_i are determined by the variational method, the basis space of the system is expanded by adding the most important excited states with a number of oscillator quanta $N \leq 2$. Thus, Maltman and Isgur introduce perturbatively other four states with $L = 1$ clusters and hidden colour, and nine states with $L = 2$ clusters.

Again, they find the new function $\psi(r)$ by the variational method, and calculate an *effective local potential* (ELP) for the *np* systems in its fundamental state (bound state $^3S_1$) by the relation

$$V^{ELP}(r) = E + \frac{\nabla^2\psi(r)}{M_N\psi(r)} \quad (\hbar = 1) \tag{4.34}$$

The potential $V^{ELP}(r)$ (whose behaviour is shown in fig.17) is characterized by a large repulsive core and appreciable ($\sim$ 25 MeV) intermediate-range attraction; moreover, it is quite similar to the potential commonly used in low-energy nuclear physics([62]).

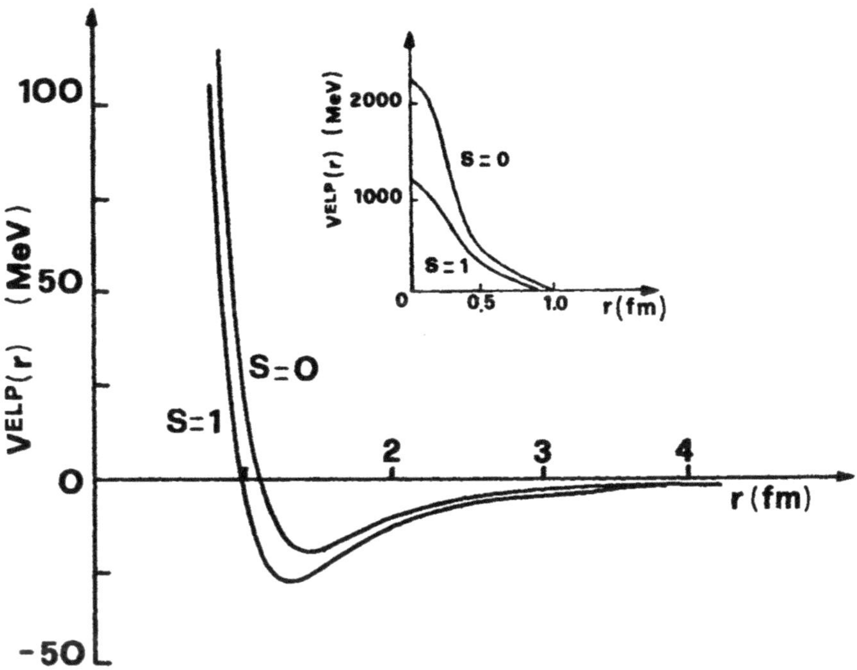

Fig. 17 – The effective NN-potential deduced in ref. (54) from quark-quark interactions for the $^3S_1$ and $^1S_0$ states. It can be compared with the *Equivalent Local Potentials* (V_{NN}^{ELP}) defined in Sec. 5.4.

As stressed by the authors, the attraction observed for $r \geq 1$ fm is critically dependent on the presence, in their calculation, of a rich spectrum of excited orbital components, whose omission — either partial or total — would weaken sensibly the attractive effects. This last remark could probably explain the conflicting results obtained in previous works.

iii) In the third phase of their calculation, the authors drastically modify the asymptotic behavior ($r \to \infty$) of the internucleon wavefunction $\psi(r)$ (4.33), in order to account (at least in a non-rigorous way) for the quark-antiquark pairs of the Fermi sea. The effect of the $q\bar{q}$-pairs is twofold: 1) they *screen* the confining potential $V^{conf}(r)$ (for increasing r), by preventing it from increasing indefinitely when $r \to \infty$. This effect could cause the elimination of the long-range component (Van der Waals force) that appears in the potential of Fig.16 and that, on the contrary, is not experimentally observed; 2) they are responsible for the meson-exchange processes that are the basic ingredient of the standard theories.

More precisely, the authors take into account the well-established experimental fact that the tail of the potential V_{NN}, for $r \geq 2$ fm, is essentially determined by one-pion exchange (OPE) processes ($\lim_{r\to\infty} V_{NN}(\underline{r}) = V^{OPE}(\underline{r})$). Therefore they *suppress* $V^{ELP}(r)$ for $r > r_0 \approx 2$ fm (so to preserve approximately its continuity) and replace it by the central part of the OPE potential. Then they calculate, by this new potential, the deuteron properties. Their results are given in Table XIV.

TABLE XIV – Deuteron properties calculated by Maltman-Isgur([59]).

quantity	exp. value	prediction
B.E. (MeV)	2.226	2.6 – 3.7
Q (mb)	2.86	2.1 ± 0.5
μ (μ_N)	0.857	0.859
r.m.s. (fm)	1.95	2.2 ± 0.5

Thus, the procedure by Maltman and Isgur represents an exemplary attempt of *hybridization* between quark and standard nuclear models. It provides encouraging results, at least in the deuterium case. However, it must be stressed that such a work is only preliminary. Many things are yet to be clarified. Among the others, it is still unclear what is the role played by hidden-colour components (CC). There is a recent claim ([63]) that they represent about the 66% of deuterium wavefunction. If so, hidden colour components would be more important than the well-known isobaric components ($\Delta\Delta$). This would probably imply a drastic revision of the conclusions drawn in a number of areas (like form factors, reaction amplitudes at intermediate energy, and so on).

4.4 – Non-relativistic quark models and nucleon-nucleon interaction.

In the last years, non-relativistic quarks models have been widely used to investigate the nucleon-nucleon (NN) interaction, by analyzing in detail the structure of the potential $V_{NN}(\underline{r})$, and directly deducing the phase shifts of NN scattering in the various *partial waves* ${}^{(2S+1)}L_J$. In particular, as we shall better see in the following, many authors have been concerned with a classical problem of nuclear physics, i.e. the determination of the V_{NN} behaviour at short internucleon distances ($r \leq 1$ fm). Indeed, it is a specially interesting task to ascertain whether or not quark models are able to reproduce those features of experimental phase shifts, usually interpreted as the evidence for a small-radius ($r_c \sim 0.5$ fm) *repulsive core* in $V_{NN}(r)$. These effects have been clearly observed in all partial waves ${}^{(2S+1)}L_J$ (except for the singlet ones).

To clarify the problem, let us answer some fondamental questions:

I) *How and why does the quark structure of the nucleons influence their reciprocal interaction?.*

It is well known that the (mean squared) charge radius of the proton is about 0.8 fm, less than (but not negligible with respect to) the nuclear radius ($\mathcal{R} \simeq 3 - 8$ fm). Moreover, the usual description of nuclear interactions by meson (π, ρ, ω, ...) exchange does not appreciably change the basic properties of nucleons. Therefore, it is likely that at least some occasional overlapping of the nucleons to occur, with unavoidable influence on the NN interaction at short internucleon distances. In other words, one expects that the short-range behaviour of $V_{NN}(\underline{r})$ is determined by the wavefunction structure of the (NN) system, described as two overlapping three-quark clusters.

II) *Is it possible, in the framework of the non-relativistic quark model, to explain the repulsive effects observed in the NN interaction at short distances?*

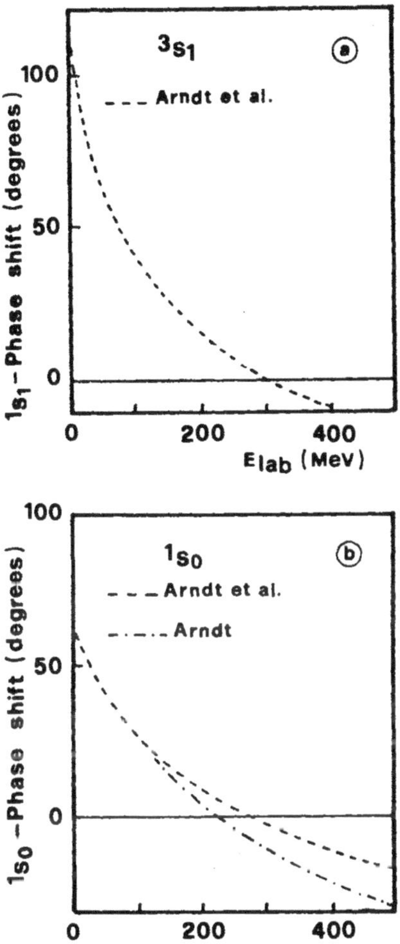

Fig. 18 – Experimental phase shifts for NN-scattering in the $^3S_1$(a) and $^1S_0$(b) waves, as a function of the laboratory energy E_{Lab}.

Such a question requires the preliminary knowledge of some notions we are going briefly to mention. One observes in NN scattering processes that some phase shifts δ_L^{SJ} for the partial waves $^{(2S+1)}L_J$, as functions of the energy E, exhibit a behaviour that cannot be interpreted by assuming a potential V_{NN} attractive for any distance r. For the wave $^3S_1$,e.g., the phase shift δ_0^{11} exhibits the behaviour shown in Fig.18. At low energies, δ_0^{11} is positive; for increasing E, it increases firstly up to a maximum, and then decreases until it becomes negative. Such a behaviour admits different explanations:

a) The potential $V_{NN}(r)$ is local, with an *hard core* (see Fig.19a);

b) $V_{NN}(r)$ is local, with a *soft core* (Fig.19b);

c) V_{NN} is a non-local potential (or a local potential singular at $r = r_c \neq 0$, as shown in Fig.19c).

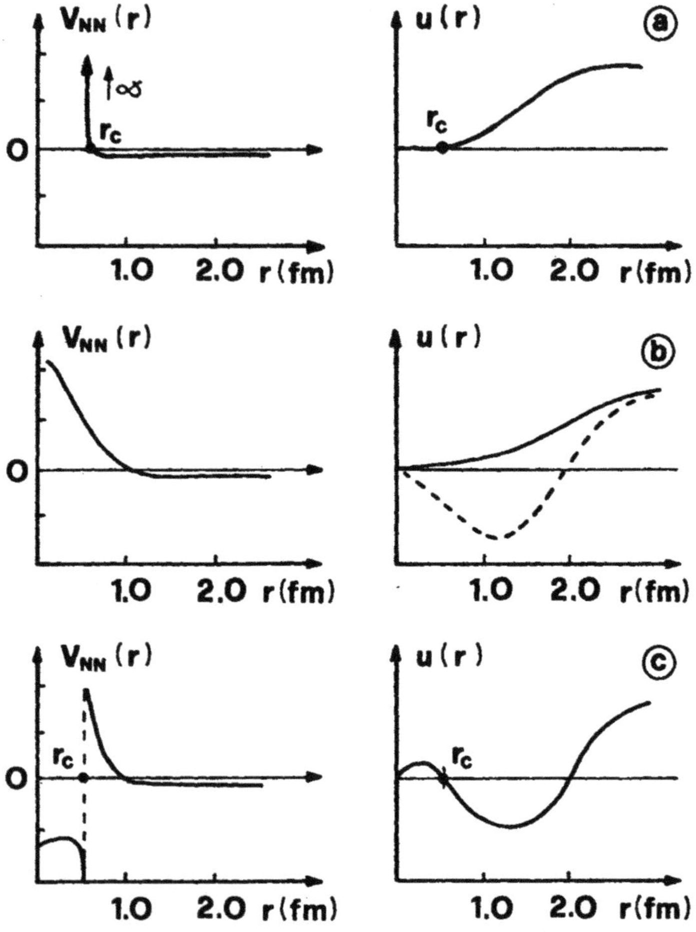

Fig. 19 – A schematic representation of the short-range behaviour of the NN-potential, $V_{NN}(r)$, and the NN-radial wave function $u(r)$ for $L = 0$. Case a): hard core; case b): soft core; case c): singular potential. In case b), the dashed and full lines can both represent the physical situation for $u(r)$.

In case b) (the most commonly used), the observed behaviour of δ_0^{11} can be explained by noting that, for attractive (repulsive) local potentials, it is $\delta_0^{11} > 0$ (< 0). At low energies, the usual S-wave state of the NN system, $u(r)$, is appreciably different from zero only at comparatively large distances, at which V_{NN} is certainly attractive. This gives positive values of δ_0^{11} for $K \ll r_c^{-1}$ ($\frac{K=(M_n c^2 E)^{\frac{1}{2}}}{\hbar c}$). On the other side, for increasing K and E, $u(r)$ is more and more squeezed towards the origin, and, therefore, $V_{NN}(r)$ becomes more and more sensitive to the values of $u(r)$ at short distances. Case a) can be considered as a limiting case of b). On the contrary, case c) is quite different: if the potential $V_{NN}(r)$ has a singularity at, say, $r = r_c, u(r)$ acquires a node at the same point. It is easy to see that, in this case, the phase shift δ_0 can have a complex behaviour and even change sign for $K \sim r_c^{-1}$. Otherwhise speaking, the phase shifts are similar to those of scattering by a hard sphere of radius r_c (i.e. $\delta \sim -r_c K$). A situation of this kind has been already observed in $\alpha - \alpha$ scattering ([64]), in which the interaction can be described in terms of a non-local potential $V_{\alpha\alpha}(r)$ (depending on L) and a wavefunction $u_{\alpha\alpha}(r)$ with a (practically) energy-independent behaviour in a neighbourhood of the node.

As already mentioned, this problem has been widely discussed in the literature. There have been a number of interesting attempts to answer our question, by using various quark models and different mathematical methods. It has been soon realized that the hard core, if any, may derive from three effects: i) the centrifugal potential barrier among quarks; ii) the Pauli principle, that forbids the spatial overlapping of quarks. By this mechanism, composite particles repel each other, by giving rise in some way to a *repulsive core* in a potential-like description; iii) specific effects due to the structure of the quark-quark potential V_{qq}. These effects (together with the Pauli principle) could be determinant in generating a V_{NN} behaviour of the type c).

Although many uncertainties still persist, it seems possible to conclude that the observed repulsive core can be indeed accounted for by quark models, probably by both the quoted effects i) and ii).

III) *Is it possible, by means of quark models, to correctly reproduce the experimentally observed behaviour of phase shifts with energy ?*. As it is well known, such a behaviour is well reproduced by standard nuclear theories. In particular, for waves with high L (≥ 3), the phase shifts are essentially affected by one-pion-exchange processes, i.e. by the OPE potential. However, the current meson theories are unable to reproduce the phases with small $L(0,2)$ without the explicit introduction of a merely phenomenological description of the V_{NN} behaviour at short distances. On the contrary, quark models can, at least in principle, describe the short-distance V_{NN} behaviour, but, at the present status of the theory, are unable to account for the asymptotic behaviour. As already emphasized in the previous section, this difficulty can be obviated by suitably connecting quark and standard models (*hybrid models*). Hopefully, one shall preserve the main advantages of both models, and achieve a good reprodution of the observed phases δ_{LS}^J (and the corresponding coupling parameters $\epsilon_{LL'}$) for the various partial waves $^{(2S+1)}L_J$. We shall come back later on hybridization methods (an example is provided by the Maltman-Isgur approach to deuterium, already discussed in sect. 4.3). For the moment, let us only stress that, among the others, hybrid models allow to fully get rid of the Van der Waals tails, predicted for V_{NN} by standard quark models.

In the remainder of this section, we shall briefly mention some works whose main aim is a critical analysis of repulsive-effect generation by means of conventional mathematical techniques. In the next section, it will be shown that the *resonating group method* (RGM) not only provides a deeper approach to the repulsive-core problem, but allows also to do explicit calculations of the phase shifts.

The subject of the V_{NN} derivation within the quark models is an old one [65,66,67]. The modern works start with the researches by Liberman (by a potential model) [66] and De Tar (by a bag model) [66]) although it is today recognized the inadequacy of their approaches. All calculations in refs.[66,67] have been done by assuming an *adiabatic approximation*, i.e. that the speed of relative motion of the two nucleon clusters is small as compared to the speed of intrinsic quark motions inside the clusters. In other words, the wavefunction of the cluster internal motion is negligibly affected by their mutual interaction (a similar approximation is exploited in the usual rotational model of the nucleus). From a mathematical point of view, this amounts to consider, in the system wavefunction, only those spin-isospin configurations corresponding to the product of two nucleon states and to neglect all spatial configurations of low symmetry (e.g. [4,2]); thus, the system wavefunction takes a form of the type $\psi([6]\{3,3\})$ (see eq.(4.27) for the notation). In these papers, Hamiltonians of the kind (4.16) have been used which, however, are not fully realistic (being unable to reproduce the hadronic spectra with good accuracy). It turns out that the mean value of $(\hat{\mathcal{H}} - \hat{\mathcal{M}})$ as a function of the internucleon distance r can be roughly identified with the central part of the potential $V_{NN}(r)$ (Born-Oppenheimer procedure). Moreover, in the adiabatic approximation $V_{NN}(r)$ always contains a strong repulsive core.

However, the adiabatic approximation is to be considered a *too drastic* one. This can be easily seen by noting, for instance, that neglecting states with low spatial symmetry is indeed a non-realistic assumption: actually, antisymmetrizing the six-quark wavefunction requires that these states are coupled to additional spin-isospin symmetries (e.g. the orbital configuration [4,2] is coupled, in the channel $(S,T) = (1,0)$, to the spin-isospin configurations $\{3,3\}$ and $\{5,1\}$), that are mixed each other by the spin-spin interaction (colour magnetic interaction).

Harvey [58] has reconsidered the problem by means of a more realistic Hamiltonian of the kind (4.7) (in which one neglects the momentum-dependent terms †). The confining potential in eq.(4.11) is taken of the form $V_{ij}^{conf} = (\mathcal{A}e^{\gamma^2 r_{ij}^2} + \mathcal{B}r_{ij}^2 + \mathcal{C})$, while the relativistic expression of the kinetic energy is adopted ††. This author includes in calculations the three components ψ_{NN}, $\psi_{\Delta\Delta}$ and ψ_{CC} (given in eqs.(4.27),(4.28) and (4.29), respectively). Then, the central potential $V_{NN}(r)$ is calculated by following the same procedure of the previous papers on the subject. The inclusion of $(\Delta\Delta)$ and (CC) components, and of lower spatial symmetry states ([4,2]), has dramatic effects, as shown in Fig.20: the repulsive core vanishes, and is replaced by a weak attraction! It reappears ($\sim$ 320 MeV) at $r = 0$ provided that one neglects the orbital components [4,2], as already observed in previous works. Analogous results have been found in channel $(S,T) = (0,1)$, i.e. for $^1S_0$ states.

† Remember that the same Hamiltonian, with the same approximation, has been considered by Maltman and Isgur in the deuterium case (see ref.[59]).

†† Clearly, this puts the problem of the C.M. kinetic energy T_{CM}, that, in this approach, cannot be subtracted in a rigorous way.

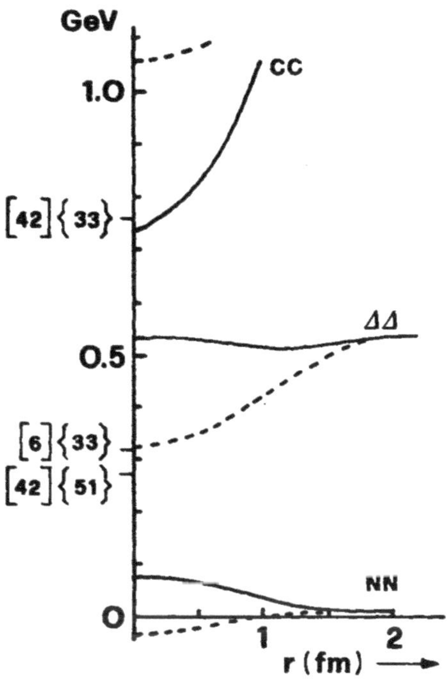

Fig. 20 – Results of the six-quark cluster calculations of ref.(58) for $(S,T) = (1,0)$ as functions of the distance, r, between clusters. The solid lines show the energies of the NN, $\Delta\Delta$ and CC configurations defined in eqs.(4.27-4.29). Dashed curves are the results of the channel mixing. Energies are given relative to the mass of non-interacting nucleons. Diagonal elements for the symmetry states at $r = 0$ are shown on the ordinate, where $[\lambda]$ and $\{\lambda'\}$ are the orbital and spin-isospin symmetries respectively.

Although Harvey's work is a not yet realistic approximation (moreover, the validity of its numerical calculations has been questioned, as we shall see later on), its main basic result is still valid: configuration mixings (that, among the others, allow to improve the adiabatic approximation) are crucial for the short-distance behaviour of $V_{NN}(r)$. Harvey's calculations do not take into account all possible states that can couple to the standard NN channel. Today, one cannot a priori exclude that any allowed configuration is important. This statement seems to be proved by the results of Maltman and Isgur ([59]) discussed in the previous section: according to these authors a further enlargement of basis states not only confirms the existence of the repulsive core, but is essential to generate an intermediate-range ($r \sim 2$ fm) attractive component in V_{NN}.

Before concluding this section, let us however quote a recent paper, apparently showing that the question is far from settled. In 1983, Storm and Watt ([63]) have found, by following Harvey's approach, that the actual existence of a repulsive core critically depends on the value of the harmonic-oscillator parameter α_h. By a mixing of states similar to the Harvey's one, but a different choice of α_h, they find a repulsive core (ranging from 175

MeV to about 800 MeV at $r = 0$, according to the assumed interaction). This shows the need to work with accurate solutions of the Schrödinger equation of the system, by determining in a fully consistent way also the non-linear parameters entering the spatial wavefunction. To our knowledge, this has not yet been done systematically in the literature ‡. Other improvements to the model may come by more consistently taking into account meson exchange processes, and by a thorough analysis of relativistic corrections to quark dynamics.

Finally, let us mention that procedures similar to those discussed above have been exploited to study other features of nuclear forces, like the structure of spin-orbit and tensor forces, in the $N - N$ potential ([68]). We shall not discuss these works for brevity's sake, and since they do not introduce any new physical concept.

‡ This point is specially delicate for those quark-quark interactions containing contact spin-spin terms, that could give rise to *collapsed* equilibrium situations.

5 – Calculation of phase shifts in NN scattering.

In this chapter, we will discuss the application of the constituent non-relativistic quark model to the calculation of phase shifts in NN scattering. After a short mathematical introduction, we will present the main results obtained up to now. Finally, we shall add further comments on the so called *hybrid models*.

5.1 – The Resonating Group Method.

The mathematical technique more commonly used for this kind of calculations is the *resonating group method (RGM)*, conceived by Wheeler at the end of Thirties ([69]) and then developed by many authors ([70,71]) and extensively applied to standard nuclear physics calculations ([72]). Such a method fully exploits the *cluster concept*, by allowing an unified treatment of bound states and of scattering and reaction processes.

The starting point of the RGM is a factorization of the physical system under study in two separate clusters I and II (see also Sect.4.2). The different cluster states α, β are represented by the antisymmetric (colourless or coloured) wavefunctions $\Phi_1^\alpha(\underline{\xi}_1)$ and $\Phi_2^\beta(\underline{\xi}_2)$, where $\underline{\xi}_1$ and $\underline{\xi}_2$ are the internal coordinates of the two clusters (for instance, the Jacobi coordinates). Then, these cluster states are coupled to antisymmetric colourless six-quark states, with definite quantum numbers (S,T), as follows:

$$\Psi_\chi^{ST,J}(\underline{\xi}_1,\underline{\xi}_2) = \sum_{\alpha\beta} C_\chi^{\alpha\beta}[\Phi_1^\alpha(\underline{\xi}_1) \otimes \Phi_2^\beta(\underline{\xi}_2)]^{ST}, \tag{5.1}$$

where the coefficiens $C_\chi^{\alpha\beta}$ provide the coupling in colour space needed to build up colour singlets. Further, let $F_\chi^L(\underline{r})$ be a set of functions describing the relative intercluster motion in a state of orbital momentum L (where $\underline{r}$ is still given by eq.(4.21)). Then, the C.M. wavefunction of the physical system in a state with quantum numbers (ST, J) can be put in the form

$$\Psi^{ST,J}(\underline{\xi}_1,\underline{\xi}_2;\underline{r}) = \hat{\mathcal{A}} \sum_{\chi=1}^{n} \sum_L [\Psi_\chi^{ST,J}(\underline{\xi}_1,\underline{\xi}_2) \otimes F_\chi^L(\underline{r})]^{JM} \tag{5.1a}$$

with obvious meaning of the symbols. The pair of clusters I and II in a state $\Psi_\chi^{ST,J}(\underline{\xi}_1,\underline{\xi}_2)$ is called a *resonating group*. In general, the theory can include an arbitrary number of resonating groups, but in its usual version one takes $n \leq 3$.

It is customary to assume that the three-quark functions $\Psi_\chi^{ST,J}$ are known *a priori* , so that one has only to determine the intercluster functions $F_\chi^L(\underline{r})$. This can be done by means of standard variational tecniques. In the following, we will denote by $\hat{\mathcal{H}}$ the usual internal motion hamiltonian, in its non-relativistic form given by eq.(4.16).

For simplicity's sake we will discuss the *single resonating group case* $(n = 1)$ thus getting rid of all superfluous indices. To further simplify the notation, we shall not take into account explicitly angular momenta and their coupling, and confine ourselves to *uncoupled waves*. Then, one has only one intercluster function $F(\underline{r})$, that is determined by the variational equation †

$$\delta \int d\underline{\xi}_1 d\underline{\xi}_2 d\underline{r}\{\Psi^+(\underline{\xi}_1,\underline{\xi}_2;\underline{r})(\hat{\mathcal{H}} - E)\Psi(\underline{\xi}_1,\underline{\xi}_2;\underline{r})\} = 0, \tag{5.2}$$

where δ denotes an arbitrary variation of the integral, generated by an arbitrary variation of Ψ (or, better, of $F(\underline{r})$), and E is the CM energy of the system.

Let us now point out that the antisymmetrizer $\hat{\mathcal{A}}$, in its most general form (including all possible permutations $\hat{P}_{ij}$ of the N particles) satisfies the conditions $\hat{\mathcal{A}} = \hat{\mathcal{A}}^+$ and $\hat{\mathcal{A}}^2 = (N!)\hat{\mathcal{A}}$. Moreover, the hamiltonian $\hat{\mathcal{H}}$ turns out to be fully symmetric. By these properties, eq.(5.2) can be put in the form:

$$\int d\underline{\xi}_1 d\underline{\xi}_2 d\underline{r}[\Psi(\underline{\xi}_1,\underline{\xi}_2)\delta F(\underline{r})]^+(\hat{\mathcal{H}} - E)\hat{\mathcal{A}}[\Psi(\underline{\xi}_1,\underline{\xi}_2)F(\underline{r})] = 0 \tag{5.3}$$

Due to the arbitrariness of the variation $\delta F(\underline{r})$, its coefficient in eq.(5.3) must identically vanish. Therefore, one gets the ***characteristic equation of RGM***

$$\int d\underline{\xi}_1 d\underline{\xi}_2 [\Psi(\underline{\xi}_1,\underline{\xi}_2)]^+(\hat{\mathcal{H}} - E)\hat{\mathcal{A}}[\Psi(\underline{\xi}_1,\underline{\xi}_2)F(\underline{r})] = 0 \tag{5.4}$$

that , in Dirac's notation, can be written as

$$\langle\Psi(\underline{\xi}_1,\underline{\xi}_2)|(\hat{\mathcal{H}} - E)\hat{\mathcal{A}}|\Psi(\underline{\xi}_1,\underline{\xi}_2)F\rangle_{\underline{r}} = 0 \tag{5.4a}$$

($\underline{r}$ fixed).

To put eq.(5.4a) in a more explicit form, it is convenient to set $\hat{\mathcal{A}} = (1 - \hat{\mathcal{A}}')$ (the operator $\hat{\mathcal{A}}'$ is defined by eq.(4.30)). Thus the integral (5.4a) becomes

$$\langle\Psi(\underline{\xi}_1,\underline{\xi}_2)|(\hat{\mathcal{H}} - E)|\Psi(\underline{\xi}_1,\underline{\xi}_2)F\rangle_{\underline{r}} - \langle\Psi(\underline{\xi}_1,\underline{\xi}_2)|(\hat{\mathcal{H}} - E)\hat{\mathcal{A}}'|\Psi(\underline{\xi}_1,\underline{\xi}_2)F\rangle_{\underline{r}} = I^D + I^{Ex} = 0, \tag{5.4b}$$

where I^D and I^{Ex} are called *direct* and *exchange* integrals, respectively. Moreover, it is better to put the hamiltonian $\hat{\mathcal{H}}$ in the form:

$$\hat{\mathcal{H}} = \{\sum_{i=1}^{N} \hat{T}_i - \hat{T}_{C.M.}\} + \sum_{i<j} \hat{V}_{ij} = (\hat{\mathcal{H}}_1 + \hat{\mathcal{H}}_2) + (\hat{T}_r + \hat{V}_{int}), \tag{5.5}$$

where:

$$\begin{cases} \hat{\mathcal{H}}_1 = \sum_{i\subset I} \hat{T}_i + \sum_{i<j(\subset I)} \hat{V}_{ij} & = \text{internal hamiltonian of cluster I} \\ \hat{\mathcal{H}}_2 = \sum_{i\subset II} \hat{T}_i + \sum_{i<j(\subset II)} \hat{V}_{ij} & = \text{internal hamiltonian of cluster II} \\ (\hat{T}_r + \hat{V}_{int}) = -(\frac{\hbar^2}{2\mu})\nabla_r^2 + \sum_{i(\subset I)} \sum_{j(\subset II)} \hat{V}_{ij} & = \text{intercluster coupling hamiltonian} \end{cases} \tag{5.6}$$

† Eq.(5.2) is nothing but a formulation of the well known ***Kohn-Hulthén variational principle*** for scattering theory([73]).

and μ is the reduced mass of the two clusters.

Let us now assume that, at least in first approximation, Ψ_1 and Ψ_2 can be considered as eigenvectors of $\widehat{\mathcal{H}}_1$ and $\widehat{\mathcal{H}}_2$, respectively, with eigenvalues E_1 and E_2. Then the direct integral $\mathcal{I}^D$ defined in eq.(5.4b) takes the form

$$\begin{aligned}\mathcal{I}^D &= \{\widehat{\mathcal{T}}_r + V^D(\underline{r}) - [E - E_1 - E_2]\}F(\underline{r}) = \\ &= \widehat{\mathcal{H}}^D - EF(\underline{r})\end{aligned} \tag{5.7}$$

where $V^D(\underline{r}) = \langle\Psi(\underline{\xi}_1, \underline{\xi}_2)|\widehat{V}_{int}|\Psi(\underline{\xi}_1, \underline{\xi}_2)\rangle_{\underline{r}}$.

To further specify the structure of the exchange integral $\mathcal{I}^{Ex}$, let us remind that $\widehat{\mathcal{A}}$ is a linear combination of all possible (non-trivial) permutation operators $\widehat{\mathcal{P}}^\alpha$. Since, in general, some permutations give the same contribution to the integral, it is convenient to take into account only those giving different contributions, by putting $\widehat{\mathcal{A}} = \sum_\alpha(-1)^{p^\alpha}N_\alpha\widehat{\mathcal{P}}^\alpha$, where $(-1)^{p^\alpha}$ is the *parity* of permutation $\widehat{\mathcal{P}}^\alpha$ and N_α is its *multiplicity*. Thus, $\mathcal{I}^{Ex}$ can be written:

$$\mathcal{I}^{Ex} = \sum_\alpha(-1)^{p^\alpha}N_\alpha\langle\Psi(\underline{\xi}_1, \underline{\xi}_2)|(\widehat{\mathcal{H}} - E)\widehat{\mathcal{P}}^\alpha|\Psi(\underline{\xi}_1, \underline{\xi}_2)F(\underline{r})\rangle_{\underline{r}}. \tag{5.8}$$

Let us do the following coordinate trasformation

$$(\underline{\xi}_1, \underline{\xi}_2) \to (\underline{u}_\alpha, \underline{r}_\alpha), \qquad \underline{r}_\alpha = \widehat{\mathcal{P}}^\alpha\underline{r} \tag{5.9}$$

under which the volume element transforms as $d\underline{\xi}_1 d\underline{\xi}_2 = |\widehat{\mathcal{J}}^\alpha| d\underline{u}_\alpha \underline{r}_\alpha$, where $\widehat{\mathcal{J}}^\alpha$ is the usual Jacobian matrix. Consequently, $\mathcal{I}^{Ex}$ takes the form

$$\mathcal{I}^{Ex} = \sum_\alpha(-1)^{p_\alpha}N_\alpha\int_{(\underline{r})}[\widehat{\mathcal{H}}(\underline{r}, \underline{r}_\alpha) - E\widehat{\mathcal{N}}(\underline{r}, \underline{r}_\alpha)]F(\underline{r}_\alpha)\,d\underline{r}_\alpha \tag{5.9a}$$

where

$$\begin{aligned}\widehat{\mathcal{H}}_\alpha(\underline{r}, \underline{r}_\alpha) &= \int_{(\underline{r}, \underline{r}_\alpha)}[\psi(\underline{\xi}_1, \underline{\xi}_2)^+]\widehat{\mathcal{H}}\widehat{\mathcal{P}}_\alpha[\psi(\underline{\xi}_1, \underline{\xi}_2)]|\widehat{\mathcal{J}}_\alpha|\,d\underline{u}_\alpha\,, \\ \widehat{\mathcal{N}}_\alpha(\underline{r}, \underline{r}_\alpha) &= \int_{(\underline{r}, \underline{r}_\alpha)}[\psi(\underline{\xi}_1, \underline{\xi}_2)^+]\widehat{\mathcal{P}}_\alpha[\psi(\underline{\xi}_1, \underline{\xi}_2)]|\widehat{\mathcal{J}}_\alpha|\,d\underline{u}_\alpha\,.\end{aligned} \tag{5.10}$$

Since $\underline{r}_\alpha$ is only a dummy integration variable, we can put $\underline{r}_\alpha = \underline{r}'$, thus giving to $\mathcal{I}^{Ex}$ the final form

$$\mathcal{I}^{Ex} = \int[\widehat{\mathcal{H}}^{Ex}(\underline{r}, \underline{r}') - E\widehat{\mathcal{N}}^{Ex}(\underline{r}, \underline{r}')]F(\underline{r}')\,d\underline{r}' \tag{5.11}$$

with:

$$\begin{aligned}\widehat{\mathcal{H}}^{Ex}(\underline{r}, \underline{r}') &= [\sum_\alpha(-1)^{p_\alpha}N_\alpha\widehat{\mathcal{H}}_\alpha(\underline{r}, \underline{r}')]_{\underline{r}_\alpha = \underline{r}'} \\ \widehat{\mathcal{N}}^{Ex}(\underline{r}, \underline{r}') &= [\sum_\alpha(-1)^{p_\alpha}N_\alpha\widehat{\mathcal{N}}_\alpha(\underline{r}, \underline{r}')]_{\underline{r}_\alpha = \underline{r}'}\end{aligned} \tag{5.12}$$

The final *RGM* equation is now obtained by adding the direct (I^D) and exchange (I^{Ex}) contributions. It can be given the compact form

$$\int [\hat{\mathcal{H}}(\underline{r},\underline{r}') - E\hat{\mathcal{N}}(\underline{r},\underline{r}')]F(\underline{r}')\,d\underline{r}' = 0 \quad , \tag{5.13}$$

where

$$\hat{\mathcal{H}}(\underline{r},\underline{r}') = \hat{\mathcal{H}}^D(\underline{r},\underline{r}')\delta(\underline{r}-\underline{r}')+\hat{\mathcal{H}}^{Ex}(\underline{r},\underline{r}') = [\hat{T}_r+V^D(\underline{r})+E_1+E_2]\delta(\underline{r}-\underline{r}')+\hat{\mathcal{H}}^{Ex}(\underline{r},\underline{r}') \tag{5.14}$$

is an *hamiltonian-like operator* †, sum of a direct local term $\hat{\mathcal{H}}^D(\underline{r},\underline{r}')\delta(\underline{r}-\underline{r}')$ and of an exchange non-local contribution $\hat{\mathcal{H}}^{Ex}(\underline{r},\underline{r}')$. Further, the operator $\hat{\mathcal{N}}(\underline{r},\underline{r}')$, is a *normalization kernel* and is given as well by the sum of a direct local term and of an exchange non-local contribution:

$$\hat{\mathcal{N}}(\underline{r},\underline{r}') = \delta(\underline{r}-\underline{r}') + \hat{\mathcal{N}}^{Ex}(\underline{r},\underline{r}') \tag{5.15}$$

Then, the resonating group method strongly suggests that the NN interactions are of non-local nature. It is easy to see that, when more resonant groups (or coupled channels) are taken into account, the final result is a *system of integro-differential equations* for the functions $F^L_\alpha(\underline{r})$, coupled by suitable pieces of the hamiltonian (for instance, by the spin-spin interaction, that couples states belonging to different irreducible representations of the SU(4) spin-isospin group).

5.2 – The spurious state problem.

The *RGM*, in its most general version, allows to calculate all functions $F^L_\chi(\underline{r})$ and to get the values of many physical observables, first of all the phase shifts of the cluster-cluster scattering in the channels labelled by χ and in the $^{(2S+1)}L_J$ wave. To explicitly construct the channel states ψ^{ST}_χ, one can take advantage of the properties of the irreducible representations of the permutation group. For instance, for $L=0$ one can build up two colourless 3-quark clusters, the nucleon N and its isobar Δ, in addition to various coloured states with the following S_3 contents:

$$\begin{aligned} [3]_O \times [111]_c \times [3]_{ST} &= \begin{cases} N(ST=1/2,1/2) \\ \Delta(ST=3/2,3/2) \end{cases} \\ [3]_O \times [21]_c \times [21]_{ST} &= \begin{cases} C_1(ST=3/2,1/2) \\ C_2(ST=1/2,3/2) \\ C_3(ST=1/2,1/2) \end{cases} \end{aligned} \tag{5.16}$$

As usual, the orbital 3-quark wavefunctions can be taken as totally symmetric states (see table VI). The total wavefunction of the tree-quark clusters is given by

$$\psi^\alpha(\underline{\xi}) = \hat{\mathcal{A}}[R_0 C^\alpha (ST)^\alpha] , \tag{5.17}$$

† Let us note that (5.14) cannot be rigorously considered as an effective hamiltonian, due to the presence of the operator $\hat{\mathcal{N}}$ in eq. (5.13).

where $\widehat{\mathcal{A}}$ is the three-particle antisymmetrizer and the index α runs over the set (N, Δ, C_1, C_2, C_3) for the color and spin-isospin functions C^α and $(ST)^\alpha$ with the symmetries given in eq. (5.16). Finally, the wavefunctions of the six-quark system, $\psi_X^{ST}(\underline{\xi}_1, \underline{\xi}_2)$, can be obtained by the product of representations of the permutation group S_6.

In many calculations the same symmetric orbital function R_0 is used for all clusters; unfortunately, this assumption leads to introduce in $F_X^L(\underline{r})$ components of *spurious* (or *forbidden) harmonic-oscillator states.*

This drawback is connected to the existence of solutions $F_{sp}^L(\underline{r})$ of the RGM equations corresponding to states that exactly vanish once antisymmetrized:

$$\psi_{sp} = \widehat{\mathcal{A}}[\psi_1 \psi_2 F_{sp}^L(\underline{r})] = 0. \tag{5.18}$$

Obviously, the spurious solutions are not physical states of the system, but — unless one is very careful — they can mix with the physical components of $F_X^L(\underline{r})$. As a consequence, in general, the solutions $F_X^L(\underline{r})$ of the RGM equation (5.13) cannot be automatically identified with the physical wavefunctions of the relative intercluster motion (although they exibit the same asymptotic behavior).

It is well beyond the aim of these lectures to give a detailed discussion of this problem, and the interested reader is referred to the literature on the subject([70,71]). Let us only note that $F_{sp}^L(\underline{r})$ satisfies the independent equations

$$\int \widehat{\mathcal{H}}(\underline{r}, \underline{r}') F_{sp}^L(\underline{r}')\, d\underline{r}' = 0\,, \qquad \int \widehat{\mathcal{N}}(\underline{r}, \underline{r}') F_{sp}^L(\underline{r}')\, d\underline{r}' = 0\,, \tag{5.19}$$

so that it is a common eigenstate to the operators $\widehat{\mathcal{H}}$ and $\widehat{\mathcal{N}}$, with zero eigenvalue.

In order to get rid of all forbidden components, it is possible, for instance, to exploit the operator $(1-\sum_{i=1}^{k} |\phi_i\rangle\langle\phi_i|)$, that is the projector out of the forbidden subspace spanned by $(|\phi_1\rangle, |\phi_2\rangle, \ldots)$. It can be shown([74]) that this amounts to replace $F_X^L(\underline{r})$ by the function

$$\tilde{F}_X^L(\underline{r}) = \int \widehat{\mathcal{N}}^{1/2}(\underline{r}, \underline{r}') F_X^L(\underline{r}')\, d\underline{r}' \tag{5.20}$$

orthogonal to the forbidden states. It is easy to realize that now $\tilde{F}_X^L(\underline{r})$ can be indeed identified with the usual wavefunction of the relative intercluster motion, and it therefore takes the meaning of a probability amplitude.

The spurious state problem has been widely discussed in many works concerned with short-range repulsive effects in the NN interaction. In fact, it is possible to prove that, whenever forbidden states satisfying eq. (4.52) exist, the radial part of $\tilde{F}^L(\underline{r})$ exibits a short-distance node, i.e. a behaviour of the type shown in fig. 18c. In this case, the physical state is actually the second state (that is, an excited state) of the NN system (the lowest is the forbidden one). Then, for instance, the radial deuteron wavefunction acquires an additional node as it usually occurs for excited states. Such an effect is responsible for the repulsive effects observed e.g. in α-α scattering([74]).

5.3 – Solution of the *RGM* equations.

The function $F^L_\chi(\underline{r})$ represents the intercluster motion in the channel χ for the partial wave $^{(2S+1)}L_J$. If χ coincides with the NN channel, then $F^L_\chi(\underline{r}) = F^L_{NN}(\underline{r})$.

Let us consider, for the sake of semplicity, the case of non-coupled waves. As usual in nuclear physics, we can write

$$F^L_{NN}(\underline{r}) = \sqrt{4\pi(2L+1)}i^l e^{i\delta_L} F^L(kr) Y^M_L(\widehat{\underline{k}}) \tag{5.21}$$

(where $\widehat{\underline{k}}$ denotes the unit vector in the $\underline{k}$-direction). In the continuum ($E > 0$), $F^L(kr)$ must exhibit the standard asymptotic behaviour

$$\lim_{r\to\infty} F^L(kr) = \frac{\sin(kr - l\frac{\pi}{2} + \delta_L)}{(kr)} \tag{5.22}$$

that determines, as in the standard theories, the phase shifts δ_L. Obviously, for coupled waves, the solution procedure is more involved and is aimed to determine either the phase shifts δ_L or the coupling parameters $\epsilon_{LL'}$.

Finally, for bound states, it is convenient to follow a similar procedure ($F^L_{NN}(\underline{r}) = F^L(r)Y^M_L(\widehat{\underline{k}})$) but, obviously, all radial functions $F^L(r)$ must asymptotically behave as a decreasing exponential.

5.4 – Recent applications of the *RGM*.

The whole list of the papers devoted to study 6-quarks systems by the *RGM* is rather long. Here we confine ourselves to discuss only some selected papers dealing with short-range repulsive effects in the NN interaction([75]). Let us only mention the existence of papers concerned with other interesting related physical problems as, for instance, the origin of spin-orbit and tensor NN forces([68,70]), or the structure of the $N\Lambda$ or $N\Sigma$ interactions([76]).

Most of these authors neglect the tensor component of the quark-quark interaction (so disregarding any partial-wave coupling), that has been taken into account only in more recent papers([77]).

Due to the practical impossibility of discussing the technical details of these studies, we shall only summarize the main results, with a particular attention to the simple case of $L = 0$ waves ($^3S_1$ and $^1S_0$).

In almost all recent papers, the proof wavefunction is constructed as the sum of NN and $\Delta\Delta$ states, in addition to a few dibarionic configurations with hidden colour. The NN interaction turns out to be ***non-local*** and ***non-adiabatic***. All information on the repulsive core (and, in some cases, on the weak medium-range attractive components) is strongly influenced by the non-local nature of the *RGM* equations. This is why many authors have attempted to solve the following basic problem: to find an ***equivalent*** NN ***local potential*** V^{ELP}_{NN}, that is a local potential able to give the same phase shift predicted by the *RGM*. For instance, Oka and Yazaki([74]) have defined the central parts of these potentials as solutions of the Schrõdinger equation

$$[-\frac{\hbar^2}{2\mu}(\frac{d^2}{dr^2} - \frac{l(l+1)}{r^2}) + V^{ELP}_{NN}(r)]\tilde{F}^L_{NN}(r) = E\tilde{F}^L_{NN}(r). \tag{5.23}$$

Such a V_{NN}^{ELP} — at least in first approximation — can be identified with a standard NN potential, but it is, evidently, state- and energy-dependent! Let us notice that different definitions of V_{NN}^{ELP} are also possible: an example is provided by the more sophisticated approach of Sukuzi and Hecht([75]), that construct V_{NN}^{ELP} in the WKB (Wentzel-Kramer-Brillouin) approximation exploiting the Wigner transforms of the non-local exchange kernels.

All these methods allow us to define an effective NN interaction (by suitably projecting the 6-quark states onto the NN channel), that, as it has been clearly proved, behaves differently at small distance depending on the orbital symmetry of the six-quarks system states. If only the 6-quark configurations with the Young orbital frame $[\lambda] = [6]$ are made allowance for †, the model gives a repulsive core in the NN-interaction at small distance; if, however, one allows for the admixture of the configuration with Young frame $[\lambda] = [42]$ ‡, the strong spin-spin colour force (colour magnetic force) turns out to be attractive, thereby resulting in a weak effective repulsion, or even an attraction, in the NN channel at small distances. If the 6-quark Young frame [42] dominates at low energy, the radial wavefunctions in S-wave NN-channels have an inner node which can simulate the conventional repulsive core (or viceversa: see the dashed curve in fig.19c). Let us remind that in our case the mixing between these two kinds of configurations is of purely dinamical nature, since both [6] and [42] orbital configurations are allowed by the Pauli principle (as previously mentioned, the spin-spin interaction plays a fundamental role in this effect).

Another possibility, already mentioned at the end of sect. 5.2, is that of a NN interaction with a forbidden state, instead of a true repulsive core. In this case $\tilde{F}_{NN}^{L}(r)$ exhibits the behavior shown in fig.19c. In order to correctly deal with this kind of occurrence, particular efforts must be devoted to assure the orthogonality of the physical states among them and to the forbidden ones.

All calculations carried out so far for $^3S_1$ and $^1S_0$ waves give negative phase shifts, thus showing the lack of any significant medium-range attraction (see our comments in sect.4.4). The phase shifts obtained by Oka and Yazaki([74]), are displaied, as an example, in fig.21. Besides, fig.22 shows the effective local potentials V^{ELP} calculated by Suzuky and Hecht ([75]) by means of the quark-quark interactions proposed by different authors. It can be seen that these potentials show a rather softcore if compared to the trivially equivalent local potential defined by eq.(5.23), but a rather hard one in comparison to the short-range terms of the phenomenological Paris potential([78]). Moreover, all of them exhibit a relevant energy-dependence. The increase in energy is almost linear, exactly as in the Paris potential.

---

† In a shell-model scheme, these correspond to s^6 configurations.

‡ In this case one includes shell-model configurations s^4p^2.

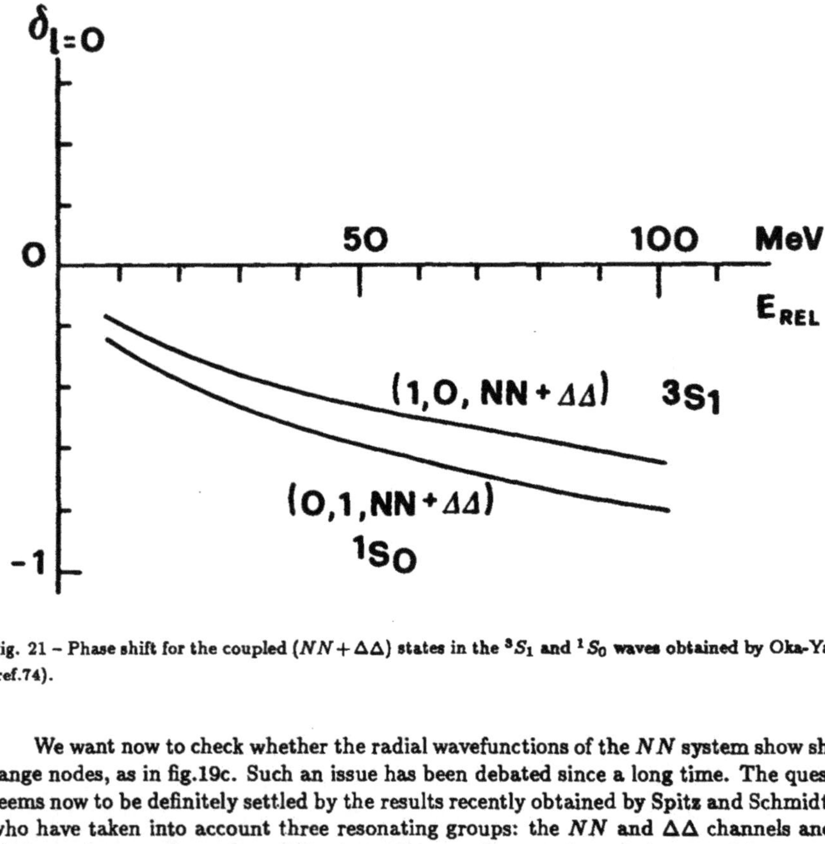

Fig. 21 – Phase shift for the coupled ($NN + \Delta\Delta$) states in the $^3S_1$ and $^1S_0$ waves obtained by Oka-Yazaki (ref.74).

We want now to check whether the radial wavefunctions of the NN system show short-range nodes, as in fig.19c. Such an issue has been debated since a long time. The question seems now to be definitely settled by the results recently obtained by Spitz and Schmidt([75]) who have taken into account three resonating groups: the NN and $\Delta\Delta$ channels and an hidden-colour configuration CC. According to these authors it is essential to exactly orthogonalize the various function spaces.

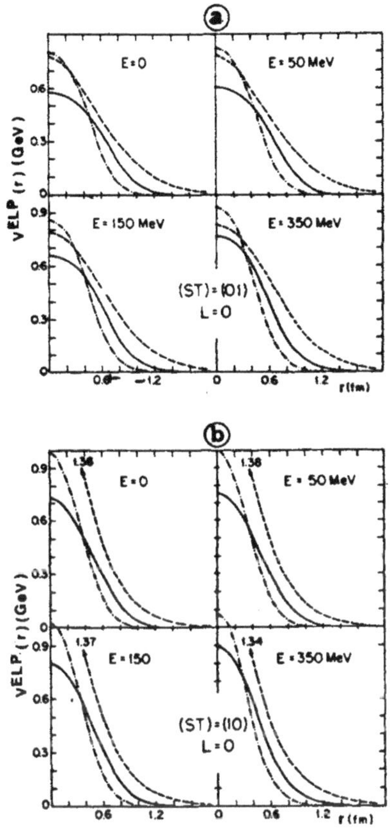

Fig. 22 – The equivalent local potentials for the $(ST) = (01)$ (a) and $(ST) = (10)$ (b) S-wave channels obtained by Suzuki-Hecht (ref. 75). Solid line: Oka-Yazaki; dashed-dotted line: Faessler, Fernandez, Lübeck and Shimizu; dashed line: Harvey.

In earlier calculations (done without ortogonalization of the channel spaces), the relative motion function of the NN-channel had a node in the S-wave at approximatively 0.5 fm. In the new treatment, in which the $\Delta\Delta$- and CC-channels are only orthogonal corrections to the NN-channel, this node is no longer present. The typical radial wavefunction obtained in ref([75]) for the wave $^3S_1$ is similar to the dashed one shown in fig.19b. Then, in the one- and three-channel approximation, the 6-quark RGM does not support the existence of a short-distance node in the NN wavefunction.

Finally, we want to stress that, according to some authors, the large cancellations of colour-Coloumbic and quark contributions indicate that the exact nature of V^{ELP} may be sensitive to the details of the parametrization of the underlying quark-quark interaction. Moreover, for waves with $L \neq 0$, only preliminary results exist; many efforts must be devoted to complete the phisical picture, although the first available results seem to be encouraging.

5.5 – Hybrid versions of the *RGM*.

A careful analysis of the existing literature permits to conclude that the quark-quark interaction described by the one-gluon approximation enables us to get a reasonable, though preliminary, reproduction of the short-distance properties of V_{NN}. However, such models are not able to give information on its long-range (and, perhaps, medium-range) behavior.

Thus, as we have previously mentioned, it seems useful to integrate the quark-quark approaches, suggested by QCD, by the usual mesonic theory of the nuclear forces, thereby generating a class of *hybrid models* (see sect. 4.4). Different hybridization techniques have been proposed. The simplest one, similar to that exploited by Maltman and Isgur([59]), and previously described in sect. 4.4, has been proposed by Miller and Kisslinger([79]). Their basic idea is to introduce a spatial separation in the description of the system considered. For intercluster separation, r, larger than a fixed value ($r_0 \approx 1\ fm$), the intercluster interaction and wavefunctions are assumed to coincide with a standard $V_{NN}(\underline{r})$ potential † and the nuclear wavefunctions $\psi_{NN}(\underline{r})$, respectively. On the contrary, for $r \leq r_0$, a 6-quark model, with wavefunctions $\psi_{6q}(\underline{r})$, is adopted. The needed matching of the two descriptions can be insured by imposing the continuity of the wavefunctions and their derivatives at $r = r_0$ (continuity of the probability amplitude and current).

Some hybridization techniques([80]) have been also used in connection with the *RGM*. For instance, Wakamatsu et al.([80]) have added to the 6-quark, one-gluon potential, $V^{(1g)}$, a long-range interaction term of the form

$$V(\underline{r}) = f(r)V^{OPE}(\underline{r}), \tag{5.24}$$

where $V^{OPE}(\underline{r})$ is the well known *one-pion NN potential* and $f(r)$ is a *cut-off function* going rapidly to zero in the region where the two nucleonic clusters overlap:

$$f(r) = \frac{(r/r_0)^{\alpha}}{[1 + (r/r_0)^{\alpha}]} \tag{5.25}$$

The results of these authors for V^{ELP} and the phase-shifts for the $^3S_1$ and $^1S_0$ waves are shown in fig.23.

---

† For instance, the Paris or Reid potentials([78]).

TABLE XV – Deuteron observables and parameters of the NN-scattering at low energy calculated by M. Wakamatsu([80]).

quantity	exp. value	prediction
B.E. (MeV)	2.226	2.226
P_D (%)	$5 \leq P_D \leq 8$	5.87
D/S	0.0271	0.0273
a(^{3}S$_1$) (fm)	5.42	5.45
r$_0$(^{3}S$_1$) (fm)	1.75	1.84
a(^{1}S$_0$) (fm)	−23.7 (pn) −18.6 (pp)	-20.0
r$_0$(^{1}S$_0$) (fm)	2.7 (pn) 2.8 (pp)	2.77

a = scattering lenght; r$_0$ = effective range.

Moreover, they have also obtained an excellent reproduction of the parameters characterizing the low-energy NN scattering (i.e. scattering lengths and effective ranges) and of the deuteron properties (see, for example, table XV and fig.24).

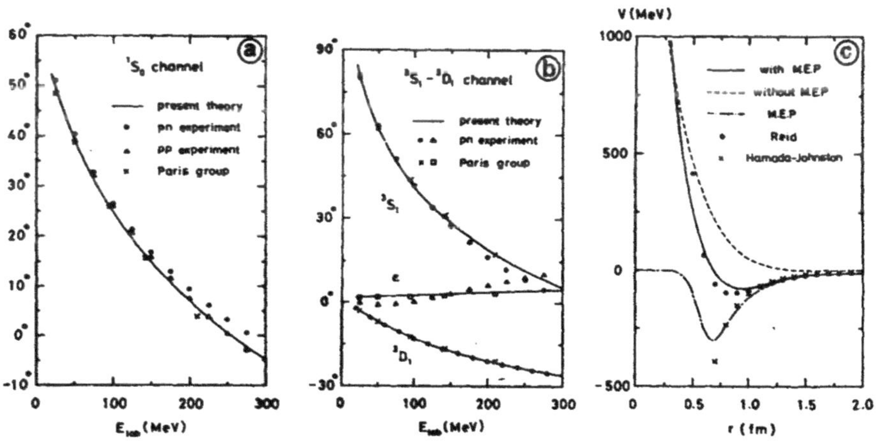

Fig. 23 – Typical results obtained by the hybrid model developed by Wakamatsu, Yakamoto and Yamauchi (ref. 80). Figs. a) and b) show phase shifts in the channels $^1S_0$ and $^3S_1 +^3 D_1$, respectively. Fig. c) shows the equivalent local potential with and without the meson-exchange potential (MEP).

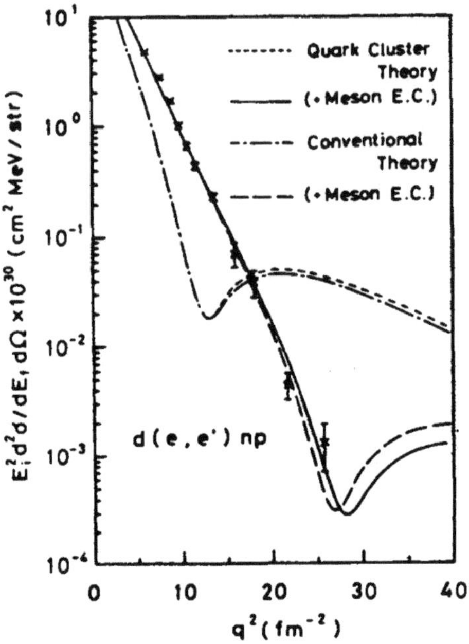

Fig. 24 – The teoretical cross sections for the threshold electrodisintegration of the deuteron at $\theta = 180°$ for $E_{np} = 1.5$ MeV, together with the experimental data. Solid and dotted curves: predictions of the cluster quark theory with and without the meson exchange currents; dashed and dash-dotted curves: conventional model with and without the meson exchange currents (ref.80).

All these results clearly show the relevant role that hybrid models can play for a quantitative and up-to-date description of the nuclear phenomena. However, let us point out that, to this aim, it is necessary to find some *natural and coherent* hybridization mechanism, fully justifiable in terms of QCD. In fact, only a model of this kind will enable us to achieve final and quantitative conclusions.

6 – Conclusions

In these lectures, we have outlined the main reasons whereby one expects that quarks - which are the basic tool of our understanding strong interactions in high energy physics - may play also a fundamental role in the description of nuclear phenomena. We have considered some of the theoretical approaches to deal with the problem of quarks in nuclei, ranging from direct applications of QCD to phenomenological quark models. As already stressed in the introduction, we have omitted, for lenght' sake, any discussion of relativistic models of the bag and soliton types. It must be understood, however, that those approaches are also promising (among the others, they directly account for confinement), expecially in the versions which try to incorporate the chiral symmetry (whose prototypes are the cloudy bag model and the skirmion model).

Clearly, a great deal of work remains to be done, both on theoretical and experimental side. Apart from a set of questions of technical nature inside any particular model, two crucial general problems are to be investigated: the description of the nucleon - nucleon interaction on the basis of the quark - quark interaction, and the consistency between the two apparently alternative pictures of nuclei valid at short distances (in terms of quarks) and at large distances (in terms of meson exchange). In spite of some partial answers to those basic questions (for instance, in the second case, the attempts to build up hybrid models), things are very far from being clarified.

Last but not least, there is presently no clear experimental evidence (apart from the EMC effect) for possible effects due to the quark substructure of nuclei. An exhaustive discussion of the experimental situation is well beyond the aim of these lectures, and would probably require a paper by itself. The status of the art can be summarized as follows: all the effects, which have been claimed as possibly ascribable to quark degrees of freedom inside nuclei, either have been experimentally disproved or admit an explanation in terms of standard mechanisms. A clear signature of a nuclear quark effect is still lacking. It is not by chance that the emphasis of our lectures was on the theoretical aspects: after all, not only good experiments are required before getting sensible ideas (as Galilei said), but also a good theoretical background is required before sensible experiments can be performed.

7 – References

1. – There exist a number of excellent textbook on nuclear physics. Let us only quote some with special emphasis on the theoretical aspects:
J.M. Blatt, V.F. Weisskopf: *Theoretical Nuclear Physics*, (Wiley, N.Y., 1952).
S. De Benedetti: *Nuclear Interactions*, (Wiley, N.Y., 1964).
A.G. Sitenko, V.K. Tartakovskij: *Lektsii poteorii iadra*, (Mir, Moskow, 1972).
A. Bohr, B.R. Mottelson: *Nuclear Structure*, vol. I (Benjamin, N.Y., 1969).
A. De Shalit, H. Feshbach: *Theoretical Nuclear Physics - vol. I: Nuclear structure*, (Wiley, N.Y., 1974).

2. –
M. Gell-Mann: *Phys. Lett.* **8**, 214(1964).
G. Zweig: *CERN report 8182/th.401* , 1964. (reprinted in *Developments in the quark theory of hadrons* vol. I, D.B. Lichtenberg and S.P. Rosen eds., Hadr. Press, Nonantum, MA, 1980)

3. – For a review of the quark model and its applications, see e.g.
D.B. Lichtenberg: *Unitary symmetry and elementary particles*, II ed.(Academic Press, N.Y., 1978).
F.E. Close: *Introduction to quarks and partons*, (Academic Press, N.Y., 1979).
A.W. Hendry and D.B. Lichtenberg: *Rep. Prog. Phys.* **41**, 1707(1978).
A.W. Hendry and D.B. Lichtenberg: *Fortschritte Phys.* **33**, 139,(1985).

4. – The original proposal is due to
M. Gell-Mann: *Acta Phys. Austriaca* suppl. **IX**, 733(1972). and to
H. Fritzsch and M.Gell-Mann: *Proc. 16th Int. Conf. on High Energy Phys.,Chicago* vol. II, 135(1972). There are a number of excellent review papers on QCD. See e.g.
W. Marciano and H. Pagels: *Phys. Rep.* **36C**, 137(1978).
A.J. Buras: *Rev. Mod. Phys.* **52**, 199(1980).
E. Reya: *Phys. Rep.* **69C**, 195(1981).
A.H. Muller: *Phys. Rep.* **73C**, 237(1981).
G. Altarelli: *Phys. Rep.* **81**, 1(1982).
M.R. Pennington: *Rep. Prog. Phys.* **46**, 393(1982).

5. –
J.J. Aubert et al.(EMC): *Phys. Lett.* **123B**, 275(1983).

6. – See e.g.
S.J. Brodsky in: *Quarks and Nuclear Forces*, D. Fries and B. Zeitnitz eds. (Springer-Verlag, Berlin, 1982), p.81.
S.J. Brodsky: *Nucl. Phys.* **A416**, 3C(1984)., and references therein.

7. – For reviews, see e.g.

N. Isgur in: *The new aspects of subnuclear physics*, A. Zichichi ed. (Plenum, N.Y., 1980),p. 107.
O.W. Greenberg: *Ann. Rev. Nucl. Part. Phys.* **28**, 327(1978).
A.J.G. Hey in: *Proc. E.P.S. Int. Conf. on high-energy Phys., Geneve, 1979*, A. Zichichi ed. (CERN, Geneve,1980).
J. Rosner in: *Tecniques and concepts of high-energy physics*, T. Ferbel ed. (Plenum, N.Y.,1981).
R.H. Dalitz: *Nucl. Phys.* **A353**, 251(1981).

8. – For general reviews, see e.g.
R.J. Jaffe in: *Acta Phys. Austriaca* suppl. **22**, 269(1980).
C. DeTar in: *Quantum flavordynamics, quantum chromodynamics and unified theories*, K.T. Mahantappa and J. Randa eds. (Plenum, N.Y., 1979)p. 393.
For reviews with special emphasis on nuclear applications, see e.g.
L. Heller in: *Quarks and nuclear forces*, D. Fries and B. Zeitnitz eds. (Springer-Verlag, Berlin, 1982),p. 145.
A.W. Thomas: *Nucl. Phys.* **A416**, 69(1984).
A.W. Thomas: *Progr. part. and nucl. phys.* **11**, 325(1984).
M. Rho: *Nucl. Phys.* **A 434**, 639(1985).
G. Ripka: *Sacs, skyrmions et solitons* in: *Proc. Ecole Joliot-Curie de Phys. Nucl.*, (1984).

9. – The Skyrme model was proposed more than twenty years ago by
T.H.R. Skyrme: *Proc. Royal Soc.* **A260**, 127(1961).
T.H.R. Skyrme: *Nucl. Phys.* **31**, 556(1962). and recently rediscovered by
N.K. Pak and H. Ch. Tze: *Ann. Phys.(N.Y.)* **117**, 164(1986).
J.M. Gipson and H.Ch. Tze: *Nucl. Phys.* **B183**, 524(1981).
E. Witten: *Nucl. Phys.* **B223**, 422,433(1983).
G.S. Adkins, C.R. Nappi, E. Witten: *Nucl. Phys.* **B228**, 552(1983). For reviews, see
A.P. Balachandran: *pp COO-3533-314 (Syracuse University, 1985)* .
V.P. Nair: *Princeton preprint (1985).*
G. Ripka: *ref. (8)* .

10. –
R. Friedberg and T.D. Lee: *Phys. Rev.* **D15**, 1694(1977).
R. Friedberg and T.D. Lee: *Phys. Rev.* **D16**, 1096(1977).
R. Goldflam and L. Wilets: *Phys. Rev.* **D25**, 1951(1982).
M.C. Birse and M.K. Banerjee: *Phys. Lett.* **136B**, 284(1984).
S. Kahana, G. Ripka and V. Soni: *Nucl. Phys.* **A415**, 351(1984).
S.Kahana and G. Ripka: *Nucl. Phys.* **A429**, 462(1984).

11. – The reader is also urged to consult the following proceedings of schools and conferences on quarks and nuclei:
Quarks and nuclear forces, D.Fries and B. Zeitnitz eds. (Springer-Verlag, Berlin, 1982).
Quarks and nuclear structure, K. Blender ed. (Springer- Verlag, Berlin, 1984).

Mesons, barions, quarks et physique nucleaire (Proc. Ecole Joliot-Curie de Phys. Nucl. 1984).

12. –
O.W. Greenberg: *Phys. Rev. Lett.* **13**, 598(1964).
O.W. Greenberg and D. Zwanziger: *Phys. Rev.* **150**, 1177(1966).
M. Gell-Mann in: *Elementary particle physics*, P. Urban ed., p. 733(Springer-Verlag, Berlin, 1972).

13. – For a review, see:
N. Isgur in: *The new aspects of subnuclear physics*, A. Zichichi ed. (Plenum, N.Y., 1980), p. 107.

14. –
S.L. Adler: *Phys. Rev.* **177**, 2426(1969).
J.S. Bell, R.Jackiw: *N.Cim.* **51**, 47(1969).

15. –
Particle data group: *Phys. Lett.* **170B**, (1986).

16. – Lie groups and related topics are discussed in a number of excellent textbooks. For a special emphasis on unitary symmetries , see first ref. (13); a more general treatment can be found in
R. Gilmore: *Lie groups, Lie algebras and some of their applications*, (Wiley, N.Y., 1974).
A.O. Barut, R. Raczka: *Theory of group representations and applications*, (PWN, Warsaw, 1980).

17. – For a review of non-abelian gauge theories, see e.g.
L.D. Faddeev, A.A. Slavnov: *Gauge fields: an introduction to quantum theory*, (Benjamin, Reading, MA, 1980).

18. –
G.H. Hooft: *Nucl.Phys.* **B33**, 173(1971).

19. –
I.B. Klapovich: *Yad. Fiz.* 10, 409(1969).

20. –
H.D. Politzer: *Phys. Rev. Lett.* **30**, 1346(1973).
D. Gross, F. Wilczek: *ibid.* **30**, 1343(1973).
D. Gross, F. Wilczek: *Phys. Rev.* **D8**, 3633(1973).
D. Gross, F. Wilczek: *Phys. Rev.* **D9**, 980(1974). For reviews, see e.g.
H.D. Politzer: *Phys. Rep.* **14C**, 129(1974).
A. Peterman: *Phys. Rep.* **53C**, 157(1979).

21. –
C. Klopfenstein et al.: *Phys. Rev.* **51**, 160(1983).
S.J. Brodsky, G.P. Lepage, P.B. Mackenzie: *Phys. Rev* **D28**, 228(1983). See also
C.L. Basham et al.: *Phys. Rev.* **D19**, 2018(1979).
P.B. Mackenzie and G.P. Lepage: *Phys. Rev. Lett.* **47**, 1244(1981).

22. – For a review of free quark searches, see e.g.
L.W. Jones: *Rev. Mod. Phys.* **49**, 717(1977).

23. –
H. Fritzsch, M. Gell-Mann, H. Leutwyler: *Phys. Lett.* **47B**, 365(1973).
S. Weinberg: *Phys. Rev. Lett.* **31**, 494(1973).
S. Weinberg: *Phys. Rev.* **D8**, 4482(1973).

24. – For reviews, see e.g. first ref. (4) and
M. Bander: *Phys. Rep.* **75C**, 205(1981).

25. – For the MIT bag, see
A. Chodos, R.L. Jaffe, C.B. Thorn, V. Weisskopf: *phys. Rev.* **D9**, 3471(1974).
A. Chodos, R.L. Jaffe, K. Johnson, C.B. Thorn: *Phys. Rev.* **D10**, 2599(1974).
For the SLAC bag, see
W.A. Bardeen, M.S. Chanowitz, S.D. Drell, M. Weinstein, T.M. Yan: *Phys. Rev.* **D11**, 1094(1975).
For the chiral bag, see
G.E. Brown, A.D. Jackson, M. Rho, V.Vento: *Phys. Lett.* **140B**, 285(1984).
V.Vento and M. Rho: *Nucl. Phys.* **A412**, 413(1984).

26. –
K. Wilson: *Phys. Rev.* **D10**, 2455(1974).
The subject is reviewed e.g. in
M. Creutz: *Quarks, gluons and lattices*, (Cambridge University Press, Cambridge,1983).
J.M. Drouffe, C. Itzykson: *Phys.Rep.* **38C**, 133(1978).
J.B. Kogut: *Rev. Mod. Phys.* **51**, 659(1979).
J.B. Kogut: *Rev. Mod. Phys.* **55**, 775(1983).

27. – See e. g.
B. Frois: *Proc. of the Ecole Joliot-Curie de Phys. Nucl.*, 1984, p. 135.

28. – Deep inelastic processes are reviewed (both on the theoretical and experimental viewpoint) in
G. Altarelli: *Riv. N. Cim.* **4**, 335(1974).

29. –
J.D. Bjorken: *Phys. Rev.* **179**, 1547(1969).

30. –
R.P. Feynman: *Phys. Rev. Lett.* **23**, 1415(1969).

31. –
E.D. Bloom et al.: *Phys. Rev. Lett.* **23**, 930(1969).
M. Breidenbach et al.: *Phys. Rev. Lett.* **23**, 935(1969).

32. –
G. Callan, D.J. Gross: *Phys. Rev. Lett.* **22**, 156(1969).

33. –
TASSO coll., R. Brandelik et al: *Phys. Lett.* **83B**, 261(1979).
TASSO coll., R. Brandelik et al: *Phys. Lett.* **86B**, 243(1979).
TASSO coll., R. Brandelik et al: *Phys. Lett.* **89B**, 418(1980).

34. –
D. Barber et al.: *Phys. Rev. Lett.* **42**, 1110,1113(1979).

35. –
J.G.H. De Groot et al.: *Phys.Lett.* **82B**, 292,456(1979).
J.G.H. De Groot et al.: *Zeitschr. Phys.* **C1**, 143(1979).

36. –
A.Bodek et al.: *Phys. Rev. Lett.* **50**, 1431(1983).
A.Bodek et al.: *Phys. Rev. Lett.* **51**, 534(1983).
A.Bodek et al.: *Phys. Rev. Lett.* **52**, 727(1984).

37. – For a review, see e.g.
K. Rith in: *Proc. Int. Europhysics Conf. on High Energy Physics*, Brighton, England, 1983.

38. –
A. Bodek, J.L. Ritchie: *Phys. Rev.* **D23**, 1070(1981).
A. Bodek, J.L. Ritchie: *Phys. Rev.* **D24**, 1400(1981).
L.L. Frankfurt, M.I. Strikman: *Nucl. Phys.* **B181**, 22(1981).
G. Berlad et al.: *Phys. Rev.* **D22**, 1547(1980).

39. –
C.H. Llewellyn Smith: *Phys. Lett.* **128B**, 107(1983).
M. Ericson, A.W. Thomas: *Phys.Lett.* **128B**, 112(1983).
J. Kubar, G. Planz, J. Szwed: *Zeitschr. Phys.* **C23**, 1985(1984).

40. –
J. Szwed: *Phys. Lett.* **128B**, 245(1983).

41. -
R.L. Jaffe: *Phys. Rev. Lett.* **50**, 228(1983).
H. Faissner, B.R. Kim: *Phys. Lett.* **130B**, 321(1983).
S. Fredriksson: *Phys. Rev. Lett.* **52**, 724(1984).

42. -
F.E. Close, R.G. Roberts, G.C. Ross: *Phys.Lett.* **129B**, 346(1983).
R.L. Jaffe,F.E. Close, R.G. Roberts, G.G. Ross: *Phys. Lett.* **134B**, 449(1984).
A.W. Hendry, D.B. Lichtenberg, E. Predazzi: *Phys. Lett.* **136B**, 433(1984).
See also
J.V. Noble: *Phys. Rev. Lett.* **46**, 412(1981).
For a recent comparison among the various theoretical models, see
R.P. Bickerstaff, M.C. Birse, G.A. Miller: *Phys. Rev.* **D33**, 3228(1986).

43. -
S.J. Brodsky, G.P. Lepage: *Phys. Rev.* **D24**, 1808(1981).

44. -
S.J. Brodsky, J.R. Hiller: *Phys. Rev.* **C28**, 475(1983).

45. -
G.P. Lepage, S.J. Brodsky: *Phys. Rev.* **D22**, 2157(1980).

46. -
S.J. Brodsky, C.R. Ji, G.P. Lepage: *Phys. Rev. Lett.* **51**, 83(1983).
S.J. Brodsky, C.R. Ji: *Phys.Rev.* **D33**, 2653(1986).

47. -
S.J. Brodsky, B.T. Chertok: *Phys. Rev. Lett.* **37**, 269(1976).
S.J. Brodsky, B.T. Chertok: *Phys. Rev.* **D14**, 3003(1976).

48. -
P.A.M. Dirac: *Rev. Mod. Phys.* **21**, 392(1949).

49. -
L. Susskind: *Phys. Rev.* **165**, 1535(1968).
K. Bardakci, M.B. Halpern: *Phys. Rev.* **176**, 1686(1968).

50. -
S. Weinberg: *Phys. Rev.* **150**, 1313(1966).

51. -
V.A. Matveev, R.M. Muradyan, A.V. Tavkhelidze: *Lett. N. Cim.* **7**, 719(1973).
S.J. Brodsky, G.R. Farrar: *Phys. Rev. Lett.* **31**, 1153(1973).
S.J. Brodsky, G.R. Farrar: *Phys. Rev.* **D11**, 1309(1975).

52. –
A. De Rujula, H. Georgi, S.L. Glashow: *Phys.Rev.* **D12**, 147(1975).

53. – For a derivation of the usual Breit-Fermi interaction, see e.g.
V.B. Berestetskii, E.M. Lifshitz, L.P. Pitaevskii: *Relativistic quantum theory*, (Pergamon, London,1971), p.280.

54. –
H.J. Lipkin: *Phys. Lett.* **74B**, 399(1978).
H.J. Lipkin: *Phys. Rev. Lett.* **41**, 1629(1978).
H.J. Lipkin: *Nucl. Phys.* **B155**, 104(1979).

55. –
N.Isgur, G. Karl: *Phys. Rev.* **D20**, 1191(1979).
L.A. Copley, N.Isgur, G. Karl: *Phys. Rev.* **D20**, 768(1979).
N.Isgur, G. Karl: *Phys. Lett.* **74B**, 353(1978).
N.Isgur, G. Karl: *Phys. Rev.* **D18**, 4187(1978).
N.Isgur, G. Karl: *Phys. Rev.* **D19**, 2653(1979).
R. Koniuk, N.Isgur: *Phys. Rev. Lett.* **44**, 845(1980).
R. Koniuk, N.Isgur: *Phys. Rev.* **D21**, 1868(1980).
R. Koniuk, N.Isgur: *Phys. Rev.* **D23**, 818(1981).
S. Godfrey, N.Isgur: *Phys. Rev.* **D32**, 189(1985).

56. –
O.W. Greenberg, H.J. Lipkin: *Nucl. Phys.* **A179**, 349(1981).

57. –
J.J. Swart: *Rev. Mod. Phys.* **35**, 916(1963).

58. –
M. Harvey: *Nucl. Phys.* **A352**, 301,326(1981).

59. –
K. Maltman, N. Isgur: *Phys. Rev. Lett.* **50**, 1827(1983).
K. Maltman, N. Isgur: *Phys. Rev.* **D29**, 952(1984).

60. – See e. g.
C.S. Warke, R. Shanker: *Phys. Rev.* **C21**, 2643(1980).

61. –
S.A. Williams, F.J. Margenta, P.D. Morley, D.L. Pursey: *Phys. Rev. Lett.* **49**, 771(1982).

62. –
M. La Combe et al.: *Phys. Rev.* **C21**, 861(1980).
K. Halinde, R. Machleidt: *Nvcl. Phys.* **A256**, 479(1976).

R.V. Reid: *Ann. Phys. (N.Y.)* **50**, 411(1963).

63. –
M.H. Storm, A. Watt: *Nucl. Phys.* **A408**, 397(1983).

64. – See e.g.
S. Sato, S. Okai, R. Tamagaki, M. Yasumo: *Progr. Theor. Phys.* **50**, 1561(1973).
S.Saito: *Progr. theor. Phys.* **41**, 705(1969).

65. – For early attempts, see e.g.
S. Otsuki, R. Tamagaki, W. Wade: *Progr. Theor. Phys.* **32**, 320(1964).
R. Tamagaki: *Rev. Mod. Phys.* **39**, 629(1967).

66. –
D.A. Liberman: *Phys. Rev.* **D16**, 1542(1977).
C. DeTar: *Phys. Rev.* **D17**, 323(1978).

67. –
Yu.F. Smirnov et al.: *Sov. J. Nucl. Phys.* **27**, 456(1978).
V.G. Nendatchin, Yu.F. Smirnov, R. Tamagaki: *Progr. Theor. Phys.* **58**, 1072(1977).
I.T. Obukhovsky et al.: *Phys. Lett.* **B83**, 321(1979).
I.T. Obukhovsky et al.: *Sov. J. Nucl. Phys.* **31**, 269(1981).

68. –
O.Morimatsu, S. Oka, K. Shimizu, K. Yazaki: *Nucl. Phys.* **A420**, 573(1984).
O. Morimatsu, K. Yazaki, M. Oka: *Nucl. Phys.* **A424**, 412(1984).

69. –
J.A. Wheeler: *Phys. Rev.* **52**, 1107(1937).

70. – For a deeper discussion of the RGM, as well as of the related *generating coordinate method* (GCM), see e.g.:
*Suppl. Progr. Theor. Phys.***62**(1977): S. Saito (p.11), H. Horiuchi (p.90), M. Kamimura (p.236).
Y.C. Tang in: *Nuclear Physics II*, T.T.S. Kuo and S.S.M. Wang eds.(Springer-Verlag, Berlin, 1981).
K. Wildemuth, Y.C. Tang: *An unified theory of the nucleus,*, (Academic Press, N.Y., 1977).

71. – For the mathematical aspects of the RGM (in particular, the role of the Pauli principle and the occurrence of spurious states), see e.g.
S. Saito: *Progr. Theor. Phys.* **41**, 705(1969).
S. Saito, S. Okai, R. Tamagaki, M. Yasumo: *Progr. Theor. Phys.* **50**, 159(1973).
T. Fliessbach, H. Walliser: *Nucl. Phys.* **A377**, 89(1982).
E.W. Schmid: *Zeitsch. Phys.* **A311**, 67(1983).

E.W. Schmid, G. Spitz: *Zeitsch. Phys.* **A321**, 581(1985).

72. – Among the old papers, of historical interest, let us quote:
R.A. Buckingham, H.S.W. Massey: *Proc. Royal Soc.* **A179**, 123(1941).
B.H. Brandsen et al.: *Proc. Royal Soc.* **A247**, 73(1978).
D.R. Thompson, Y.C. Tang: *Phys. Rev.* **159**, 806(1967).
S. Hachberg et al.: *Proc. Royal Soc.* **A68**, 746(1955).

73. –
L. Hulthen: *Math. Rev.* **6**, 111(1944).
Compt. Rend. Dizieme Congr. Math. Scandinaves 1946, Copenhagen, p.201(1947)
*Arkiv. Math. Astron. Fys.*35A25(1948)
W. Kohn: *Phys. Rev.* **74**, 1763(1948).
W. Kohn: *Phys. Rev.* **84**, 495(1951).
See also
R.G. Newton: *Scattering theory of waves and particles*, II ed. (Springer-Verlag, Berlin, 1982), chapt. 11.3.

74. –
M. Oka, K. Yazaki: *Progr. Theor. Phys.* **66**, 556,572(1981).
M. Oka, K. Yazaki: *Nucl. Phys.* **A402**, 477(1983) and references therein.

75. –
J.E.F.T. Ribeiro: *Zeitsch. Phys.* C5, 27(1980).
A. Faessler, F. Fernandez, G. Lubeck, K. Shimizu: *Nucl. Phys.* **A402**, 555(1983).
K. Suzuki, K.T. Hecht: *Phys. Rev.* **C27**, 299(1983).
K. Suzuki, K.T. Hecht: *Phys. Rev.* **C28**, 1458(1983).
K. Suzuki, K.T. Hecht: *Phys. Rev.* **C29**, 1586(1984).
G. Spitz, W. Schmidt: *Few Body Systems* **1**, 37(1986).

76. –
M. Harvey, J. Letourneur, R. Lorazo: *Nucl. Phys.* **A424**, 428(1984).

77. – See e.g.
J. Burger, N.H. Hofmann: *Phys. Lett.* **148B**, 25(1984).

78. – For the Paris Potential, see
M. Lacombe et al.: *Phys. Rev.* **C21**, 861(1980).
R. Vink Mau: *Nucl. Phys.* **A374**, 3(1982).
The Reid potential was proposed by
R. V. Reid jr.: *Ann. Phys.* 50, 411(1968).

79. –
G.A. Miller, L.S. Kisslinger: *Phys. Rev.* **C27**, 1602,1669(1983).
E.M. Hessley, L.S. Kisslinger, G.A. Miller: *Phys. Rev.* **C28**, 1277(1983).

P.A.M. Guichon, G.A. Miller: *Phys. Lett.* **134B**, 15(1984).

80. -
A. Faessler, F. Fernandez: *Phys. Lett.* **124B**, 145(1983).
M. Oka, K.Yazaki: *Nucl. Phys.* **A402**, 477(1983).
V.I. Kukulm, V.N. Pomerantsev, V.M. Krasnopolsky, P.B. Sazonov: *Phys. Lett.* **135B**, 20(1984).
M. Berger, H.J. Weber: *Int. Rep. UVA - MKPH - T - 84 - 6* , (1984).
M. Wakamatsu, R. Yamamoto, Y. Yamaguchi: *Phys. Lett.* **146B**, 148,153(1984).
K. Brones, A. Faessler, F. Fernandez, K. Shimizu: *Zeitsch. Phys.* **A320**, 609(1985).

CORRELATED BASIS FUNCTION THEORY FOR FERMION SYSTEMS

by
A.Fabrocini[(*)]
Physics Department, University of Pisa
Istituto Nazionale di Fisica Nucleare - Sezione di Pisa (Italy)
S.Fantoni[(**)]
Physics Department, University of Lecce
Istituto Nazionale di Fisica Nucleare - Sezione di Lecce (Italy)

[(*)]Dipartimento di Fisica and INFN, Sezione di Pisa,
Piazza Torricelli 2, I-56100 Pisa (Italia)
[(**)] Dipartimento di Fisica,Università di Lecce,
Via Arnesano, I-73100 Lecce (Italia)

CORRELATED BASIS FUNCTION THEORY FOR FERMION SYSTEMS

by A. Fabrocini and S. Fantoni

Introduction.

Experimental and theoretical investigations on strongly interacting Fermi systems have been in the past and, most problably, will also be in the future an exciting field of basic research. Here, we will limit our attention to macroscopic quantum systems, which are constituted by a *very large* number A of fermions ($A \to \infty$) kept at a finite value of the density ρ and are described by means of a non relativistic hamiltonian of the form:

$$H = -\frac{\hbar^2}{2m} \sum_{i=1,A} \nabla_i^2 + \sum_{j>i=1,A} v_{ij} + \sum_{k>j>i=1,A} v_{ijk}. \tag{1.1}$$

Liquid ^3He, nuclear matter, neutron matter, electron plasma, etc. are examples of such systems. In all these systems the interactions are characterized by a strong repulsion at short distances which induces strong correlation effects and accurate solutions of the many-body Schroedinger equation

$$H\psi_n = E\psi_n, \tag{1.2}$$

are required to fully understand the experimental data available for these systems in terms of the properties of the free constituent particles and of their bare interaction.

The Green Function MonteCarlo (GFMC) method developed by *Kalos et al.*[1] provides an exact solution for the ground state wave eigenfunction ψ_0 and the corresponding eigenvalue E_0. Basically the method consists in solving the equation

$$\psi_0(R) = E_0 \int G(R,R')\psi_0(R')dR', \tag{1.3}$$

where $G(R,R')$ is the Green Function defined as

$$G(R,R') = \sum_n \frac{|\psi_n(R)\rangle\langle\psi_n(R')|}{E_n}, \tag{1.4}$$

R represents the set of coordinates $\mathbf{r}_1 \ldots \mathbf{r}_A$ and $\psi_n(R)$ is the n-th eigenstate corresponding to the eigenvalue E_n. The actual calculation[2] is performed by solving the following iterative scheme

$$\bar{\psi}_k(R) = E_k \int \bar{G}(R,R')\bar{\psi}_{k-1}(R')dR', \tag{1.5}$$

where E_k is a constant which ensures that ψ_k, as well as ψ_{k-1} is properly normalized and

$$\bar{G}(R,R') = \psi_v(R)G(R,R')\psi_v(R'), \tag{1.6}$$

$$\bar{\psi}(R) = \psi(R)\psi_v(R), \tag{1.7}$$

with $\psi_v(R)$ being an approximate solution which generally is taken from variational calculations. The modified Green function $G(R,R')$ is sampled by using MonteCarlo techniques.

In the case of Bose systems, like liquid ^4He, very accurate estimates of E_0 are obtained. In the case of Fermi systems the presence of nodes in $\psi_v(R)$ causes the appearance of divergencies that can not be handled numerically. The problem is partially overcome[3] by keeping the nodes of $\psi(R)$ to be the same as in $\psi_v(R)$. The fixing of the nodes allows for numerical solutions of $\bar{G}(R,R')$, which however furnish only upperbounds of the ground state energy. To release the fixed node constraint unfortunately brings to a rapid increase of the statistical error in the iterative solution of eq.(1.5).

Conventional many-body theories rely on perturbative schemes based on uncorrelated basis functions $|i]$, given by the eigenfunctions of an independent particle hamiltonian of the type:

$$H_0 = \sum_{i=1,A} (-\frac{\hbar^2}{2m}\nabla_i^2 + U(k_i)). \tag{1.8}$$

The interaction hamiltonian

$$H_I = \sum_{j>i=1,A} v_{ij} + \sum_{k>j>i=1,A} v_{ijk} - \sum_{i=1,A} U(k_i), \tag{1.9}$$

is then treated perturbatively. The *ad hoc* perturbative scheme required by the presence of a repulsive core in the interaction H_I was originally derived by *Brueckner*[4], *Bethe*[5] and *Goldstone*[6] and called BBG expansion. The main idea, here, is to sum all the perturbative terms coming from particle-particle excitations, namely from intermediate states of the type $|h_1h_2p_1p_2]$, $|h_1h_2p_1'p_2']$.... As a result, the singular matrix elements of the bare potential v are replaced by the well behaved matrix elements of the reaction matrix G given by:

$$G = v - v(\frac{Q}{E})G, \tag{1.10}$$

where Q is the Pauli operator which constraints the intermediate states to be particle states and E is the energy corresponding to the intermediate state.

The expansion in power of G does not converge, but there are formal arguments[7], as well as numerical calculations[8,9] indicating that an expansion on the number of hole lines may converge. Still the hole-line expansion should be considered as a low density expansion whose convergence properties are not fully understood and whose application to dense system, like liquid ^3He, does not appear to be justified. Moreover, certain class of perturbative diagrams, like e.g. the ring diagrams may need to be summed at all orders[10] also in the case of less dense systems, like nuclear matter.

Other expansion schemes have been devised to sum a larger class of perturbative diagrams which could include the effects due to both short and long range correlations. The Coupled Cluster theory[11], Parquet theory[12], massive summations of p-p, p-h and h-h excitations[13] are interesting examples, which however have not yet shown their full applicability and usefulness for realistic calculations in nuclear matter and liquid ^3He.

Correlated basis theories of strongly interacting Fermi systems, firstly proposed by *Feenberg* and coworkers[14,15], naturally arise from the variational theories in which the trial wave function is taken of the form

$$|0) = \frac{G|0]}{[0|G^{\dagger}G|0]^{1/2}}, \tag{1.11}$$

where $|0]$ is the ground state wave function of a Fermi gas having the same density ρ of the liquid and G is a suitable chosen correlation operator, and

$$E_0 = (0|H|0), \tag{1.12}$$

is an upperbound to the binding energy of the liquid.

Originally, the operator G was taken to be of the so called *Jastrow* type, namely a product of pair correlation functions,

$$G^{(J)} = \prod_{j>i=1,A} f(r_{ij}), \tag{1.13}$$

and $f(r_{ij})$ was determined by minimizing the expectation value E_0. Typically, $f(r_{ij})$ is very small at short distances ($r_{ij} \to 0$) and heals to 1 at large distance ($r_{ij} \to \infty$).

The *Jastrow* ansatz, in spite of its semplicity, has revealed itself to be quite realistic indeed, since it gives reasonable upperbounds to the binding energy of liquids where the interaction is purely central, like e.g. ^3He, v_2 – *model* of neutron matter[16], OMY-model of nuclear matter[17] etc., and it also capable of explaining the long wavelength limit behaviour of the static structure function :

$$S(k) = (0|\rho_{\mathbf{k}}{}^{\dagger}\rho_{\mathbf{k}}|0)/A, \tag{1.14}$$

$$\rho_{\mathbf{k}} = \sum_{i=1,A} \exp i\mathbf{k}\cdot\mathbf{r}_i. \tag{1.15}$$

The following two main indications come out from these earlier calculations: (i) *the product form for the operator G is capable to provide a reasonable description of both the short and long range correlations.* In fact the short range correlations are very effective in providing good estimates of the binding energy, whereas the long range correlations are crucial for a correct behaviour of $S(k)$ at small values of k and therefore of the phonon type of excitations of the liquid; (ii) *configurations in which many particles are simultaneously correlated may well be very important* and one has to take them properly into account to calculate the expectation value of the hamiltonian or of other interesting operators. The Fermi Hyper Netted Chain (FHNC) theory[18-20] and the MonteCarlo technique[2] can provide for the required accuracy in the above calculations.

More recently, correlation operators G containing two-body state dependent correlations $F(i,j)$, as well as three-body correlations $f_3(\mathbf{r}_{ij},\mathbf{r}_{ik})$ have been used. Such operators have the following form

$$G = G^{(J)} \prod_{k>j>i=1,A} f_3(\mathbf{r}_{ij},\mathbf{r}_{ik}) S \prod_{j>i=1,A} F(i,j), \tag{1.16}$$

where S is the operator which symmetrizes $\prod_{i<j} F(i,j)$ (notice that $F(i,j)$ an $F(i,k)$ do not commute in general).

State dependent correlations are particularly important in nuclear matter, since the N-N interaction is strongly state dependent. The expectation value of the tensor components of the interaction on the wave function of eq.(1.11) with G given by $G^{(J)}$ is zero. As a result, the *Jastrow* ansatz gives an unbound nuclear matter if realistic N-N potentials are employed.

The FHNC theory has been extended[21,22] to treat correlation operators of the type given in eq.(1.16). It is believed that the Single Operator Chain (FHNC/SOC) approximation[23] leads to accurate results for the expectation value of two- and three-body operators.

Triplet correlations are increasingly important as the density of the system gets higher and higher. For instance, they give a not negligeable contribution to the binding energy of liquid ^3He. A correlation operator of the type given in eq.(1.16) gives a very realistic description of the ground state wave function of liquid ^3He: in fact the equilibrium energy and density obtained[24,25] with such a trial function are $E_0 =\sim -2.5K$ and $\rho_0 = 0.28\sigma^{-3}$ ($\sigma = 2.556\mathring{A}$) which is very close to the GFMC estimates[26] and to the experimental values $-2.47K$ and $0.277\sigma^{-3}$ respectively.

Althought the variational theory seems to do very well into describing several properties of the ground state of strongly interacting Fermi systems, one still needs a perturbative scheme to get a completely realistic understanding of the true eigenstates of the hamiltonian. Especially the calculation of quantities which are more related to the dynamics of the systems, like e.g. the momentum distribution, the Landau parameters, the response functions, requires to go beyond the variational theory.

The correlated basis (CB) states are defined as

$$|\mathbf{h}_1 \dots \mathbf{p}_n) = \frac{G|\mathbf{h}_1 \dots \mathbf{p}_n]}{[\mathbf{h}_1 \dots \mathbf{p}_n|G^{\dagger}G|\mathbf{h}_1 \dots \mathbf{p}_n]^{1/2}}, \tag{1.17}$$

where $|\mathbf{h}_1 \dots \mathbf{p}_n]$ denotes an n-particle n-hole Fermi gas state and $|\mathbf{h}_1 \dots \mathbf{p}_n)$ is the *corresponding correlated* CB state.

The CB states are not orthogonal to each other. They have been used in not orthogonal perturbation theory[27,28] (CBPT) to study various properties of liquid helium[29,30] and nuclear matter[31,32].

It is possible to orthogonalize the CB states. It seems preferable to use an orthogonalization procedure which preserves the diagonal matrix elements of the hamiltonian. Recently, a two-step orthogonalization procedure has been devised[33] which uses a combination of the *Schmidt* and the *Löwdin* orthogonalizations[34], such that the energies of the orthogonal states are equal to those of the CB states.

The CBF theory, the general properties of the matrix elements of the unity and the hamiltonian operators and the FHNC summation procedures are described in Section 2. Sections 3,4 and 5 discuss few applications for hard-spheres, liquid ^3He and nuclear matter respectively. The last Section is devoted to conclusions.

. Correlated basis function theory

A CBF theory depends upon the correlation operator G chosen to define the CB states of eq.(1.17). There is freedom into choosing G, similarly to the freedom one has for the single particle potential $U(k_i)$ in BBG theory. In practice, G is selected so that the ground state wave function of eq.(1.11) already provides a good estimate for the binding energy per particle of the system. This is achieved by performing a preliminary variational calculation of E_0 by using a properly chosen form for the correlation operator. Few calculations have been performed with $G^{(J)}$, some others with a more complex correlation operator like that given in eq.(1.16). It is likely that the more realistic is the correlation operator, the more rapid is the convergence of the perturbative series.

Let us write the hamiltonian in the form:

$$H = H_0 + H_I, \tag{2.1}$$

where the *unperturbed* part H_0 is defined in terms of the CB states, by means of the following equation

$$(i|H_0|j) = E_i^v \delta_{ij}, \tag{2.2}$$

where the *variational estimates* E_i^v are given by the diagonal matrix elements:

$$E_i^v = (i|H|i). \tag{2.3}$$

The energies E_i^v are called variational estimates since they depend on the correlation operator G used to define the CB states and G has been determined variationally, as previously discussed.

The CB states form a complete but not orthogonal set of basis functions. They can be used in a not orthogonal perturbative scheme or they can be orthogonalized amongst themselves and then used in standard perturbation theory.

The interaction hamiltonian is given by

$$(i|H_I|j) = (1 - \delta_{ij})(i|H - E_n|j), \tag{2.4}$$

with E_n being the eigenvalue of the energy to be calculated. The presence of E_n in the r.h.s. of the above equation is necessary in not orthogonal CBF theory and is due to the fact that the two kets $|i)$ and $|j)$ are in general not orthogonal. If the CB states are orthogonalized the presence of E_n in eq.(2.4) is unnecessary.

In what follows we give an outline of the cluster properties of the CB matrix elements

$$H_{ij} = (i|H_0|j), \tag{2.5}$$

$$N_{ij} = (i|j), \tag{2.6}$$

review the theories to calculate them, and, finally, discuss both the not orthogonal and the orthogonal perturbative schemes.

2.1 CB matrix elements

In this section we review the calculation and the properties of the overlap matrix elements $(\mathbf{h'}_1 \dots \mathbf{p'}_m | \mathbf{h}_1 \dots \mathbf{p}_n)$ and the interaction matrix elements $(\mathbf{h'}_1 \dots \mathbf{p'}_m | H | \mathbf{h}_1 \dots \mathbf{p}_n)$ by using the Power Series cluster expansion[35] and diagrammatic methods. From the definition of the CB states given in eq. (1.17) it follows that the CB matrix elements are constructed from the matrix elements of $G^\dagger G$ and of $G^\dagger H G$ on the Fermi gas states $|\mathbf{h}_1 \dots \mathbf{p}_n]$

The operators $G^\dagger G$ and $G^\dagger H G$ are expanded by making the following substitutions

$$f^2(r_{ij}) = 1 + h(r_{ij}), \tag{2.7}$$

$$f_3^2(\mathbf{r}_{ij}, \mathbf{r}_{ik}) = 1 + h_3(\mathbf{r}_{ij}, \mathbf{r}_{ik}), \tag{2.8}$$

$$F(i,j) = 1 + \hat{h}(i,j). \tag{2.9}$$

In doing that one has to take care of commuting G with the hamiltonian and the state dependent correlations between themselves.

The various terms of the expansion are characterized by integrals that contain a small groups of particles, denoted also as *cluster terms*. They are conveniently represented by diagrams, few examples of which are displayed in Figs. 1 and 2. They obey the following diagrammatical rules:

(i) The diagrams are built with points, oriented lines, dashed lines, black triangles, wavy lines and thick lines.

The points or vertices of a diagram represent coordinates $\mathbf{r}_i$. The solid points (internal points) imply an integration over the corresponding coordinate times ρ, the circles (external points) represent coordinates over which there is no integration. In the final calculation of a matrix elements there are no circles; however in the process of summing up all the cluster terms contributing to a given matrix elements it is usefull to compute particular *n-body distribution functions*

$$g_n(1,\cdots,n) = \frac{A!}{n!\rho^n} \frac{\int d\mathbf{r}_{n+1}\cdots d\mathbf{r}_A \Psi_0^\dagger(1,\cdots,A)\Psi_0(1,\cdots,A)}{\int d\mathbf{r}_1\cdots d\mathbf{r}_A \Psi_0^\dagger(1,\cdots,A)\Psi_0(1,\cdots,A)}, \tag{2.10}$$

which have n circles. For instance, the expectation value of a two-body operator is given by

$$(0|\sum_{i\le j} v(r_{ij})|0) = \frac{\rho}{2}\int d\mathbf{r}_{ij} g_2(r_{ij}) v(r_{ij}). \tag{2.11}$$

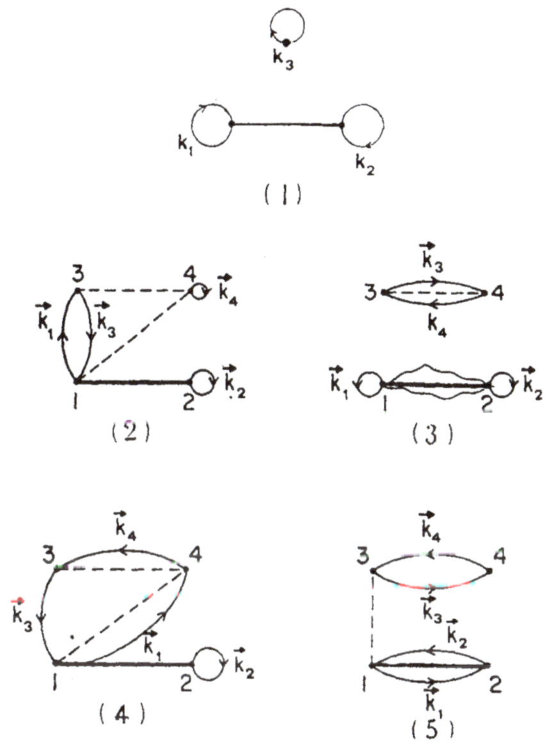

Fig. 1- *Examples of cluster diagrams in the expansion of the diagonal matrix elements of the hamiltonian.*

The dashed lines, black triangles and wavy lines represent the correlations h, h_3 and $2f^2\hat{h}$ respectively. One can also have two superimposed wavy lines, representing two state dependent correlations of the same couple of particles coming respectively from $G^\dagger$ and G and given by $f^2\hat{h}^\dagger\hat{h}$.

An oriented line going from i to j denotes a factor $\exp[i\mathbf{k}\cdot\mathbf{r}_{ij}]$ which is obtained when the state $\mathbf{k}$ is occupied in the bra $[\mathbf{h}'_1 \ldots \mathbf{p}'_m|$ by the particle j and in the ket $|\mathbf{h}_1 \ldots \mathbf{p}_n]$ by the particle i. In calculating the non diagonal matrix elements there are states occupied in the bra which are not occupied in the ket. In this case one has to choose a correspondence between the occupied states in the bra and those in the ket. Correspondingly, the oriented lines are denoted as *transition lines* and have two labels $\mathbf{a}_k$ and $\mathbf{b}_{k'}$ indicating that particle i is in the state $\mathbf{a}_k$ in the ket $|\mathbf{h}_1 \ldots \mathbf{p}_n]$ and $\mathbf{b}_{k'}$ in the bra $[\mathbf{h'}_1 \ldots \mathbf{p'}_m|$. For instance in the diagrams of Fig. 2 the correspondence is $\mathbf{h}_i$ with $\mathbf{p}_i$ and $\mathbf{h}_j$ with $\mathbf{p}_j$.

(ii) The points must be connected by either correlation or exchange lines. In principle all diagrams should have A points. However, diagrams in which uncorrelated particle are exchanged give zero contribution. One is then left with diagrams in which a given number A_c of points is correlated, namely is reached by any of the correlation lines, and each of the remaining $A - A_c$ points is only reached by an isolated loop of a state line. For instance the diagram (1.1) of Fig. 1 has $A_c = 2$ and the remaining $A-2$ points are unlinked points like $\mathbf{r}_3$. Isolated unconnected points give a unity factor and therefore they are omitted in the diagrams. As a consequence, *each point of a diagram must be reached by at least one correlation line* (diagram (1.1) of Fig. 1 is not allowed).

In the cluster diagrams contributing to a n-body distribution function, the n external points (circles) do not need to be reached by correlation lines, since, in the final calculation the distribution function is multiplied either by correlation or interaction functions depending on the variable represented by the external points.

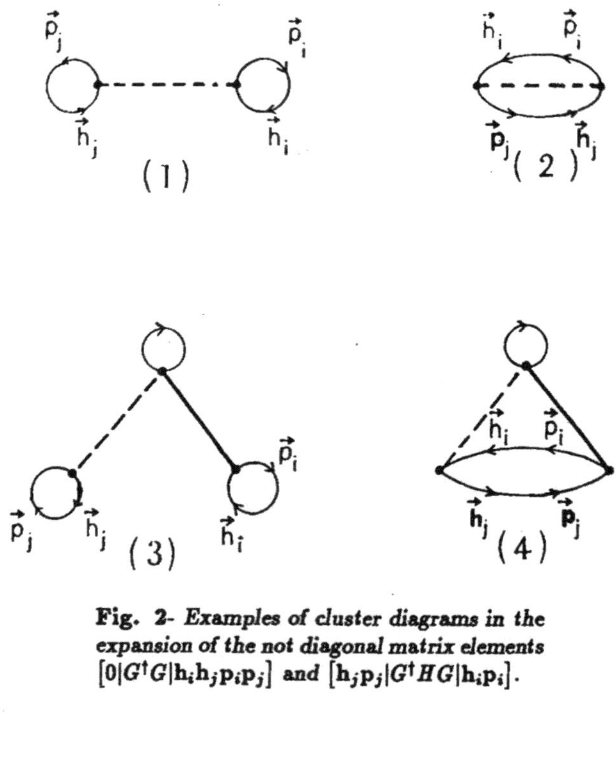

Fig. 2- ***Examples of cluster diagrams in the expansion of the not diagonal matrix elements*** $[0|G^\dagger G|\mathbf{h}_i\mathbf{h}_j\mathbf{p}_i\mathbf{p}_j]$ ***and*** $[\mathbf{h}_j\mathbf{p}_j|G^\dagger H G|\mathbf{h}_i\mathbf{p}_i]$.

The exchange lines must form closed loops and the various loops must be unconnected. The contribution of a diagram is given by:

$$\frac{(-1)^{N_c-N_l}}{s\Omega^{N_c}}\sum_{\mathbf{k}_1,\mathbf{k}_2\cdots}\int d\mathbf{r}_1\cdots d\mathbf{r}_{N_c} < \eta(1)\cdots\eta(N_c)|(\text{product of all the}$$

$$\text{correlation and exchange operators})|\eta(1)\cdots\eta(N_c)>, \tag{2.12}$$

where $\eta(i)$ generically stands for the spin-isospin state of the particle i, Ω is the normalization volume, N_c is the number of points in the diagrams, which equals the number of state-lines and N_l is the number of closed loops. The sum is extended to all the states $\mathbf{k}_i$ which are occupied in both the the ket $|\mathbf{h}_1 \ldots \mathbf{p}_n]$ and in the bra $[\mathbf{h}'_1 \ldots \mathbf{p}'_m|$ and are not explicitely singled out in the calculation. For instance in evaluating diagonal matrix elements one is interested to select those diagrams having one given hole or particle line (see the discussion on the next section).

The factor Ω^{-N_c} comes from the normalization of the plane waves states and the factor $(-1)^{N_c-N_l}$ takes into account the sign changes due to the permutation operators in the Slater wave function. Only one way of labelling the particles is considered in the diagrams. For instance all the diagrams that can be obtained by labelling the points 1,2,3 and 4 of diagram (1.2) in a different way are ignored. There are $N_c!/s$ ways of labelling the particles, where the symmetry factor s equals the number of exchanges which leave the diagram unchanged. The factor $N_c!$ cancels the two $\frac{1}{\sqrt{N_c!}}$ factors due to the normalization of the Fermi gas states.

r) In principle all the labels $\mathbf{k}_1, \mathbf{k}_2 \cdots$ in a diagram are different from each other, however the restriction $\mathbf{k}_1 \neq \mathbf{k}_2 \neq \cdots$ can be ignored because of the antisymmetry of the Slater function. Fig. 3 illustrates that diagrams with two or more identical state lines cancel each other. In fact it is usefull to remove such a restriction, since the summation in eq. (2.10) can be performed for each $\mathbf{k}_i$ independently, leading to

$$A^{N_h} \times (\text{product of exchange functions} \quad l(k_F r_{ij})), \tag{2.13}$$

where

$$l(k_F r) = \frac{\nu\Omega}{(2\pi)^3}\int d\mathbf{k}\Theta(k-k_F)\exp(i\mathbf{k}\cdot\mathbf{r}) = \frac{3}{x^3}(x\cos(x) - \sin(x)), \tag{2.14}$$

with $x = k_F r$ and ν the degeneracy of the system.

By substistuting the factor given in eq. (2.13) into eq. (2.12) one gets the *important* result that the various cluster diagrams have the following simple dependence on A

$$(\text{any cluster diagram}) \propto A^{N_U - N_D}, \tag{2.15}$$

where N_U is the number of unconnected pieces of the diagram (for instance $N_U = 1$ for the diagrams (1.2),(1.4),(1.5) and (2.1-4), $N_U = 2$ for the diagram (1.3) and $N_U = 3$ for

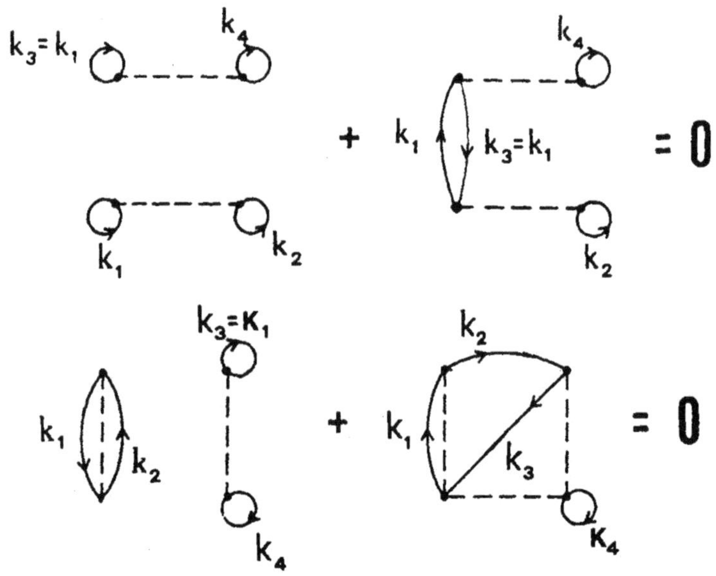

Fig. 3- *Cancellation of diagrams having two or more identical state lines.*

diagram of Fig. 6) and $N_D = N_c - N_k$ is the number of singled out states, which must be lesser than the number of states which are occupied in the ket and not occupied in the bra. The factor A^{N_U} in eq. (2.15) is due to the fact that each unconnected piece of a cluster diagram brings a factor Ω after spatial integration because of the traslational invariance of the system.

2.1a CB Matrix Elements of the Identity Operator.

Let us first calculate the ratio

$$R_{ab} = \frac{[\mathbf{h}_1 \cdots \mathbf{p}_n| G^\dagger G |\mathbf{h}_1 \cdots \mathbf{p}_n]}{[\mathbf{h}'_1 \cdots \mathbf{p}'_m| G^\dagger G |\mathbf{h}'_1 \cdots \mathbf{p}'_m]}, \tag{2.16}$$

which occurs in many expressions. Let $\mathbf{k}_i$ be the single particle states occupied in both the states $|\mathbf{h}_1 \cdots \mathbf{p}_n]$ and $|\mathbf{h}'_1 \cdots \mathbf{p}'_m]$; let $\mathbf{a}_i$ be the N_D states which are occupied in $|\mathbf{h}_1 \cdots \mathbf{p}_n]$ and not occupied in $|\mathbf{h}'_1 \cdots \mathbf{p}'_m]$ and $\mathbf{b}_i$ be the N_D states which are occupied in $|\mathbf{h}'_1 \cdots \mathbf{p}'_m]$ and not occupied in $|\mathbf{h}_1 \cdots \mathbf{p}_n]$. In the following we assume that N_k is proportional to A, whereas N_D is finite.

The numerator of eq. (2.16) contains both connected and unconnected diagrams and, in virtue of the A dependence of the cluster diagrams given in eq. (2.15), can be written in the following factorized form

$$[\mathbf{h}_1 \cdots \mathbf{p}_n| G^\dagger G |\mathbf{h}_1 \cdots \mathbf{p}_n] = \big(1 + \Sigma(\mathbf{a}_1, \cdots, \mathbf{a}_{N_D})\big)\big(1 + \Sigma\big), \tag{2.17}$$

where $\Sigma(\mathbf{a}_1, \cdots, \mathbf{a}_{N_D})$ is the sum of diagrams which contain one or more of the N_D lines $\mathbf{a}_i$ and any number of the lines $\mathbf{k}_j$, and in which a line $\mathbf{a}_i$ occurs at most once, whereas a line $\mathbf{k}_j$ may occur any number of times. Similarly, Σ is the sum of diagrams having only $\mathbf{k}_j$-lines each of which may occur any number of times. It is convenient to simplify the above equation as follows

$$\Sigma(\mathbf{a}_1, \cdots, \mathbf{a}_{N_D}) = \sum_{i=1,N_D} \Sigma_C(\mathbf{a}_i) + \sum_{j>i=1,N_D} (\Sigma_C(\mathbf{a}_i + \mathbf{a}_j) + \Sigma_C(\mathbf{a}_i)\Sigma_C(\mathbf{a}_j)) + \cdots, \tag{2.18}$$

where Σ_C means sum of *connected* diagrams. $\Sigma_C(\mathbf{a}_i)$ sums the connected diagrams having one $\mathbf{a}_i$-line and any number of $\mathbf{k}_j$-lines, $\Sigma_C(\mathbf{a}_i + \mathbf{a}_j)$ sums connected diagrams having one $\mathbf{a}_i$-line and one $\mathbf{a}_j$-line etc.

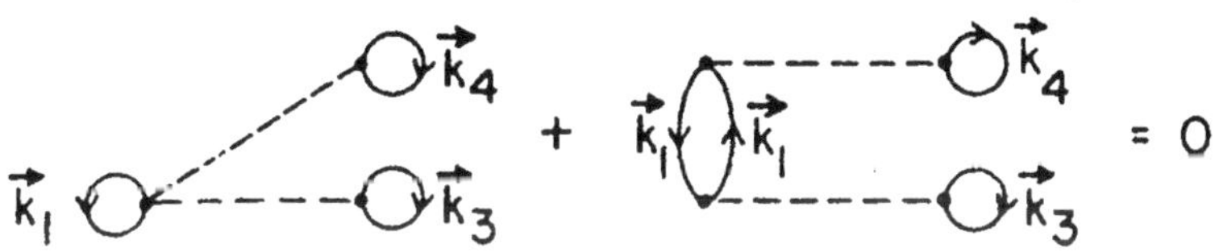

Fig. 4- *Cancellation of reducible diagrams.*

A generic diagram contributing to $\Sigma_C(\mathbf{a}_i)$ is of order 1, since $N_U = 1$ and $N_D = 1$. Similarly, $\Sigma_C(\mathbf{a}_i + \mathbf{a}_j)$ are of order A^{-1}, $\Sigma_C(\mathbf{a}_i + \mathbf{a}_j + \mathbf{a}_k)$ is of order A^{-2} etc., and we neglect them in the thermodynamic limit with the following result

$$[\mathbf{h}_1 \cdots \mathbf{p}_n| \, G^\dagger G \, |\mathbf{h}_1 \cdots \mathbf{p}_n] = \prod_{i=1,N_D} (1 + \Sigma_C(\mathbf{a}_i))(1 + \Sigma_C). \tag{2.19}$$

A similar expression can be obtained for the denominator of eq. (2.16), leading to

$$R_{ab} = \prod_{i=1,N_D} \frac{1 + \Sigma_C(\mathbf{a}_i)}{1 + \Sigma_C(\mathbf{b}_i)}. \tag{2.20}$$

In the case of the *Jastrow* ansatz the sum $\Sigma_C(\mathbf{a}_i)$ is further simplified to give only irreducible diagrams[35]. In fact reducible diagrams cancel each other as illustrated in Fig. 4. Such a cancellation is not complete if state-dependent correlations are present in G. That would have been the case if the correlations in the diagrams of Fig. 4 were of the type $F^\dagger(i,j)F(i,j)$, since $F(1,3)$ does not commute with $F(1,4)$.

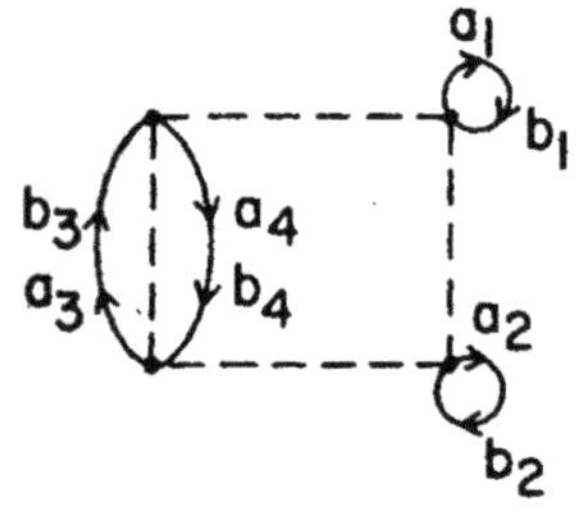

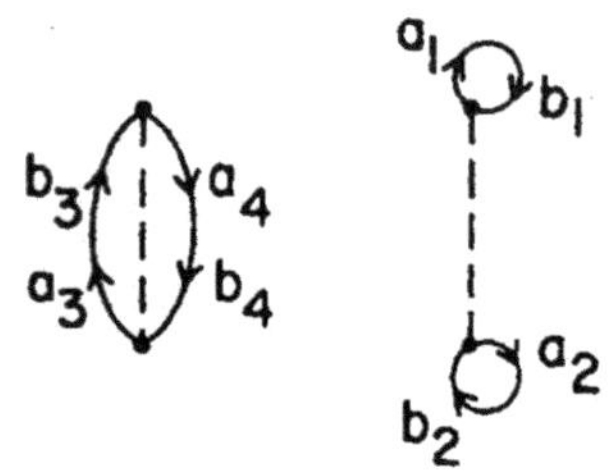

Fig. 5- *Connected and disconnected diagrams in the sum* $\Sigma(\mathbf{a}_1\mathbf{b}_1+\cdots+\mathbf{a}_4\mathbf{b}_4)$.

Non diagonal matrix elements of the identity operator are in general different from zero since the CB state are not orthogonal. The diagrams occurring in the expansion of

$$[\mathbf{h}_1'\cdots\mathbf{p}_m'|\,G^\dagger G\,|\mathbf{h}_1\cdots\mathbf{p}_n]\,, \tag{2.21}$$

have N_D transition lines, namely oriented lines with two state labels. By using the same procedure discussed above one easily gets the result

$$[\mathbf{h}_1'\cdots\mathbf{p}_m'|\,G^\dagger G\,|\mathbf{h}_1\cdots\mathbf{p}_n] = \Sigma(\mathbf{a}_1\mathbf{b}_1+\cdots+\mathbf{a}_{N_D}\mathbf{b}_{N_D})(1+\Sigma), \tag{2.22}$$

where $\Sigma(\mathbf{a}_1\mathbf{b}_1+\cdots+\mathbf{a}_{N_D}\mathbf{b}_{N_D})$ is the sum of diagrams containing one each of the N_D transition lines and any number of $\mathbf{k}_j$-lines. The CB matrix element is then given by

$$\begin{aligned}(\mathbf{h}_1'\cdots\mathbf{p}_m'|\mathbf{h}_1\cdots\mathbf{p}_n) &= \frac{[\mathbf{h}_1'\cdots\mathbf{p}_m'|\,G^\dagger G\,|\mathbf{h}_1\cdots\mathbf{p}_n]}{[\mathbf{h}_1\cdots\mathbf{p}_n|\,G^\dagger G\,|\mathbf{h}_1\cdots\mathbf{p}_n]}\times\left(\frac{[\mathbf{h}_1\cdots\mathbf{p}_n|\,G^\dagger G\,|\mathbf{h}_1\cdots\mathbf{p}_n]}{[\mathbf{h}_1'\cdots\mathbf{p}_m'|\,G^\dagger G\,|\mathbf{h}_1'\cdots\mathbf{p}_m']}\right)^{\frac{1}{2}}\\ &= \frac{\Sigma(\mathbf{a}_1\mathbf{b}_1+\cdots+\mathbf{a}_{N_D}\mathbf{b}_{N_D})}{\prod_{i=1,N_D}(1+\Sigma(\mathbf{a}_i))^{\frac{1}{2}}(1+\Sigma(\mathbf{b}_i))^{\frac{1}{2}}}.\end{aligned} \tag{2.23}$$

Momentum conservation requires that

$$\sum_{i=1,N_D}\mathbf{a}_i=\sum_{i=1,N_D}\mathbf{b}_i\quad. \tag{2.24}$$

If there are no subsets containing a number $d < N_D$ of $\mathbf{a}_i$ and $\mathbf{b}_i$ states, such that

$$\sum_{i\in d} \mathbf{a}_i = \sum_{i\in d} \mathbf{b}_i, \tag{2.25}$$

all the transition lines must occur in a single connected diagram, and the order of magnitude of the matrix element (2.23) is A^{-N_D+1}. It has to be noticed that, since the momentum conservation requires a change in at least two states, then $N_D \geq 2$. Let N_U be the number of subsets satisfying eq. (2.25) with $N_U = 1$ if there are no such subsets. Since each subset must have changes in two or more momentum states, then $2N_U < N_D$. If $N_U \geq 2$ we can have disconnected diagrams, as illustrated in diagram (5.2), contributing to the numerator of eq. (2.23). In this case (2.23) is of the order A^{-N_D+2}. In general one has

$$[\mathbf{h}'_1 \cdots \mathbf{p}'_m|\mathbf{h}_1 \cdots \mathbf{p}_n] \quad \propto A^{-N_D+N_U}, \tag{2.26}$$

with $N_U \leq N_D/2$ and $N_U \geq 1$. The leading terms in A correspond to diagrams which have a number N_U of disconnected pieces and can therefore be factorized into the product of N_U overlap matrix elements.

b CB Matrix Elements of the Hamiltonian.

Let us first consider the diagonal matrix elements of the hamiltonian given in eq. (1.1). The terms of $[\mathbf{h}_1 \ldots \mathbf{p}_n|G^\dagger H\, G|\mathbf{h}_1 \ldots \mathbf{p}_n]$ in which the kinetic energy operator ∇_i^2 operates on the uncorrelated ket $|\mathbf{h}_1 \ldots \mathbf{p}_n]$ give

$$T_F(\mathbf{h}_1 \ldots \mathbf{p}_n) \quad [\mathbf{h}_1 \ldots \mathbf{p}_n|G^\dagger G|\mathbf{h}_1 \ldots \mathbf{p}_n], \tag{2.27}$$

where

$$T_F(\mathbf{h}_1 \ldots \mathbf{p}_n) = \sum_{i=1,A} \frac{\hbar^2}{2m} k_i^2. \tag{2.28}$$

The above sum is extended to all the occupied states in $|\mathbf{h}_1 \ldots \mathbf{p}_n]$, denoted as $\mathbf{k}_i, i = 1, A$. The other terms of $[\mathbf{h}_1 \ldots \mathbf{p}_n|G^\dagger H\, G|\mathbf{h}_1 \ldots \mathbf{p}_n]$ contain either one of the potentials v_{ij}, v_{ijk} or gradients of correlations $\nabla_i^2 f_{ij}, \nabla_i f_{ij} \cdot \nabla_i f_{ik}, \nabla_i f_{ij} \cdot \nabla_i |\mathbf{h}_1 \ldots \mathbf{p}_n]$, and they are evaluated with cluster expansions. In the diagrammatic notation each of the *interaction* functions $f_{ij} v_{ij} f_{ij}, f_{ij} \nabla_i^2 f_{ij}$ etc. is represented by a thick line, and it is straightforward to show that

$$[\mathbf{h}_1 \ldots \mathbf{p}_n|G^\dagger H\, G|\mathbf{h}_1 \ldots \mathbf{p}_n] = \big(T_F(\mathbf{h}_1 \ldots \mathbf{p}_n) + \Sigma_C(I)\big)(1 + \Sigma), \tag{2.29}$$

where $\Sigma_C(I)$ is the sum of all connected diagrams that have one interaction line and any number of lines $\mathbf{k}_i, i = 1, A$. The diagonal matrix elements are then given by

$$(\mathbf{h}_1 \ldots \mathbf{p}_n|H|\mathbf{h}_1 \ldots \mathbf{p}_n) = T_F(\mathbf{h}_1 \ldots \mathbf{p}_n) + \Sigma_C(I) \quad , \tag{2.30}$$

and they are of order A. One is typically interested in the difference of the above diagonal matrix elements with that evaluated on the ground state $|0)$, which furnishes the excitation energy of the state $|\mathbf{h}_1 \ldots \mathbf{p}_n)$. This is given by

$$(\mathbf{h}_1 \ldots \mathbf{p}_n|H|\mathbf{h}_1 \ldots \mathbf{p}_n) - (0|H|0) = \sum_i \Sigma_C(I + \mathbf{p}_i) - \sum_j \Sigma_C(I + \mathbf{h}_j) + O(\frac{1}{A})$$
$$= \sum_i e(p_i) - \sum_j e(h_j). \tag{2.31}$$

where $\Sigma_C(I + \mathbf{a}_i)$ is the sum of all connected diagrams that have one interaction line and one $\mathbf{a}_i$ line and any number of lines $\mathbf{k}_i, i = 1, A$. The excitation energies are of order 1 if the number of single particle excited states is finite as we are assuming here. Note that in the case of a system at a given temperature $T \neq 0$ the number of single particle states is of the order of A.

The non diagonal matrix element $[\mathbf{h}'_1 \ldots \mathbf{p}'_m|G^\dagger H\, G|\mathbf{h}_1 \ldots \mathbf{p}_n]$ has one term containing $T_F(\mathbf{h}_1 \ldots \mathbf{p}_n)$ and others containing an interaction line and it can be expressed as follows

$$[\mathbf{h}'_1 \ldots \mathbf{p}'_m|G^\dagger H\, G|\mathbf{h}_1 \ldots \mathbf{p}_n] = T_F(\mathbf{h}_1 \ldots \mathbf{p}_n)[\mathbf{h}'_1 \ldots \mathbf{p}'_m|G^\dagger G|\mathbf{h}_1 \ldots \mathbf{p}_n]$$
$$+ \Sigma(I + \mathbf{a}_1\mathbf{b}_1 + \ldots \mathbf{a}_{N_D}\mathbf{b}_{N_D})(1 + \Sigma), \tag{2.32}$$

with the result

$$(\mathbf{h}'_1 \ldots \mathbf{p}'_m|H|\mathbf{h}_1 \ldots \mathbf{p}_n) = \Big(T_F(\mathbf{h}_1 \ldots \mathbf{p}_n)\Sigma(\mathbf{a}_1\mathbf{b}_1 + \ldots \mathbf{a}_{N_D}\mathbf{b}_{N_D}) + \Sigma(I + \mathbf{a}_1\mathbf{b}_1 +$$
$$\ldots \mathbf{a}_{N_D}\mathbf{b}_{N_D})\Big) \Big/ \prod_{i=1,N_D} (1 + \Sigma_C(\mathbf{a}_i))^{\frac{1}{2}} (1 + \Sigma_C(\mathbf{b}_i))^{\frac{1}{2}}. \tag{2.33}$$

If there are N_U subsets in the transition, then the quantity $\Sigma(\mathbf{a}_1\mathbf{b}_1 + \ldots \mathbf{a}_{N_D}\mathbf{b}_{N_D})$ is of order $A^{-N_D+N_U}$ and the leading diagrams of $\Sigma(I+\mathbf{a}_1\mathbf{b}_1 + \ldots \mathbf{a}_{N_D}\mathbf{b}_{N_D})$, illustrated in Fig. 6, are of order $A^{-N_D+N_U+1}$. The Fermi kinetic energy term T_F is of order A, therefore one gets the result

$$(\mathbf{h}'_1 \ldots \mathbf{p}'_m|H|\mathbf{h}_1 \ldots \mathbf{p}_n) \propto A^{-N_D+N_U+1}. \tag{2.34}$$

The largest terms have $N_D = 2, N_U = 1$ and they are of order 1.

The non orthogonality of CB states raises the order of magnitude of the nondiagonal matrix elements of the hamiltonian by a factor A. In the following, we will show that the nondiagonal matrix elements of $H - E_0$ are reduced in order of magnitude of a factor A with respect to those of H. As a consequence they are of the same order of magnitude of those of the identity operator. From eqs. (2.23) and (2.33) it follows that

$$(\mathbf{h}'_1 \ldots \mathbf{p}'_m|H - E_0|\mathbf{h}_1 \ldots \mathbf{p}_n) = \big((T_F(\mathbf{h}_1 \ldots \mathbf{p}_n) - T_F(0) - \Sigma^0_C(I))$$
$$\times\, \Sigma(\mathbf{a}_1\mathbf{b}_1 + \ldots \mathbf{a}_{N_D}\mathbf{b}_{N_D}) + \Sigma(I + \mathbf{a}_1\mathbf{b}_1 + \ldots \mathbf{a}_{N_D}\mathbf{b}_{N_D})\big)$$
$$\Big/ \prod_{i=1,N_D} (1 + \Sigma_C(\mathbf{a}_i))^{\frac{1}{2}} (1 + \Sigma_C(\mathbf{b}_i))^{\frac{1}{2}}, \tag{2.35}$$

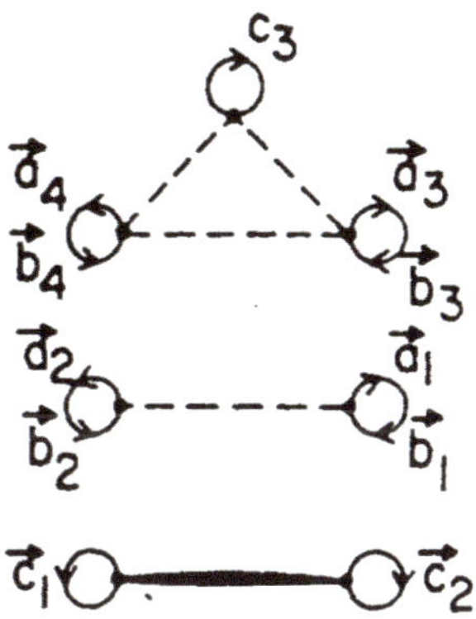

Fig. 6-*Examples of disconnected diagrams in the expansion of* $\Sigma(I + \mathbf{a}_1\mathbf{b}_1 + \ldots \mathbf{a}_{N_D}\mathbf{b}_{N_D})$.

where $\Sigma_C^0(I)$ is the sum of connected diagrams contributing to the expectation value of the hamiltonian on the CB ground state E_0. The difference

$$T_F(\mathbf{h}_1 \ldots \mathbf{p}_n) - T_F(0) = \frac{\hbar^2}{2m} \sum_{i=1,n} (p_i^2 - h_i^2), \tag{2.36}$$

is of order 1, and the quantity $\Sigma(I + \mathbf{a}_1\mathbf{b}_1 + \ldots \mathbf{a}_{N_D}\mathbf{b}_{N_D})$, as illustrated in Fig. 6 can be separated into a factorized part and a connected one, namely

$$\Sigma(I + \mathbf{a}_1\mathbf{b}_1 + \ldots \mathbf{a}_{N_D}\mathbf{b}_{N_D}) = \Sigma(\mathbf{a}_1\mathbf{b}_1 + \ldots \mathbf{a}_{N_D}\mathbf{b}_{N_D})\Sigma_C(I)$$

$$+ \Sigma(\text{con.} \quad I + \mathbf{a}_1\mathbf{b}_1 + \ldots \mathbf{a}_{N_D}\mathbf{b}_{N_D}), \tag{2.37}$$

where $\Sigma_C(I)$ is the sum of connected diagrams with an interaction line and any number of $\mathbf{k}_i$ lines and $\Sigma(\text{con.} \quad I + \mathbf{a}_1\mathbf{b}_1 + \ldots \mathbf{a}_{N_D}\mathbf{b}_{N_D})$ is the subset of $\Sigma(I + \mathbf{a}_1\mathbf{b}_1 + \ldots \mathbf{a}_{N_D}\mathbf{b}_{N_D})$ diagrams in which the interaction line occurs together with a transition line in a connected part of a diagram. The states $\mathbf{k}_i$ include the states $\mathbf{x}_i$ which are not occupied in the ground state $|0)$ and the states $\mathbf{y}_i$ which are occupied in $|0)$ only. We then have

$$\Sigma_C(I) - \Sigma_C^0(I) = \sum_i (\Sigma(I + \mathbf{x}_i) - \Sigma(I + \mathbf{y}_i)) + O(\frac{1}{A}). \tag{2.38}$$

Here, $\Sigma(I + \mathbf{a}_i)$ denotes the subset of Σ_C that has one $\mathbf{a}_i$ and these terms are of order 1. Finally we get the result

$$(\mathbf{h}_1' \ldots \mathbf{p}_m' | H - E_0 | \mathbf{h}_1 \ldots \mathbf{p}_n) = \Big(\Sigma(\text{con.} \quad I + \mathbf{a}_1\mathbf{b}_1 + \ldots \mathbf{a}_{N_D}\mathbf{b}_{N_D})$$

$$+\Sigma(\mathbf{a}_1\mathbf{b}_1+\ldots\mathbf{a}_{N_D}\mathbf{b}_{N_D})\Big(\frac{\hbar^2}{2m}\sum_{j=1,n}(p_j^2-h_j^2)+\sum_i(\Sigma_C(I+\mathbf{x}_i)$$

$$-\Sigma_C(I+\mathbf{y}_i)\Big)\Big/\left(1+\Sigma_C(\mathbf{a}_i)\right)^{\frac{1}{2}}\left(1+\Sigma_C(\mathbf{b}_i)\right)^{\frac{1}{2}}, \tag{2.39}$$

which is of order $A^{-N_D+N_U}$ whereas the matrix element of H is of the order $A^{-N_D+N_U+1}$. If $N_D < 4$ the cluster diagrams are necessarily linked and they do not lead to any divergency in the perturbation theory (see section 2.3). On the contraray, if $N_D \geq 4$ one may have disconnected diagrams which do lead to divergencies. Such divergencies need to be cancelled in order the perturbation theory to be applicable.

2.2 Summations of Cluster Diagrams: FHNC Theory

In most cases of interest large classes of cluster diagrams need to be summed up. Especially for calculations of CB matrix elements in dense systems, like liquid ^3He, it is necessary to perform massive summations of cluster diagrams in order to reach the required accuracy.

A cluster diagrams with a number N_c of vertices has a density factor of ρ^{N_c}. However the density ρ is not necessarily a good expansion parameter, but rather is the combination of both the density and the range of the correlation function $h(r_{ij})$ to give rise to the effective expansion parameter of the cluster term series.

To understand that it is usefull to consider the *chain* diagrams (also denoted as *nodal* diagrams since they have at least one node, namely a point such that any pattern joining the two external points of the diagram must necessarily pass through it) displayed in Fig. 7. It is easy to verify that the Fourier transform of the function $N^{(n)}(r)$ corresponding to a nodal diagram with $n-1$ nodal points is given by

$$\rho\tilde{N}^{(n)}(k)=(\rho\tilde{h}(k))^n, \tag{2.40}$$

where

$$\tilde{f}(k)=\int d\mathbf{r}\exp(i\mathbf{k}\cdot\mathbf{r})f(r). \tag{2.41}$$

In the long wavelength limit $k\to 0$ one has

$$\rho\tilde{N}^{(n)}(k\to 0)=\left(\rho\int d\mathbf{r}h(r)\right)^n, \tag{2.42}$$

and the integral $\int d\mathbf{r}h(r)$ is of order ρ^{-1} in realistic cases. The quantity $\rho\int d\mathbf{r}h(r)$ can be quite large irrespective to the density of the system. For instance, the *optimal Jastrow* factor $h(r)$ in liquid helium has been found[36,37] to have a *long range behaviour*, namely $h(r)\to\beta/r^2$ as $r\to\infty$. The Fourier transform of such a long ranged function diverges

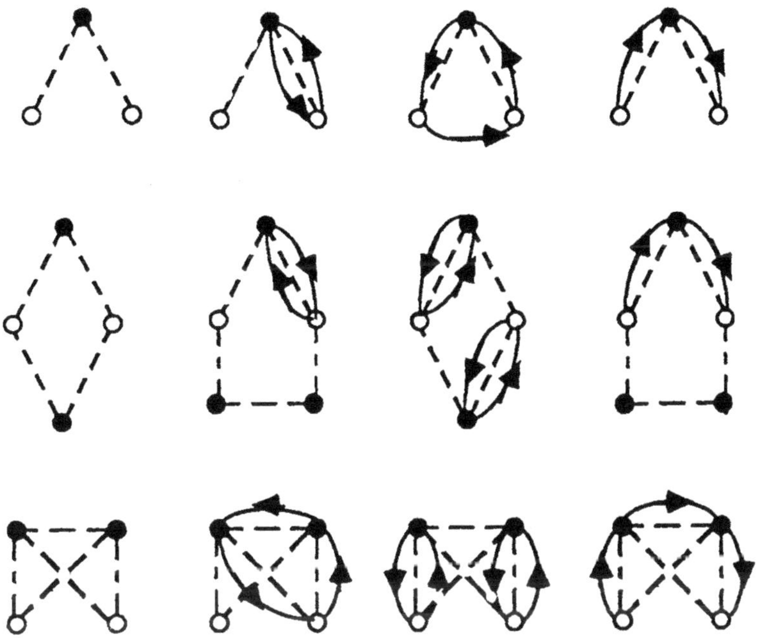

Fig. 7-*Examples of cluster diagrams contributing to the two-body distribution function. The single vertex loops are omitted for simplicity. The diagrams in the first row are nodal, those in the second row are composite and the diagrams in the bottom row are elementary. The diagrams in the first column are of the dd-type; de-, ee- and cc-diagrams are given in the subsequent columns.*

for $k \to 0$ as $-\beta/(2\pi k)$. As a consequence, a chain diagram like the first of Fig. 7, with n correlations diverges as $(1/k)^n$, whereas the total sum of the chain diagrams

$$\sum_{n=2}^{\infty} \tilde{N}^{(n)}(k) = \frac{\rho(\tilde{h}(k))^2}{1-\rho\tilde{h}(k)}, \tag{2.43}$$

has the proper behaviour $\beta/(2\pi k)$. A similar feature is encountered in the electron plasma also, for which the optimal *Jastrow* correlation heals to 1 as $1/r$. It is important to notice that such long range order of the correlation functions is strictly related to the long wavelength limit of the structure function $S(k)$ of the system. In particular they explain the linear behaviour of $S(k)$ for liquid ^3He and the quadratic one for e-plasma as $k \to 0$ which are related to the phonon and plasmon collective excitations respectively.

Not nodal diagrams, like the *composite diagrams* shown in the second row of Fig. 7 and the *elementary* or *bridge* diagrams, displayed in the third row, give important contributions to the short range part of the distribution function and, therefore they have to be taken into account to obtain accurate evaluation of the binding energy of the system.

In practice it seems preferable to perform massive summations of the above diagrams rather than to make use of some expansion parameter. The wound parameter (or correlation volume)

$$K = \rho \int dr (1 - f(r))^2 (1 - \frac{1}{\nu} l^2(k_F r)), \tag{2.44}$$

is often considered to give an indication for the convergence properties of the massive summations of cluster diagrams. In less dense condensed systems, like nuclear matter, the many body contributions are not as important as in liquid helium. However, in doing a variational calculation to get both an upperbound of the ground state energy and the correlation operator G, one has *necessarily* to include the many body contributions, otherwise no minimum is found for the energy.

Massive summations of cluster diagrams can be performed by using the FHNC theory[18,19] and its generalizations to include state dependent correlations[21,22] or to calculate density matrices[38], non diagonal matrix elements[39,40], etc. They are extensively described in review papers[15,20,21,40,41], which the reader is referred to. In the following, we limit ourselves to summarize some of their important features to calculate the quantities discussed in the previous section.

2.2a Diagonal Matrix Elements. Ground State

Let us first consider the diagonal matrix element of the hamiltonian on the ground state $|0)$, which is given by $T_F(0)+\Sigma_C^0$. As discussed in section 2.1b, two- and three-body interaction operators enter into the calculation of Σ_c^0. The terms having two-body interaction central operators (say $f^2(r_{ij})v^c(r_{ij})$, with v^c being the central component of the interparticle potential) are calculated by using eq.(2.11).

The diagrams contributing to the two-body distribution function can be summed up by means of the FHNC theory, which is exemplified in Fig. 8 in the case of correlation bounds of the Bose type. The nodal diagrams $N(r)$ built up with a given bound $X(r)$ are easily summed by using the convolution property and eq. (2.43). Richer and richer bounds are then constructed, by using the property that, in the r space, the composite diagrams can be expressed as products of nodal and elementary diagrams functions, as follows

$$X(r) = f^2(r) \exp(N(r) + E(r)) - N(r) - 1, \tag{2.45}$$

where $E(r)$ represents the sum of elementary diagrams. In principle the above scheme *exactly* sums all the Σ_C^0 diagrams. However the elementary diagrams cannot be summed all, since they involve irreducible multidimensional integrals. The approximation which includes four-body elementary diagrams E_4 is denoted as FHNC/4, that including both E_4 and E_5 is denoted as FHNC/5, etc.

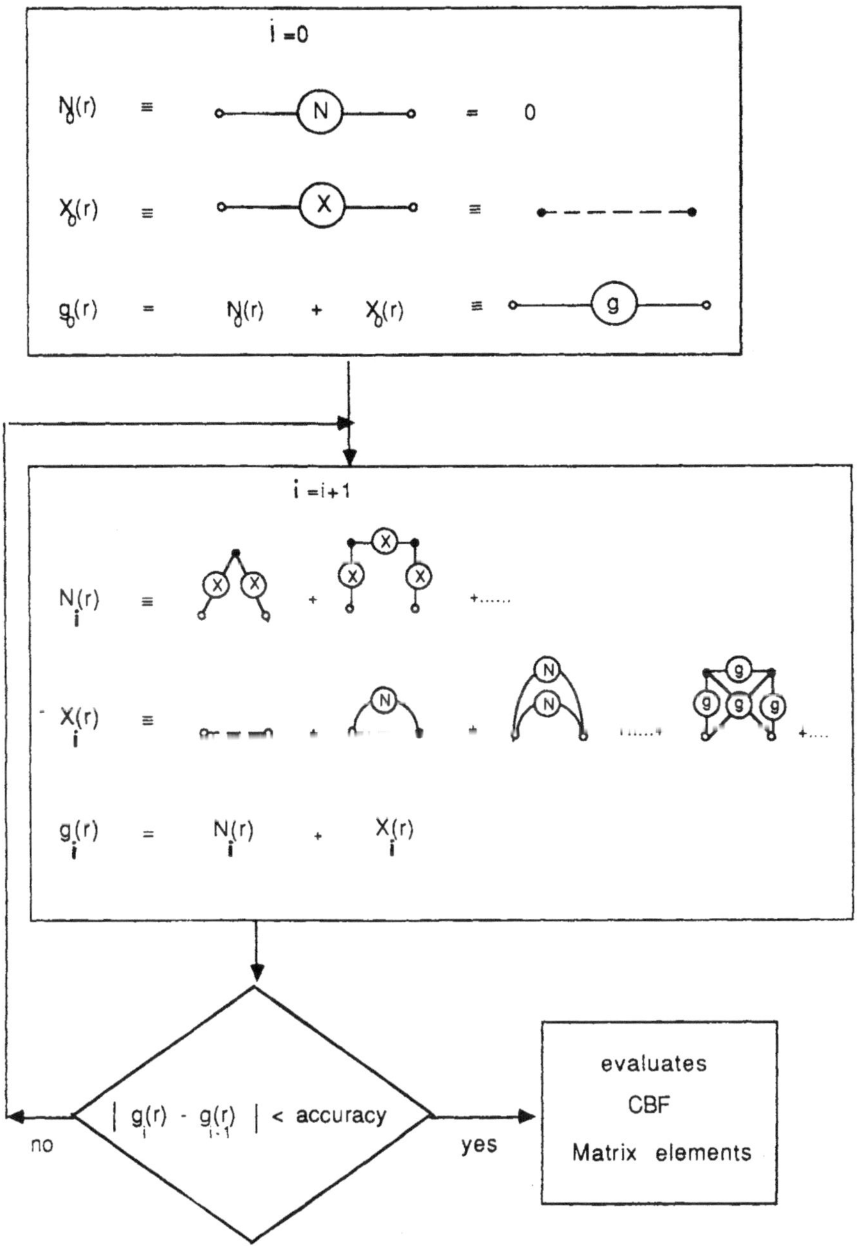

Fig. 8-*Diagrammatical scheme of FHNC summations.*

The actual FHNC equations must take into account that there are statistical correlations in addition to the Bose-type correlations. The scheme discussed above has to be generalized to include $l^2(k_F r_{ij})$ and $l(k_F r_{ij})$ correlations, with the constraints that (i) two any $l^2(k_F r_{ij})$ correlations cannot have a point in common and (ii) the $l(k_F r_{ij})$ correlations must form closed loops (which follow from the diagrammatical rules discussed in section 2.1). This implies that it is necessary to distinguish four types of nodal,composite and elementary diagrams, which are labelled with the subscripts *dd*, *de*, *ee* and *cc*. The subscript *dd* means that the corresponding diagrams have only Bose-type correlations reaching both the external points. The subscript *e* means that two exchange correlations of the type $l(k_F r_{ij})$ reach the corresponding external point and, finally, *c* stands for circular exchange correlations . Such a classification of diagrams is illustrated in Fig. 7. The functions $N_{mn}(r_{ij})$ summing the different types of nodal diagrams denoted as *mn-nodal distribution functions* are obtained by solving the following FHNC equations

$$N_{mn}(r_{ij}) = \rho \sum_{l,l'} \int d\mathbf{r}_k \theta_{ll'} (X_{ml}(r_{ik}) + N_{ml}(r_{ik}))X_{l'n}(r_{kj}), \tag{2.46}$$

where $mn = dd, \quad de, \quad ee, \quad cc$ and $l(l')$ is d or e,

$$\theta_{ll'} = (1 + \delta_{ll',dd})(1 + \delta_{ll',de})(1 + \delta_{ll',ed}) - 1, \tag{2.47}$$

and $\delta_{ll',mn}$ is 1 or 0 wether $ll' = mn$ or not. The nodal function $N_{cc}(r_{ij})$ is given by

$$N_{cc}(r_{ij}) = \rho \int d\mathbf{r}_k (X_{cc}(r_{ik}) + N_{cc}(r_{ik}) - \frac{1}{\nu} l(k_F r_{ik}))X_{cc}(r_{kj}). \tag{2.48}$$

The *composite distribution functions* $X_{mn}(r_{ij})$ are given by

$$X_{dd}(r_{ij}) = g_{dd}(r_{ij}) - N_{dd}(r_{ij}) - 1, \tag{2.49}$$

$$X_{de}(r_{ij}) = g_{dd}(r_{ij})T_{de}(r_{ij}) - N_{de}(r_{ij}), \tag{2.50}$$

$$\begin{aligned} X_{ee}(r_{ij}) = {} & g_{dd}(r_{ij})\{T_{ee}(r_{ij}) - \nu[l(k_F r_{ij})/\nu - T_{cc}(r_{ij})]^2 + T_{de}^2(r_{ij})\} \\ & - N_{ee}(r_{ij}) + l(k_F r_{ij})/\nu, \end{aligned} \tag{2.51}$$

$$X_{cc}(r_{ij}) = g_{dd}(r_{ij})(T_{cc}(r_{ij}) - l(k_F r_{ij})/\nu) - N_{cc}(r_{ij}), \tag{2.52}$$

where

$$g_{dd}(r_{ij}) = f^2(r_{ij}) \exp(T_{dd}(r_{ij})), \tag{2.53}$$

$$T_{mn}(r_{ij}) = N_{mn}(r_{ij}) + E_{mn}(r_{ij}) + C_{mn}(r_{ij}). \tag{2.54}$$

The *mn*-distribution functions $g_{mn}(r_{ij})$ are simply obtained by summing the corresponding *nodal* and *composite* distribution functions

$$g_{mn}(r_{ij}) = N_{mn}(r_{ij}) + X_{mn}(r_{ij}) - \delta_{mn,dd}, \tag{2.55}$$

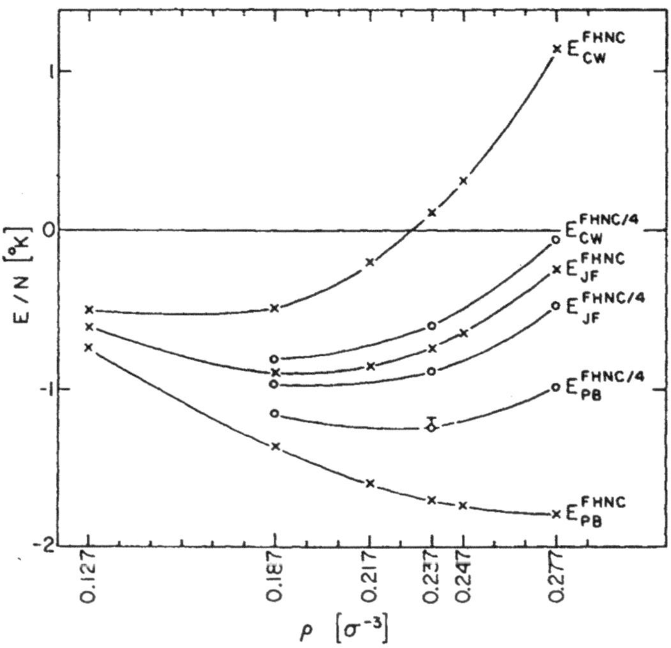

Fig. 9-Taken from ref. (44). Comparison of FHNC and FHNC/4 approximations in liquid ^3He.

and the full two-body distribution function is given by

$$g_2(r_{ij}) = g_{dd}(r_{ij}) + 2g_{de}(r_{ij}) + g_{ee}(r_{ij}), \tag{2.56}$$

The functions $C_{mn}(r_{ij})$ appearing in eq. (2.54) are due to triplet correlations and are obtained by integrating $f_3(\mathbf{r}_{ij}, \mathbf{r}_{ik})$, after being properly dressed with two-body bounds $g_{mn}(r_{ij})$, on the coordinate $\mathbf{r}_k$, namely

$$C_{mn}(r_{ij}) = \rho \sum_{l,l'} \int d\mathbf{r}_k \left[f_3^2(\mathbf{r}_{ij}, \mathbf{r}_{ik}) - 1\right] \theta_{ll'} g_{ml}(r_{ik}) g_{l'n}(r_{kj}), \tag{2.57}$$

for $mn = dd, de, ee$ and

$$C_{cc}(r_{ij}) = \rho \int d\mathbf{r}_k \left[f_3^2(\mathbf{r}_{ij}, \mathbf{r}_{ik}) - 1\right] g_{cc}(r_{ik}) g_{cc}(r_{kj}). \tag{2.58}$$

The three-body distribution function is then given by

$$g_3(\mathbf{r}_{ij}, \mathbf{r}_{ik}) = f_3^2(\mathbf{r}_{ij}, \mathbf{r}_{ik}) \sum_{exch} g_{lm}(r_{ij}) g_{m'n}(r_{jk}) g_{n'l'}(r_{ki}) [1 + A_{l''m''n''}(\mathbf{r}_{ij}, \mathbf{r}_{ik})], \tag{2.59}$$

where the subscripts (l, l', l''), (m, m', m'') and (n, n', n'') follow the exchange rules and the functions A_{lmn} sum the Abe diagrams[43,24,37].

The FHNC equations can be solved numerically by means of iterative procedures similar to that illustrated by Fig. 8. For dense systems like liquid helium the convergence may be difficult to reach. In such cases it is usefull to smooth out the iterative procedure by introducing a reduction parameter x multiplying the nodal functions in the following way

$$N = xN_{new} + (1 - x)N_{old}, \tag{2.60}$$

where N_{new} is obtained from eqs. (2.46-48) if N_{old} and X_{old} are used on their r.h.s. Other methods of numerical solutions of the FHNC equations are discussed in ref. (44).

A crucial point for the accuracy of the FHNC theory in the calculation of the diagonal matrix elements of the hamiltonian is the treatment of the *elementary* and *Abe* diagrams. They are very important in liquid ^3He , as shown in Figs.9-10 but they do not give a large contribution in nuclear matter calculations. In fact most of the calculations in nuclear matter have been performed in FHNC approximation. Figs.9-10 display the energy per particle of the *Jastrow* model of liquid ^3He as a function of the density ρ, calculated by using the *McMillan* correlation function

$$f(r) = \exp\Big(- \frac{1}{2}(b/r)^5\Big), \tag{2.61}$$

with $b = 1.13\sigma$ (Fig. 9) or $b = 1.15\sigma$ (Fig. 10). The expectation value of the kinetic energy can be written in different ways by properly playing with integrations by part. The three commonly used forms[44] are the *Jackson-Feenberg* T_{JF} in the form generalized[45] to infinite Fermi systems by *Fantoni-Rosati*, the *Clark-Westhaus* T_{CW} and the *Pandharipande-Bethe* T_{PB}. Their explicit expressions can be found in refs. (24) and (44). Here we limit ourselves to briefly discuss the relative merits of the three expressions. T_{JF} involves the low order distribution functions only. For instance in the *Jastrow* model the energy expectation value E_{JF} is given by[45,24]

$$E_{JF} = T_F + W_B + W_\phi + U_\phi + U_T, \tag{2.62}$$

where $T_F = 3(\hbar k_F)^2/(10m)$ is the Fermi kinetic energy and the two-body terms W_B and W_ϕ

$$\begin{aligned} W_B = \frac{1}{2}\rho \int d\mathbf{r} g_2(r) v_{JF}(r) = \\ \frac{1}{2}\rho \int d\mathbf{r} g(r)\Big(v(r) - \frac{\hbar^2}{2m}\frac{\nabla^2 f^2 - (f')^2}{f^2}\Big), \end{aligned} \tag{2.63}$$

$$W_\phi = -\frac{\rho}{\nu}\int d\mathbf{r}\Big((g_{dd}(r) - 1)((l - \nu T_{cc})\hat{l} + \frac{\hbar^2}{4m}(l')^2) - \nu g_{dd} E_{cc}\hat{l}\Big), \tag{2.64}$$

$$\hat{l} = \frac{\hbar^2}{4m}\nabla^2 l \quad . \tag{2.65}$$

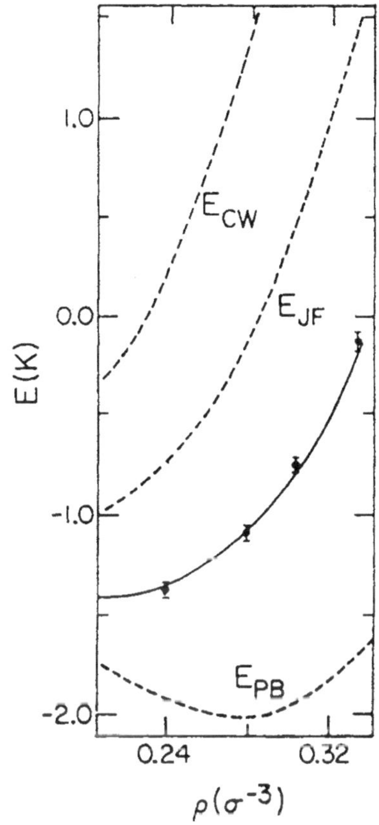

Fig. 10-*Comparison of JF, PB and CW energies in FHNC from ref.(24). The solid curves represent the results in FHNC/s approximation .The FHNC/α results practically coincide with the FHNC/s ones. The data points are the MC results.*

They provide the main contribution. The three-body term U_ϕ is generally very small . The other three-body term U_T is due to triplets and three-body interactions. Their expressions are given in refs. (24) and (45) and are not reported here for sake of brevity.

T_{PB} is obtained without doing any integration by parts and, as a consequence, its two-body operator has a smoother behaviour at small interparticle distances with respect to T_{JF}. It results that a not too great accuracy is required in the calculation of the two-body distribution function. However T_{PB} has terms involving three- and more-body operators[24] which are not at all negligeable and, therefore, it becomes impracticable if triplet correlations are used. Finally, T_{CW} has neither a smooth two-body operator nor has negligeable three- and more-body terms. However the operators involved in it have gradients acting on the correlation operator only and, therefore it may be usefull if one has some good prescription for the distribution functions like for instance in the *Lado* scheme[46,47].

The three kinetic expressions should give exactly the same results if the elementary and Abe terms are correctly included. Fig. 9 shows that FHNC and FHNC/4 are not still satisfactory in the case of liquid helium.

Recently two approximate methods have been devised to include the important contributions from elementary diagrams and they give results which are in very good agreement with the available MonteCarlo calculations, as it can be seen from Fig. 10. Both the *scaling*[48] (FHNC/s) and the *interpolation equation*[49] (FHNC/α) methods were originally derived for the case of Bose systems and the reader is referred to the original papers and to ref. (37) for a complete discussion. Their extensions to Fermi systems are fully described in refs. (24), (25) and are not repeated here.

The above discussion has been limited to the case of central correlations for pedagogical purposes. In fact state dependent correlations have been found to be extremely important for almost all the strongly interacting systems considered up to now. The treatment for summing the cluster diagrams containing wavy lines is technically complicated by the operatorial structure of the correlations and by their non commutativity, which forces to take into account the various orderings of the correlations in both the bra and the ket. An approximation which has been found to be adequate in all the calculations performed is the FHNC/SOC which is described in ref.(23). Higher order corrections[50,51,25] to this approximation have been found to be small both in nuclear matter and in liquid ^3He. A few examples of diagrams included in this approximation are displayed in Fig. 11.

All the operatorial chains and the operatorial rings used as vertex corrections in the separable diagrams are built with single operator correlations, namely wavy or exchange lines. The interaction line is always dressed with the full operatorial correlation $F^\dagger(mn)F(mn)$. Therefore one may have up to five operatorial links on the interaction points and they are treated exactly. The chain-separable terms are not very large and are generally approximated. Operatorial rings are also used to vertex correct the internal points not involved by the interaction and also added to the the central bounds used in the FHNC summation. The diagrams in which two or more operatorial rings are connected each other are treated in an approximate way.

2.2b- Diagonal Matrix Elements: excited states.

In this section we briefly review the methods to calculate the expectation value of the hamiltonian on the CB excited states, whose cluster properties have been discussed in section 2.1b. The quantities to be evaluated are the sums $\Sigma_C(I + \mathbf{a}_i)$ of all connected diagrams having one interaction line I and one state line $\mathbf{a}_i$ and any number of state lines

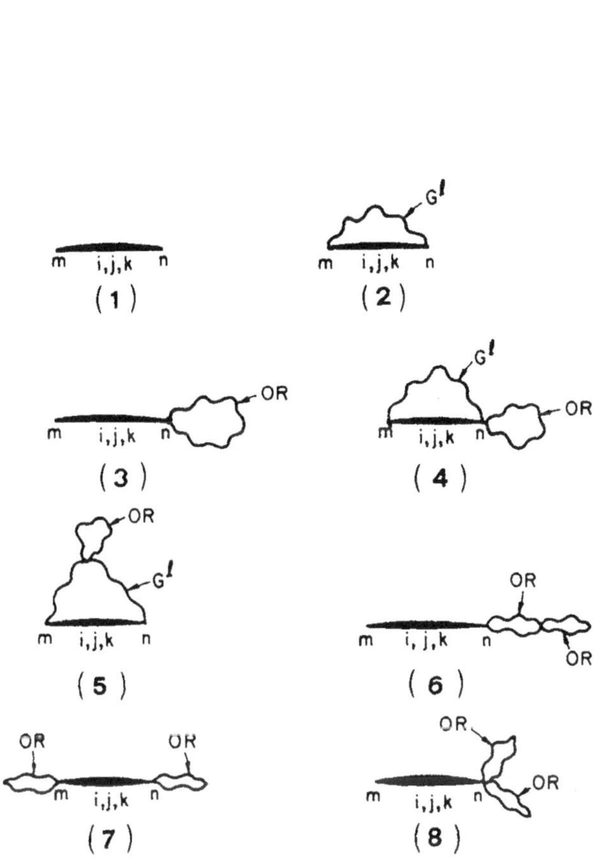

Fig. 11 *Diagrams included in the FHNC/SOC approximation. The diagrams may contain any number of hypernetted chains of central correlations that are not shown as well as the exchanges. All the operatorial chains or rings displayed are built with single operators, namely with single wavy lines.*

$\mathbf{k}_i, i = 1, A$ below the Fermi sea, given in eq. (2.31).

The diagrams to be summed up may be considered as two-body distribution function diagrams, namely diagrams having two external points, from one of which the state line $\mathbf{a}_i$ leaves. Such a state line either it is closed on itself, as in the case of the direct terms which therefore do not depend explicitely on the variable $\mathbf{a}_i$, or goes into the other external point in the case of the exchange terms, which then are in the form of Fourier transforms of proper distribution functions with $\mathbf{a}_i$ as argument. In calculating the energy of n particle-hole CB excited states only the difference of Σ_C quantities enter (see eq. (2.31)), therefore only the exchange terms need to be calculated.

One of the correlation lines of the diagrams has necessarily to be an interaction line I. For the *Jastrow* model and the E_{JF} energy prescription such a line can either come from the term W_B of eq. (2.63) and therefore be $f^2 v_{JF}$ or come from the W_Φ term and be $\hat{l}$ or $(l')^2$.

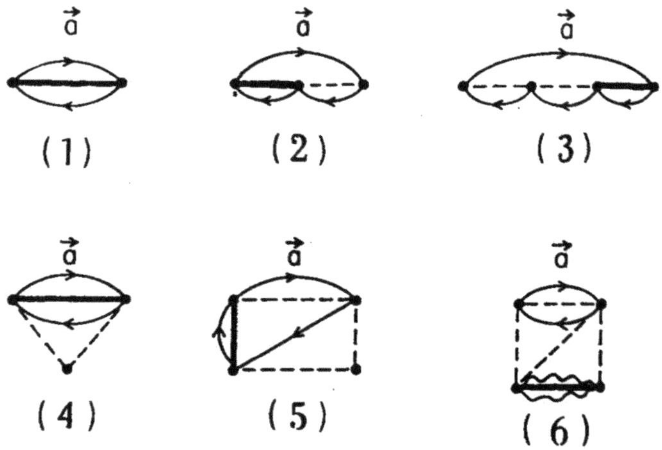

Fig. 12-*Cluster diagrams contributing to $\Sigma_C(I+\mathbf{a}_i)$ and summed up by FHNC' or FHNC$^\lambda$ equations.*

A few examples of such diagrams are displayed in Fig. 12.

The FHNC equations to sum these diagrams are usually denoted as FHNC' equations. An easy way to derive them is to linearize the FHNC equations using the various interaction terms as driving terms as discussed below. In the case of the Jastrow + triplet model and the E_{JF} prescription one has to consider the following composite functions

$$X_{dd} = g_{dd} - N_{dd} - 1, \tag{2.66}$$

$$X_{de} = g_{dd}T_{de} - N_{de}, \tag{2.67}$$

$$X_{ee} = g_{dd}\{T_{ee} - \nu[(l + \lambda\hat{l})/\nu - T_{cc}]^2 + T_{de}^2 - \frac{\hbar^2}{2m}\lambda(l')^2/\nu\} - N_{ee}, \tag{2.68}$$

$$X_{cc} = g_{dd}[T_{cc} - (l + \lambda\hat{l})/\nu] - N_{cc} + (l + \lambda\hat{l})/\nu, \tag{2.69}$$

where

$$g_{dd} = f^2(1 + \lambda v_{JF})\exp(T_{dd}), \tag{2.70}$$

Similarly the triplet functions C become

$$C_{mn}(r_{ij}) = \rho \sum_{exch} \int d\mathbf{r}_k \left[f_3^2(\mathbf{r}_{ij}, \mathbf{r}_{ik}) \left(1 - \lambda\frac{\hbar^2}{8m} \sum_{cycl} \nabla_i^2 \ln f_3^2(\mathbf{r}_{ij}, \mathbf{r}_{ik}) \right) - 1 \right]$$
$$\times\, g_{ml}(r_{ik}) g_{l'n}(r_{kj}), \tag{2.71}$$

where the interaction term multiplied by λ in the last equation comes from U_T. The chain equations (2.46)-(2.48) are left unchanged. The various quantities entering into the FHNC' equations are then simply given by

$$Y'(r) = \left(\frac{\partial Y(r,\lambda)}{\partial \lambda}\right)_{\lambda=0}, \tag{2.72}$$

where Y stands for any of the FHNC functions N, X, E or g.

The partial derivative in the above equation can be performed numerically, by solving the FHNC$^\lambda$ equations (2.66)-(2.71) keeping λ as a smallness parameter. This method is particularly usefull since it mantains all the approximations used in the ground state calculation for the elementary diagrams. Its extension to include state dependent correlations and three-body forces is straightforward.

The formal application of eq. (2.72) leads to the FHNC' equations [39,15,41,52] whose derivation is left as an exercise to the interested reader.

The excitation energy of a n particle-n hole CB state given in eq. (2.31) consists of a Fermi kinetic energy term and of a set of cc-type diagrams . The single particle energy is of the form

$$e(k) = \frac{\hbar^2}{2m}k^2 + U_0 + U(k), \tag{2.73}$$

where the term U_0 is a constant originating from those diagrams in which the state line k is not exchanged. The momentum dependent part is then given by

$$U(k) = \frac{\rho \tilde{X}'_{cc}(k)}{D(k)} + \frac{\hbar^2}{4m}\left(k^2 \rho \tilde{X}_{cc}(k) + 2 \quad Y(k)\right), \tag{2.74}$$

where $D(k) = 1 - \rho\tilde{X}_{cc}(k)$ and the function $Y(k)$ comes from U_ϕ and the $(l')^2$ part of W_ϕ of eq.(2.64). Its explicit expression may be found in ref.(41).

The single particle energy $e(k)$ can also be calculated by infinitesimally changing the uncorrelated momentum distribution of the ground state. This can be performed numerically[53] by removing a small fraction x of particles from a thin spherical shell at $k = k_F$ $(k = h$ in case of hole states) and putting them in a thin spherical shell at $k = p$ $(k = k_F)$. Up to terms linear in x we have

$$\frac{E(x,k)}{A} = \frac{E_0}{A} + x[\pm e(k) \mp e(k_F)], \tag{2.75}$$

where the upper and lower signs refer to particles ($k > k_F$) and holes ($k < k_F$) respectively. $E(x,k)$ is computed as E_0 by using the modified Slater function

$$l(x,k,r) = l(k_F r) \pm \left(\frac{\sin(kr)}{kr} - \frac{\sin(k_F r)}{k_F r}\right), \tag{2.76}$$

and consequently modifying the k-dependent terms.

2.2c- Non diagonal matrix elements.

Let us first analyze the overlap matrix element between the ground state $|0)$ and the *2-ph* CB state $|\mathbf{h}_1\mathbf{h}_2\mathbf{p}_1\mathbf{p}_2)$, whose general expression is given in eq. (2.23). We use the phase convention that $\mathbf{h}_1$ goes into $\mathbf{p}_1$. The diagrams contributing to $\Sigma(\mathbf{h}_1\mathbf{p}_1+\mathbf{h}_2\mathbf{p}_2)$ have two transition lines and any number of $\mathbf{k}_i$ lines.

Because of momentum conservation both the transition lines belong to the same linked diagram ($N_U=1$), therefore the diagrams contributing to $\Sigma(\mathbf{h}_1\mathbf{p}_1+\mathbf{h}_2\mathbf{p}_2)$ are linked.

It is convenient to separate the set of diagrams into two subsets: (i) the set of direct diagrams, in which the two transition lines belong to two separate exchange loops ; (ii) the set of exchange diagrams, in which the two transition lines belong to the same exchange loop

$$\Sigma(\mathbf{h}_1\mathbf{p}_1+\mathbf{h}_2\mathbf{p}_2)=\Sigma_{dir}(\mathbf{h}_1\mathbf{p}_1+\mathbf{h}_2\mathbf{p}_2)+\Sigma_{exch}(\mathbf{h}_1\mathbf{p}_1+\mathbf{h}_2\mathbf{p}_2). \tag{2.77}$$

Diagrams (13.1), (13.4) and (13.6) of Fig. 13 are direct whereas diagrams (13.2) and (13.7) are of the exchange type (in the Figure $i=1$ and $j=2$).

Let us briefly discuss the structure of the direct term Σ_{dir}. In ref. (40) it is shown that, in the case of state independent correlation, they can be factorized into the product of irreducible quantities. For each of the two transition lines one can use a folding procedure which shows such a factorizability property. The folding procedure of the transition lines is illustrated in Fig. 14 and can be applied to diagrams (13.4,7). As a consequence of this important property the diagrams can be recast in such a way as to have a common separable part given by

$$V(\mathbf{h}_1\mathbf{h}_2\mathbf{p}_1\mathbf{p}_2)=(1+\Sigma_C(\mathbf{h}_1))(1+\Sigma_C(\mathbf{h}_2))(1+\Sigma_C(\mathbf{p}_1))(1+\Sigma_C(\mathbf{p}_2)). \tag{2.78}$$

After the diagrammatical exclusion of such a separable part one is left with a two- plus a three- plus a four-body irreducible structures, in which the separabilty points can be considered as external points. Therefore one has

$$\Sigma_{dir}(\mathbf{h}_1\mathbf{p}_1+\mathbf{h}_2\mathbf{p}_2)=V(\mathbf{h}_1\mathbf{h}_2\mathbf{p}_1\mathbf{p}_2)\sum_{i=2,4}\Sigma_{dir}^{Irr,i}(\mathbf{h}_1\mathbf{p}_1+\mathbf{h}_2\mathbf{p}_2), \tag{2.79}$$

where the label i indicates the number of external points of the corresponding diagrams. For instance, diagrams (13.1-4) and (13.6,7) have $i=2$ and the diagram (13.5) has $i=3$. There is a partial cancellation of the quantities at the denominator of eq. (2.23) and the separable terms given in eq. (2.78) with the result

$$\begin{aligned}(0|\mathbf{h}_1\mathbf{h}_2\mathbf{p}_1\mathbf{p}_2)=&(1+\Sigma_C(\mathbf{h}_1))^{\frac{1}{2}}(1+\Sigma_C(\mathbf{h}_2))^{\frac{1}{2}}(1+\Sigma_C(\mathbf{p}_1))^{\frac{1}{2}}(1+\Sigma_C(\mathbf{p}_2))^{\frac{1}{2}}\\&\times\sum_{i=2,4}\left(\Sigma_{dir}^{Irr,i}(\mathbf{h}_1\mathbf{p}_1+\mathbf{h}_2\mathbf{p}_2)+\Sigma_{exch}^{Irr,i}(\mathbf{h}_1\mathbf{p}_1+\mathbf{h}_2\mathbf{p}_2)\right).\end{aligned} \tag{2.80}$$

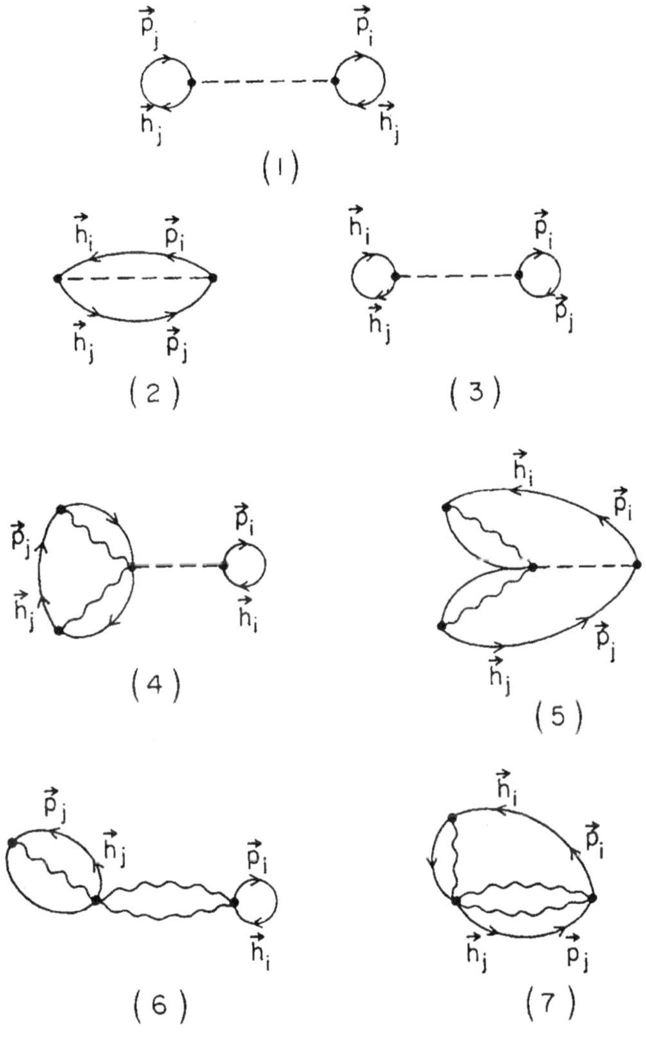

Fig. 13- *Examples of diagrams contributing to* $\Sigma(\mathbf{h}_i\mathbf{p}_i + \mathbf{h}_j\mathbf{p}_j)$.

In the *Jastrow model* and FHNC approximation one has

$$(1+\Sigma_C(\mathbf{k})) = \frac{1}{1-\rho\tilde{X}_{cc}(k)}, \tag{2.81}$$

$$\Sigma_{dir}^{Irr,2}(\mathbf{h}_1\mathbf{p}_1 + \mathbf{h}_2\mathbf{p}_2) = \Gamma_{dd}(k), \tag{2.82}$$

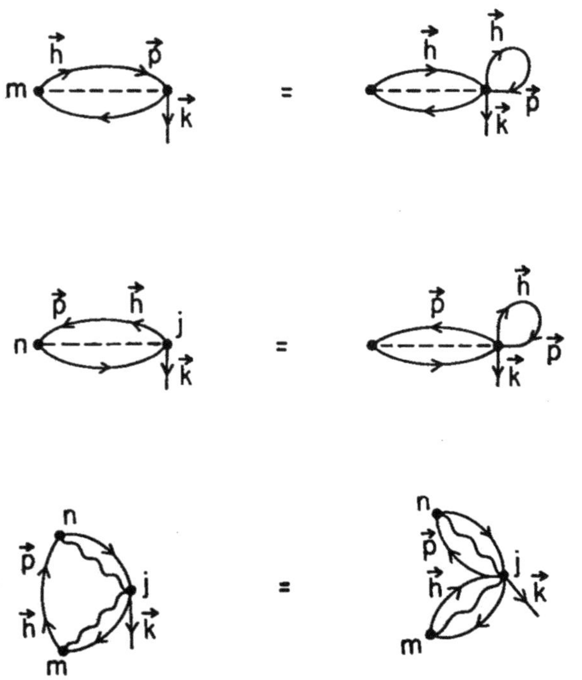

Fig. 14- *Diagrammatical equations showing the folding procedure discussed in the text.*

$$\Gamma_{dd}(k) = \rho(\widetilde{g_{dd} - 1})(k), \tag{2.83}$$

where $\mathbf{k} = \mathbf{p}_1 - \mathbf{h}_1 = \mathbf{h}_2 - \mathbf{p}_2$ is the momentum transfer.

The diagrams contributing to the exchange part Σ_{exch} have the same diagrammatical representation of $\Sigma_{dir}(\mathbf{h}_1\mathbf{p}_2 + \mathbf{h}_2\mathbf{p}_1)$. In fact, by changing the phase convention the direct terms become the exchange terms and viceversa. Therefore the exchange terms can be formally obtained from the direct ones by changing the momentum transfer $\mathbf{k}$ into the exchanged momentum transfer $\mathbf{s} = \mathbf{p}_2 - \mathbf{h}_1$ and by including an extra exchange operator P_{ij}, where i and j are the particles being in the states $\mathbf{p}_1$ and $\mathbf{p}_2$ in the ket. For instance, both the diagram (13.2) and (13.3) correspond to the same exchange term. In the *Jastrow* model the exchange operator carries a factor $-1/\nu$ only, therefore one gets the result

$$\Sigma_{exch}^{Irr,2}(\mathbf{h}_1\mathbf{p}_1 + \mathbf{h}_2\mathbf{p}_2) = -\frac{\Gamma_{dd}(s)}{\nu}. \tag{2.84}$$

In the case of the *Correlation Operator* model the factorizations discussed above are not valid anymore and one has extra terms which take into account of the non-commutativity of the state dependent correlations. The explicit expressions for the overlap matrix elements in this more realistic case are based on the FHNC/SOC algebra and are reported in ref. (40).

A complete similar discussion can be done for the non diagonal matrix elements of the hamiltonian, given in eq. (2.39), which, in the case of a matrix element between the ground state and a $2 - ph$ state becomes

$$(0|H - E_0|\mathbf{h}_1\mathbf{h}_2\mathbf{p}_1\mathbf{p}_2) = (0|\mathbf{h}_1\mathbf{h}_2\mathbf{p}_1\mathbf{p}_2)\times$$

$$\Big(\frac{\hbar^2}{2m}(p_1^2 + p_2^2 - h_1^2 - h_2^2) - \Sigma_C(\text{con.}\ \ I + \mathbf{h}_1) - \Sigma_C(I + \mathbf{h}_2)\Big)$$

$$+ \frac{\Sigma(I + \mathbf{h}_1\mathbf{p}_1 + \mathbf{h}_2\mathbf{p}_2)}{(1 + \Sigma_C(\mathbf{h}_1))^{\frac{1}{2}}(1 + \Sigma_C(\mathbf{h}_2))^{\frac{1}{2}}(1 + \Sigma_C(\mathbf{p}_1))^{\frac{1}{2}}(1 + \Sigma_C(\mathbf{p}_2))^{\frac{1}{2}}}, \tag{2.85}$$

In fact the quantity $\Sigma(\text{con.}\ \ I + \mathbf{h}_1\mathbf{p}_1 + \mathbf{h}_2\mathbf{p}_2)$ has similar cluster properties as the corresponding quantity of the overlap matrix, since momentum conservation implies that the interaction line and the two transition lines belong to the same linked diagram. With respect to the preceding discussion here one has to use the FHNC' quantities instead of the FHNC ones. There are terms in which the interaction line I is contained in the Σ^{irr} part and, in the case of the simple *Jastrow* model they give

$$\Sigma_{dir}^{Irr,2}(I + \mathbf{h}_1\mathbf{p}_1 + \mathbf{h}_2\mathbf{p}_2) = \rho\tilde{g}'_{dd}(k), \tag{2.86}$$

and similarly for the exchange term. Other terms have the interaction line contained in the separable part given by eq.(2.78) with $1 + \Sigma_C(a)$ going to $\Sigma_C(I + a)$ and they partially cancel the separable terms in eq.(2.85). Explicit expressions for the non diagonal matrix elements will be given in sections 3 and 4 in connection with the discussion of the calculations performed for the *homework* hard-sphere problem and in liquid ^3He.

In nuclear matter calculations the state dependent correlations technically complicate the massive summation of many body cluster terms. However, once the correlation operator is given, at the equilibrium density, the many body diagrams containing operatorial correlations are not giving large contributions and therefore they can be neglected without loosing too much of accuracy. Massive summations of many body cluster diagrams can be technically performed by using the FHNC/SOC approximation. This has been done in the calculation of the response functions of nuclear matter[40,54] and work is in progress to evaluate at the same degree of accuracy the non diagonal matrix elements of the hamiltonian, which crucially enters in several calculations reviewed in section 5, in the form of an effective interaction.

In the low order cluster calculations of the matrix elements of the hamiltonian it is necessary to include the two-body terms as well as the three-body separable terms because of the cancellation with normalization factors as previously discussed.

At every order of the Power Series cluster expansion[25] these cancellations are preserved. Such cluster diagrams are displayed in Fig. 15. In eq. (2.85) the quantity $(0|\mathbf{h}_1\mathbf{h}_2\mathbf{p}_1\mathbf{p}_2)$ has not zeroth order contributions. In first order it is given by diagrams (13.1) and (13.2).

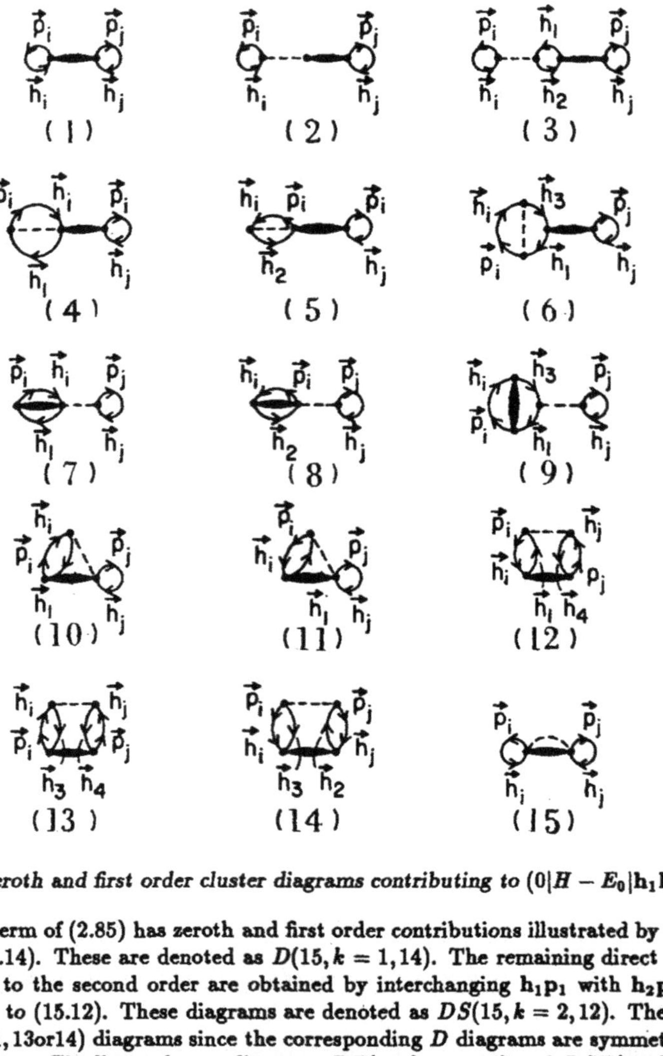

Fig. 15- *Zeroth and first order cluster diagrams contributing to* $(0|H - E_0|\mathbf{h}_1\mathbf{h}_2\mathbf{p}_1\mathbf{p}_2)$.

The second term of (2.85) has zeroth and first order contributions illustrated by diagrams (15.1) to (15.14). These are denoted as $D(15, k = 1, 14)$. The remaining direct diagrams contributing to the second order are obtained by interchanging $\mathbf{h}_1\mathbf{p}_1$ with $\mathbf{h}_2\mathbf{p}_2$ in diagrams (15.2) to (15.12). These diagrams are denoted as $DS(15, k = 2, 12)$. There are no $DS(15, k = 1, 13\text{or}14)$ diagrams since the corresponding D diagrams are symmetric under this interchange. Finally, exchange diagrams $DE(15, k = 1, 14)$ and $DSE(15, k = 2, 12)$ are obtained by exchanging $\mathbf{h}_1$ with $\mathbf{h}_2$. We obtain

$$(0|H - E_0|\mathbf{h}_1\mathbf{h}_2\mathbf{p}_1\mathbf{p}_2) = T_1 + \sum_{k=1,14} \big(D(15,k) + DE(15,k)\big)\big(1 - \frac{1}{2}(\delta_{k4} + \delta_{k5})\big)$$

$$+ \sum_{k=2,12} \big(DS(15,k) + DSE(15,k)\big)\big(1 - \frac{1}{2}(\delta_{k4} + \delta_{k5})\big), \tag{2.87}$$

where T_1 is the first term in eq. (2.85). Note that there is a partial cancellation of diagrams having $k = 8$ with the terms of T_1. Diagrams with $k = 9, 13, 14$ have not been included in the *two- plus three-body separable* approximation used to calculate the effective interaction in CO model of nuclear matter[31] and discussed in section 5.

Pertubative correction treatments in correlated theory.

In this section we summarize the main features of the two mostly used perturbative schemes in CB theory, namely the not orthogonal scheme CBPT and a recently developed orthogonal one. It is known that if all CB-states are orthogonalized together the resulting orthogonal states are not as good as the original CB states under several respects. For instance, the expectation value of the hamiltonian in the ground state of the orthogonalized set is higher than E_0 of the CB ground state[14,55] whereas the ground state eigenvalue must be below E_0 in virtue of the variational principle. Therefore a two-step process, in which the CB states are first orthogonalized by using *Löwdin* transformations and then used in perturbation theory is not recommended. The *Löwdin* orthogonalization moves the energies of low lying states up, and the perturbative corrections move them down, both effects being larger than the net displacement[14,55]. In not orthogonal correlated theory only the net displacement is calculated.

However it is not clear how important are the spuriousities of the orthogonality corrections which are present at every order of CBPT. Moreover, in the calculation of other quantities than the perturbative corrections to the energy eigenvalues, like, for instance, that of the response functions or the spectral function, one needs to have sets of excited states. It has been found that in order to avoid double counting and to properly satisfy the sum rule properties it is necessary to have orthogonal states. For this reason we outline in section 2.3b an orthogonal correlated basis (OCB) theory which preserves the CB diagonal matrix elements in the thermodynamic limit. Therefore conventional perturbation theory with these orthogonal states is expected to have a convergence as good as that of CBPT.

a- Not orthogonal CBF perturbation theory.

In not orthogonal CBF theory one treats perturbatively the interaction hamiltonian of eq.(2.4). The correction to the variational estimate E_n^v, given by the diagonal matrix element of the hamiltonian on the CB state $|n)$, is obtained by the non orthogonal perturbation theory expression[56]

$$\Delta E_n = E_n - E_n^v = \sum_{p\neq n} \frac{|H_{np} - E_n N_{np}|^2}{E_n - E_p^v} + \sum_{p\neq q\neq n} \frac{(H_{np} - E_n N_{np})(H_{pq} - E_n N_{pq})(H_{qn} - E_n N_{qn})}{(E_n - E_p^v)(E_n - E_q^v)} + \ldots, \qquad (2.88)$$

where the matrix elements H_{ij} and N_{ij} are defined in eqs. (2.5) and (2.6) respectively.

The various orders $\Delta E_n^{(i)}$ of the perturbative series are obtained by expanding eq.(2.88) around $E_n = E_n^v$ and then inserting ΔE_n back on the r.h.s. repeatedly.

The second order perturbative correction is given by

$$\Delta E_n^{(2)} = \sum_{p \neq n} \frac{|H_{np} - E_n^v N_{np}|^2}{E_n^v - E_p^v}, \tag{2.89}$$

and it has been applied to study nuclear matter[31,32] and liquid ^3He[29,30] within the set of $2p-2h$ CB states.

A poorly adapted feature of this series is that each $\Delta E_n^{(i)}$ has a catastrophic dependence on the particle number A. We have seen in the section 2.1 that the quantity $H_{ij} - E_0^v N_{ij}$ is given in general by the sum of well behaved *linked* terms and one or more *unlinked* terms which are written as products of linked quantities. These unlinked terms arise if the CB states $|i)$ and $|j)$ differ in more than three orbitals $d_{ij} \geq 4$ and diverge in the thermodynamic limit $A, \Omega \to \infty$ with ρ kept constant. Similarly, all the terms coming out from insertions of ΔE_n on the r.h.s. of eq.(2.88) are divergent in the large A-limit, since ΔE_n is of the order of A.

It has been rigorously proved[28] that in the case of the ground state all the divergent terms of the perturbative series exactly cancel out, the net result being that ΔE_0 *is given by the sum of all linked Goldstone-type diagrams* having the following diagrammatical rules

the diagrams are constructed with up- and down-going solid lines representing particle and hole states and with orizonthal, not oriented thick solid and thick dashed lines denoting the *linked parts* of the interaction matrix element

$$V_{ij} = H_{ij} - E_i^v N_{ij}, \tag{2.90}$$

and of the overlap matrix element N_{ij} respectively; the particle (hole) lines either enter or outgo the thick lines and they belong to the ket $|j)$ or to the bra $(i|$ whether they stay over or below the thick line;

- every interaction and overlap line must have an equal number $d \geq 2$ of entering and outgoing oriented lines because the particle number A is conserved;

- an energy denominator $E_0^v - E_i^v$ is associated with any V_{ij} line (except for the case $i = 0$);

- the topmost orizonthal line is always a V-line; every N-line must be directly connected with the orizonthal line which lies immediately over it;

- the sign of a diagram is given by $(-)^{n_L + n_N}$, where n_L is the number of closed loops and n_N is the number of N-lines.

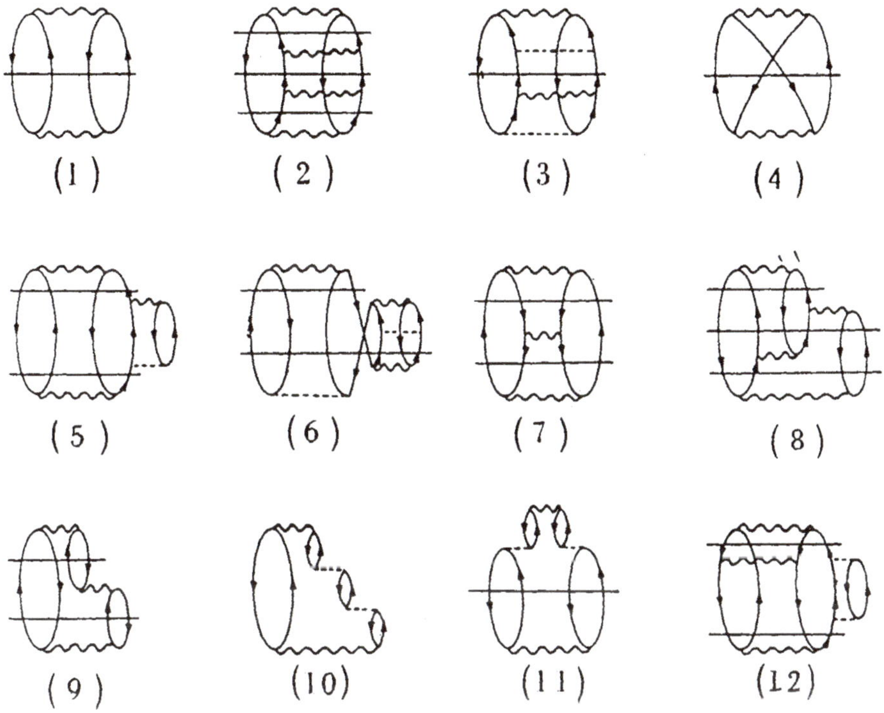

Fig. 16- *Linked Goldstone-like diagrams occuring in CBPT theory for ΔE_0. The last two diagrams are not allowed by the found diagrammatical rules.*

Few examples of CBF diagrams contributing to ΔE_0 are shown in Fig.16.

For instance one has that

$$\text{Diagram (16.1)} = \sum_{\alpha \in (h_1 h_2 p_1 p_2)} \frac{v_{0\alpha} v_{\alpha 0}}{E_0^v - E_\alpha^v}. \tag{2.91}$$

Summing to the diagram (16.1) a similar diagram having a dashed line in place of the bottom solid line on gets the second order perturbative correction to the ground state energy ΔE_0 due to $2 - ph$ admixtures to the CB $|0\rangle$.

There are few differencies between the CBF diagrams and the standard Goldstone diagrams of BBG theory which need to be pointed out. (i) The thick dashed line is a typical diagrammatic element of the CBF diagrams: it represents the overlap matrix element and is entirely due to the fact that CB states are not orthogonal amongst themselves. (ii) The thick lines in CBF diagrams may be reached by $2d_{ij}$ oriented lines with $d \geq 2$. Matrix elements of a two-body operator between CB states having $d_{ij} > 2$ may well be different from zero. In BBG theory the basis states are uncorrelated, therefore matrix elements of a two-body operator for $d_{ij} > 2$ necessarily vanish. (iii) The interaction line in CBF theory corresponds to the linked portion of a matrix element of the interaction between correlated states and it is well behaved even in the case of strongly repulsive potentials. This implies that massive summations, like that of the ladder diagrams, which must be done in BBG theory, is unnecessary in CBF theory expecially if realistic correlation operators G are used.

2.3b- Orthogonal Perturbation Theory.

The orthogonal perturbation theory outlined in this section is simply the standard perturbation theory based on correlated states which are obtained from the CB states after having orthogonalized them. The CB states are not orthogonalized all together because we want to mantain the diagonal matrix elements of the hamiltonian calculated with CB states. This is achieved by a two step orthogonalization process[33] such that a given $n-hp$ CB state is not orthogonalized with the $m-hp$ CB states with $m > n$, but only with those having $m \leq n$.

An intermediate, partially orthogonal (PO) set of states, denoted as $|\mathbf{h}_1 \ldots \mathbf{p}_n\}$ is defined by

$$|\mathbf{h}_1 \ldots \mathbf{p}_n\} = |\mathbf{h}_1 \ldots \mathbf{p}_n) - \sum_{m<n} \sum_{\mathbf{h}'_1 \ldots \mathbf{p}'_m} |\mathbf{h}'_1 \ldots \mathbf{p}'_m)(\mathbf{h}'_1 \ldots \mathbf{p}'_m|\mathbf{h}_1 \ldots \mathbf{p}_n). \tag{2.92}$$

The PO $n-ph$ states are obtained by orthogonalizing the CB $n-ph$ states to all $m-ph$ states with $m < n$ by means of the *Schmidt* procedure. Since any CB $n-ph$ state is directly coupled to few states with $m < n$, the PO states are not too different from the CB states. In ref. (33) it is proved that the diagonal matrix elements do not change in going from the CB to PO states in the thermodynamic limit, and therefore the PO states are normalized.

A PO $n-ph$ state is orthogonal to all PO $m \neq n$ states (in the thermodinamic limit) by construction, but not to other $n-ph$ PO states. A complete orthogonalization is obtained by making separate *Löwdin* transformations to orthogonalize the $n-ph$ states among themselves

$$\begin{aligned} |\mathbf{h}_1 \ldots \mathbf{p}_n> = {}& |\mathbf{h}_1 \ldots \mathbf{p}_n\} - \frac{1}{2} \sum_{\mathbf{h}'_1 \ldots \mathbf{p}'_n} |\mathbf{h}'_1 \ldots \mathbf{p}'_n\}\overline{\{\mathbf{h}'_1 \ldots \mathbf{p}'_n|\mathbf{h}_1 \ldots \mathbf{p}_n\}} \\ & + \frac{3}{8} \sum_{\mathbf{h}'_1 \ldots \mathbf{p}'_n} \sum_{\mathbf{h}''_1 \ldots \mathbf{p}''_n} \overline{\{\mathbf{h}''_1 \ldots \mathbf{p}''_n|\mathbf{h}'_1 \ldots \mathbf{p}'_n\}} \;\; \overline{\{\mathbf{h}'_1 \ldots \mathbf{p}'_n|\mathbf{h}_1 \ldots \mathbf{p}_n\}}. \\ & \ldots \end{aligned} \tag{2.93}$$

The states $|\mathbf{h}_1 \ldots \mathbf{p}_n>$ are the desired correlated orthonormal states (CO). The coefficients $+1, -1/2, +3/8, \ldots$ of the various terms in the *Löwdin* transformation are those that occur in the expansion of $(1+x)^{-1/2}$. The bar over any matrix element denotes that diagonal elements must be omitted, namely

$$\overline{(a|b)} = (a|b)(1-\delta_{ab}), \tag{2.94}$$

where $|a)$ and $|b)$ are two states in any basis.

In ref. (33) it is proved that

$$<\mathbf{h}_1 \ldots \mathbf{p}_n|\mathbf{h}_1 \ldots \mathbf{p}_n> = (\mathbf{h}_1 \ldots \mathbf{p}_n|\mathbf{h}_1 \ldots \mathbf{p}_n) \quad + \quad 0\left(\frac{1}{A}\right). \tag{2.95}$$

A practical scheme of orthogonalization is the following: *Löwdin* transform the $1-ph$ CB states among themselves; partially orthogonalize the $2-ph$ CB states to the CO states $|0>$ and $|\mathbf{h}_i\mathbf{p}_i>$ states, etc. Such an orthogonalization makes use of the overlap matrix elements N_{ij} in an iterative scheme. It has been applied to the calculation of the response functions of nuclear matter[40,54], which will be discussed in section 6 and few iterations have always been sufficient to reach convergence.

A similar procedure has been used[57] to orthogonalize the states of few-body systems. The states were classified according to the number of particles in the continuum. A significative contribution from the orthogonalization correction were found in the calculations of the charge response.

Standard perturbation theory is used with CO states. Therefore the perturbative unlinked diagrams cancel in virtue of the *Goldstone* theorem[58]. The only new types of divergencies may come from the matrix elements. It may be proved that the cluster diagrams contributing to the matrix elements with CO states are linked, being the *Schmidt* orthogonalization to cause the cancellation of the unlinked diagrams contributing to the matrix elements with CB states.

c- Comparison of OCB with CBPT

Orthogonal and not orthogonal CBF theories make use of different perturbative schemes to calculate various quantities of interest. In CBPT theory one typically takes the number of interaction lines (solid or dashed) as an expansion parameter. One may perform ladder summations[28,59], but they are expected to be not important because the important correlations are supposed to be already contained into the correlation operator G. It follows that each order of this scheme contains spuriousities due to the lack of orthogonality corrections. For instance, the following sum rule for the static structure function

$$\begin{aligned} S(k) &= <0|\rho_{\mathbf{k}}^{\dagger}\rho_{\mathbf{k}}|0> \\ &= \sum_{\mathbf{h},\mathbf{p}} <0|\rho_{\mathbf{k}}^{\dagger}|\mathbf{hp}><\mathbf{hp}|\rho_{\mathbf{k}}^{\dagger}\rho_{\mathbf{k}}|0>, \end{aligned} \tag{2.96}$$

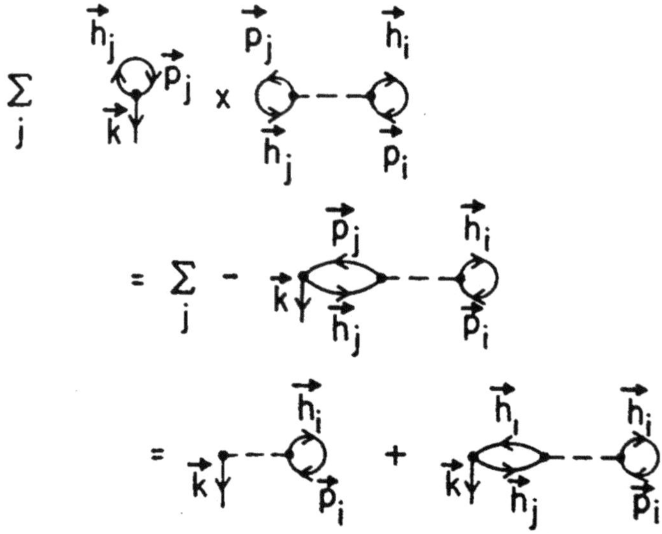

Fig. 17- *Illustration of the diagrammatic multiplication. The second equality is obtained by writing the sum over* $\mathbf{p}_2$ *as a sum over all states minus the sum over hole states.*

holds in virtue of the fact that $\rho_{\mathbf{k}}|0>$ is a combination of $1-ph$ states if $\rho_{\mathbf{k}}$ commutes with the correlation operator G (the above sum rule is then violated if the state dependent correlations contain gradient operators, like for instance back-flow or $\mathbf{L}\cdot\mathbf{S}$ correlations). If CB states are used in eq. (2.96) the sum rule is violated since the $1-ph$ CB states are not orthogonal amongst themselves.

In orthogonal theory there are no explicit interaction lines of the dashed type, since the states are orthogonal. However they implicitely come into the theory in the calculation of the orthogonal states. For instance, diagram (16.10) is an orthogonality correction of diagram (16.1). In OCB theory diagram (16.10) is absent, but it is implicitely included in the calculation of diagram (16.1). Actually, the calculation of diagram (16.1) implies the sum of the whole series of ring diagrams of that type with any number of dashed lines.

A comparison of the two perturbative schemes can be easily done analysing the cluster diagrams coming at the first order of the Power Series expansion, namely those having only one dynamical correlation amongst the particles which are not directly connected by the hamiltonian operator. In order to do that it is necessary to convert a given product of terms as a series of cluster terms. For instance, the quantity

$$\sum_{\mathbf{h}'_1\mathbf{p}'_1}(0|H-E_0|\mathbf{h}'_1\mathbf{h}_2\mathbf{p}'_1\mathbf{p}_2)(\mathbf{h}'_1\mathbf{h}_2\mathbf{p}'_1\mathbf{p}_2|\mathbf{h}_1\mathbf{h}_2\mathbf{p}_1\mathbf{p}_2), \tag{2.97}$$

enters in the definition of the CO state $|\mathbf{h}_1\mathbf{h}_2\mathbf{p}_1\mathbf{p}_2>$ if one use OCB theory or in the CBPT diagram (16.9). It is possible to convert the above quantity in a sum of cluster diagrams as illustrated in Fig. 17.

In the following, we discuss the corrections to the CBPT non diagonal matrix elements of the hamiltonian given in eq. (2.87) due to the orthogonalizations of the $2-ph$ CB states among themselves. In OCB theory the PO states $|\mathbf{h}_1\mathbf{h}_2\mathbf{p}_1\mathbf{p}_2\}$ are given by

$$|\mathbf{h}_1\mathbf{h}_2\mathbf{p}_1\mathbf{p}_2\} = |\mathbf{h}_1\mathbf{h}_2\mathbf{p}_1\mathbf{p}_2) - |0)(0|\mathbf{h}_1\mathbf{h}_2\mathbf{p}_1\mathbf{p}_2), \tag{2.98}$$

since the $1-ph$ states are orthogonal to the $2-ph$ having zero total momentum. It follows that

$$\{0|H|\mathbf{h}_1\mathbf{h}_2\mathbf{p}_1\mathbf{p}_2\} = (0|H - E_0|\mathbf{h}_1\mathbf{h}_2\mathbf{p}_1\mathbf{p}_2), \tag{2.99}$$

which shows that CBPT theory takes into account the non orthogonality of $|\mathbf{h}_1\mathbf{h}_2\mathbf{p}_1\mathbf{p}_2)$ and the ground state. In order to analyze the effect of the non orthogonality of the $2-ph$ states among themselves in first order of the Power Series expansion, we calculate the first order *Löwdin* correction to this matrix element:

$$\Delta H = -\frac{1}{2}\sum_{2'}(0|H - E_0|2')(2'|\mathbf{h}_1\mathbf{h}_2\mathbf{p}_1\mathbf{p}_2), \tag{2.100}$$

where the $2-ph$ states $|2')$ have only two momenta different from $|\mathbf{h}_1\mathbf{h}_2\mathbf{p}_1\mathbf{p}_2)$, namely

$$\begin{aligned} |2') = &|\mathbf{h}_1'\mathbf{h}_2\mathbf{p}_1'\mathbf{p}_2), \quad |\mathbf{h}_1\mathbf{h}_2\mathbf{p}_1'\mathbf{p}_2'), \quad |\mathbf{h}_1\mathbf{h}_2'\mathbf{p}_1'\mathbf{p}_2), \\ &|\mathbf{h}_1\mathbf{h}_2'\mathbf{p}_1\mathbf{p}_2'), \quad |\mathbf{h}_1'\mathbf{h}_2'\mathbf{p}_1\mathbf{p}_2), \quad |\mathbf{h}_1'\mathbf{h}_2\mathbf{p}_1\mathbf{p}_2'). \end{aligned} \tag{2.101}$$

States with three or four momenta different from $|\mathbf{h}_1\mathbf{h}_2\mathbf{p}_1\mathbf{p}_2)$ do not contribute to ΔH in first order of the PS expansion, which is given by

$$\begin{aligned} H_{OCB} = &H_{CBPT} + \Delta H = T_1 + D(15.1) + DE(15.1) \\ &+ \frac{1}{2}\sum_{k=2,14}(D(15,k) + DE(15,k))(1 - \delta_{k5} + \delta_{k8}) \\ &+ \frac{1}{2}\sum_{k=2,12}(DS(15,k) + DSE(15,k))(1 - \delta_{k5} + \delta_{k8}) \\ &- \frac{1}{2}(D(15.15) + DE(15.15)). \end{aligned} \tag{2.102}$$

One should notice that the orthogonality corrections cancel the diagram (15.5) along with its DS, DE and DSE counterparts, and they reduce the contribution of all other first order diagrams, except for D(15,8) and its counterparts, by a factor of two. However the new diagrams D(15,15) and DE(15.15) are added. It follows that it is not consistent to calculate the first order PS corrections to the non diagonal matrix elements without taking into account the orthogonality corrections. Particularly in dense systems, like liquid ^3He diagrams of the type D(15,5) give large contributions[29,60] in CBPT and these may be entirely spurious.

However there are also higher order perturbative diagrams, for instance those having three non diagonal matrix elements with $|\mathbf{h}_1\mathbf{h}_2\mathbf{p}_1\mathbf{p}_2)$ and $|\mathbf{h}_1'\mathbf{h}_2\mathbf{p}_1'\mathbf{p}_2)$ as intermediate states, which furnish cluster diagrams of the same order in the PS expansion as those with two non diagonal matrix elements. Therefore a more detailed cluster analysis is needed to really understand the convergence properties of the perturbative series.

3. Fermi Hard Sphere systems.

Much effort has been devoted to the study of the Hard Sphere (HS) dilute Fermi gas, i.e. a low-density Fermi gas whose particles interact via a repulsive hard core potential[61]. The interest in this system is twofold: a) a number of physical systems is characterized by a strong two-body repulsion at small interparticle distances; b) because of its low density, the perturbative approach applied to the HS Fermi gas provides reliable results that can be used as tests of other many-body techniques.

The expansion parameter in the low-density (LD) series is $x = k_F c$ where c is the hard core radius. Quantities whose values have been computed using this approach are: the ground-state energy[62], the momentum distribution $n(k)$[63,64] and the self-energy $\Sigma(k,E)$[63].

We will concentrate on E/A, whose LD expansion, up to terms in x^3, is[62]:

$$\frac{E}{A} = \frac{3}{5}\frac{\hbar^2}{2m}k_F^2\{1 + \frac{5}{3}(\nu - 1)[\frac{2x}{3\pi} + \frac{4}{35\pi^2}(11 - ln4)x^2 + (\frac{1}{15\pi} + 0.0755$$

$$+ 0.059 \quad (\nu - 3)x^3] + (\nu + 1)\frac{x^3}{3\pi}\}. \tag{3.1}$$

The first term in eq.(3.1) is the Fermi kinetic energy, the x^2 term is entirely due to the two-particle interaction in $l = 0$ waves whilst $l = 1$ waves mainly give a term proportional to x^3. For $x < 0.5$, eq.(3.1) is certainly correct by more than 1%, and it is presumably reliable up to $x \leq 0.8$.

The HS Fermi gas has been also studied using the variational approach. We will examine its results when the Jastrow correlation operator $G^{(J)}$ of eq.(1.13) is adopted and we will compare them with the CBF and the LD estimates . The two-body correlation function $f(r_{ij})$ has the form:

$$f(r_{ij} \leq c) = 1 - e^{-\mu(r-c)},$$

$$f(r_{ij} \geq c) = 0, \tag{3.2}$$

where μ is a parameter whose value is fixed by minimizing the energy.

In Table I the variational estimates of z, defined as

$$z = \frac{E}{A} \quad {T_F}^{-1} \quad , \tag{3.3}$$

x	z^{JF}	z^{PB}	z^{CW}	z^{av}	z^{SD}	z^{LD}	ν	μ
0.2	1.271	1.269	1.272	1.27	1.24	1.245	4	4.2
	1.090	1.089	1.092	1.09	1.08	1.081	2	4.3
0.4	1.637	1.633	1.639	1.63	1.59	1.597	4	2.6
	1.218	1.217	1.218	1.22	1.20	1.196	2	2.7
0.6	2.182	2.163	2.196	2.17	2.12	2.120	4	2.0
	1.400	1.395	1.404	1.40	1.37	1.361	2	2.1
0.8	3.025	2.956	3.083	2.99	2.95	2.875	4	1.7
	1.673	1.655	1.684	1.66	1.63	1.597	2	1.8
1.0	4.376	4.146	4.580	4.26	4.25	3.927	4	1.4
	2.073	2.024	2.110	2.05	2.02	1.920	2	1.6

Table I. *Variational estimates of z. See text.*

are presented. Three different forms for the kinetic energy have been used: the Jackson-Feenberg (JF), the Pandharipande-Bethe (PB) and the Clark-Westhaus (CW). All of them give identical results provided the two- and three-body distribution functions are exactly computed (see section 2.2a). The z-values of Table I have been obtained within the FHNC/0 approximation for g_2 and the Kirkwood Superposition approximation (KSA) for g_3. The difference between the three estimates is a measure of the importance of the elementary and Abe diagrams. The Table shows also the average value of z

$$z^{av} = \frac{z^{JF} + z^{PB}}{2}, \tag{3.4}$$

the low-density estimate z^{LD} of eq.(3.1) and the value z^{SD} obtained by using a state-dependent correlation function, namely

$$f_{SD}(ij) = f_E(r_{ij})\frac{(1 - P^{\sigma\tau})}{2} + f_O(r_{ij})\frac{(1 + P^{\sigma\tau})}{2}, \tag{3.5}$$

where $P^{\sigma\tau}$ is the spin-isospin exchange operator and $f_{E(O)}$ is the even(odd) part of the correlation function. The Table reports the z values for $\nu = 4, 2$;the variational parameters μ are also given.

The difference between z^{av} and z^{LD} is less than 2% up to $x = 0.6$. The state dependent results practically coincide with the LD ones up to that x value. For higher x's the accuracy of the LD expansion is questionable and z^{av} and z^{SD} are quite close.

CBF perturbative theory provides a tool to improve on the accuracy of the Jastrow results. We have computed the second order perturbative correction $\Delta E_0^{(2)}$ of eq.(2.89) to the energy per particle both in the non orthogonal and in the orthogonal scheme.In CBPT, the $2-ph$ states considered are the $|\mathbf{h}_1\mathbf{h}_2\mathbf{p}_1\mathbf{p}_2)$ states of eq.(1.17) ; the orthogonalized $2-ph$ states $|\mathbf{h}_1\mathbf{h}_2\mathbf{p}_1\mathbf{p}_2>$ are given by eq.(2.93) after truncating the *Löwdin* series at the second order.

One of the major problems in applying CBF theory is the calculation of the Hamiltonian and overlap matrix elements whose cluster expansion has been discussed in the previous section of the paper. The simplest approximation is to retain only two-body cluster diagrams and it is expected to be reliable only at very low densities. The two-body dressed (TBD) approximation sums an infinite set of terms and consists into keeping the diagrams prescribed by CBF theory where only two points are reached by the ph transition lines. TBD approximation consists into assuming $\Sigma_C(a) = \Sigma_C(I+a) = 0$ and keeping only the $\Sigma^{Irr,2}$ terms. The direct part of the non-diagonal matrix element of the Hamiltonian in TBD reads:

$$(0|H-E_0|\mathbf{h}_1\mathbf{h}_2\mathbf{p}_1\mathbf{p}_2)^{TBD}_{dir} = [\Gamma'_{dd}(k) + \frac{\Gamma_{dd}(k)}{4}\frac{\hbar^2}{2m}(p_1^2+p_2^2-3h_1^2-3h_2^2)]$$

$$\times \quad \delta(\mathbf{p}_1+\mathbf{p}_2-\mathbf{h}_1-\mathbf{h}_2), \tag{3.6}$$

where $\mathbf{k} = \mathbf{p}_1 - \mathbf{h}_1 = \mathbf{h}_2 - \mathbf{p}_2$ is the direct momentum transfer and $\rho\Gamma'_{dd} = \rho\bar{g}'_{dd}$. The exchange contribution is given by $\mathbf{p}_1 \leftrightarrow \mathbf{p}_2$. To obtain eq.(3.6) we have used the JF identity:

$$<\Psi^\dagger\nabla^2\Psi> = \frac{1}{4}\left(<\Psi^\dagger\nabla^2\Psi> + <(\nabla^2\Psi^\dagger)\Psi) - 2<\nabla\Psi^\dagger\cdot\nabla\Psi>\right). \tag{3.7}$$

Three- and more-body diagrams with transition lines involving more than two particles originate both separable and non separable diagrams. We will consider only those separable diagrams that can be written as products of two-body dressed terms, namely those having the irreducible part of the form $\Sigma^{Irr,2}$:

$$(0|H-E_0|\mathbf{h}_1\mathbf{h}_2\mathbf{p}_1\mathbf{p}_2)^{SBP}_{dir} = \{(0|H-E_0|\mathbf{h}_1\mathbf{h}_2\mathbf{p}_1\mathbf{p}_2)^{TBD}_{dir}$$

$$+[\frac{\rho\tilde{X}'_{cc}(p_1)}{D(p_1)} + \frac{\hbar^2}{4m}\Big(Z(p_1) - Z(h_1) + h_1^2\rho\tilde{X}_{cc}(h_1) + \mathbf{h}_1\cdot\mathbf{p}_1(D(h_1) + \frac{Z(h_1)}{h_1^2})$$

$$(D(p_1) + \frac{Z(p_1)}{p_1^2})\Big) \quad +1\to 2 \quad] \quad \delta(\mathbf{p}_1+\mathbf{p}_2-\mathbf{h}_1-\mathbf{h}_2)\}$$

$$\times\frac{1}{\sqrt{D(p_1)D(p_2)D(h_1)D(h_2)}}, \tag{3.8}$$

where:

$$Z(a) = -a^2\left(\rho\tilde{X}_{cc}(a) - Y(a)\right). \tag{3.9}$$

A third approximation, denoted here as 3BS approximation is obtained by retaining, in eq.(3.8), only terms linear in $\tilde{X}_{cc}$, $\tilde{X}'_{cc}$ and Z.

A pathology of the SEP approximation is immediately recognized: in systems with $\nu = 1$ (e.g. polarized liquid ^3He), for some value of the momentum lesser than k_F it may happen that $\rho\tilde{X}_{cc} = 1$. In this case, the vertex correction $1/D$ diverges. In systems with $\nu \neq 1$, the effect of the vertex corrections is to strongly enhance the value of the matrix element providing a large perturbative correction. This behavior is partially corrected by the*Löwdin* orthogonalization procedure. Let's consider in details the Γ'_{dd} part of the 3BS approximation. In CBPT it contributes as:

$$\Gamma'_{dd}(k)\left(1 + \frac{\rho}{2}[\tilde{X}_{cc}(h_1) + \tilde{X}_{cc}(p_1) + 1 \to 2]\right); \tag{3.10}$$

the second order of the sl Löwdin series cancels exactly $X_{cc}(h_i)$ to give, in OCB:

$$\Gamma'_{dd}(k)\left(1 + \frac{\rho}{2}[\tilde{X}_{cc}(p_1) + \tilde{X}_{cc}(p_2)]\right). \tag{3.11}$$

Consistent corrections to the 3BS approximation due to the orthogonalization are obtained by retaining the 3BS-like diagrams coming from the *Löwdin* corrected matrix elements. This is a lengthy procedure which will be described elsewhere[52] , whose result is:

$$\begin{aligned} <0|H - E_0|\mathbf{h}_1\mathbf{h}_2\mathbf{p}_1\mathbf{p}_2>^{O3BS}_{dir} = (0|H - E_0|\mathbf{h}_1\mathbf{h}_2\mathbf{p}_1\mathbf{p}_2)^{TBD}_{dir}\{1 + \frac{\rho}{2}[\tilde{X}_{cc}(p_1) + \tilde{X}_{cc}(p_2)]\} \\ + \Big(\frac{\Gamma_{dd}(k)}{2}\{\rho\tilde{X}'_{cc}(p_1) + \frac{\hbar^2}{4m}[\ h_1^2\rho\tilde{X}_{cc}(h_1) - (k^2 - h_1^2 - 2h_2^2)\frac{\rho\tilde{X}_{cc}(p_1)}{2} \\ - 2\mathbf{p}_1 \cdot \mathbf{h}_1\rho X_{cc}(p_1) + Z(p_1)\ \ ] + 1 \leftrightarrow 2\ \ \} - R_{dir}\ \Big) \\ \times \delta(\mathbf{p}_1 + \mathbf{p}_2 - \mathbf{h}_1 - \mathbf{h}_2), \end{aligned} \tag{3.12}$$

where

$$R_{dir} = \frac{\Gamma_{dd}(k)S_F(k)}{2}\{2\Gamma'_{dd}(k) + \frac{\hbar^2}{2m}\frac{\Gamma_{dd}(k)}{2}[p_1^2 - 3h_1^2 + 1 \to 2]\} \tag{3.13}$$

and $S_F(k)$ is the free Fermi gas structure function.

Table II presents the results for x in the above approximations. Up to $x = 0.8$ the TBD, 3BS, O3BS and SEP CBF corrections are almost indistinguishable and they bring the Jastrow results closer to both the LD and SD estimates. The case $x = 1$ clearly shows the reduction of the CBF correction due to the orthogonalization: O3BS correction is smaller than 3BS. The SEP correction is the largest, as expected from the previous discussion. An interesting feature of the results at $x = 0.8, 1$ is that the CBF and SD results tend to agree between themselves and they are both slightly different from the LD ones. This is consistent with the fact that in this region the LD expansion starts to loose reliability.

x	z^{av}	z^{TBD}	z^{3BS}	z^{O3BS}	z^{SEP}	z^{SD}	z^{LD}	ν
0.2	1.27	1.26	1.25	1.26	1.26	1.24	1.24	4
	1.09	1.08	1.08	1.08	1.08	1.08	1.08	2
0.4	1.63	1.61	1.61	1.61	1.61	1.59	1.60	4
	1.22	1.21	1.21	1.21	1.21	1.20	1.20	2
0.6	2.17	2.15	2.14	2.14	2.14	2.12	2.12	4
	1.40	1.39	1.38	1.39	1.38	1.37	1.36	2
0.8	2.99	2.96	2.94	2.94	2.94	2.95	2.87	4
	1.66	1.65	1.64	1.64	1.63	1.63	1.60	2
1.0	4.26	4.21	4.18	4.20	4.17	4.25	3.93	4
	2.05	2.03	2.02	2.03	2.01	2.02	1.92	2

Table II. *Second order CBF values of of z. See text.*

4. Liquid ^3He.

Atomic liquid ^3He at $T = 0$ has been a challenging problem for the many-bodys physicists for a long period. Standard Brueckner perturbative theory does not work here because the hard-core of the interparticle potential is even larger than the average distance between particles $r_0 = (4\pi\rho/3)^{-1/3}$ and the strong correlations, not only of the Jastrow type, cannot be treated perturbatively. The high density of the system makes also difficult the convergence of clusters summation techniques as FHNC and the accuracy of the results strongly depends on the approximations of the calculation. The latest[26] GFMC result for the ground state energy per particle has been obtained within the *fixed node* (FN) approximation and gives an upperbound of $E_0 = -2.37 \pm .01K$ at the equilibrium density $\rho_0 = 0.277\sigma^{-3}(\sigma = 2.556\AA)$, to be compared with the experimental value $-2.47K$. The variational approach faces with the above mentioned problem arising from the need of using realistic wave functions; this makes difficult the evaluation of the expectation value of the hamiltonian.

Recent variational calculations[(24,25)] have employed the correlation operator G of eq.(1.16) containing Jastrow, three-body and state-dependent correlations. The state dependent part of G may contain either back-flow correlations:

$$F_B(i,j) = \exp\left(\imath\eta_B(r_{ij})\mathbf{r}_{ij} \cdot \hat{\mathbf{K}}_{ij}\right), \tag{4.1}$$

where the operator $\hat{\mathbf{K}}$ acts on $|0]$ as follows

$$\hat{\mathbf{K}}_{mn} \exp\left(\imath \sum_{i=1}^{A} \mathbf{k}_{\alpha_i} \cdot \mathbf{r}_i\right) = (\mathbf{k}_m - \mathbf{k}_n) \exp\left(\imath \sum_{i=1}^{A} \mathbf{k}_{\alpha_i} \cdot \mathbf{r}_i\right), \tag{4.2}$$

or spin-dependent correlations:

$$F_S(i,j) = 1 + \eta_S(r_{ij})\sigma_i \cdot \sigma_j. \tag{4.3}$$

According to the models used we will talk of J,JT,JTB,JTS and JTSB wave functions. The last one contains both back-flow and spin correlations.

A semioptimized pair correlation function $f(r)$ having the proper long- range behavior

$$f(r \longrightarrow \infty) = 1 - \frac{\beta}{r^2} \tag{4.4}$$

has been used in all the variational calculations presented in this section. A fully optimized pair correlation function is extremely difficult to obtain if triplet and state dependent correlations are present . For this reason the function $f(r)$ has been chosen as the *optimal* Jastrow factor of the underlying mass-3 boson system. The triplet function $f_3(\mathbf{r}_{ij}, \mathbf{r}_{ik})$ and the back-flow function $\eta_B(r)$ have the forms given in ref.(24). The spin function $\eta_S(r)$ is evaluated by minimizing the energy expectation value calculated at the second order in the spin correlations, under the constraints that both $\eta_S(r)$ and $d\eta_S(r)/dr$ heal to zero at some distance d_S[25].

An exact evaluation of the integrals defining the two- and three-body distribution functions for a sufficient large number of particles could in principle be done using the MonteCarlo method . Unfortunately, the integrals must be sampled in a finite size box, with all the functions going to zero at the borders. This limitation does not allow the MC technique to handle long ranged correlation functions as the one employed here. Integral equations method, as FHNC, do not have this problem, but ,because of the large density of the system, the elementary diagrams contributions must be somehow taken into account. We will give results obtained within the FHNC/α approximation of ref.(49) and the scaling approximation (FHNC/s) of ref.(48). The realistic HFDHE2 two-body potential of *Aziz et al.*[65,66] has been adopted. Triplet and back-flow contributions have been evaulated as in ref.(24), moreover three-body kinetic energy terms due to back-flow correlations have been added[52]. More details about the variational calculations may be found in ref.(25).

The variational results for the g.s. energy are given in table III and the comparison with the experimental data is shown in Fig.18. FHNC/α and FHNC/s approximations agree closely at $\rho = \rho_0$, even if the total energy comes out from strong cancellations between the kinetic and potential terms, that in turn do not agree as well. The difference is due to the fact that in ref.(24) the elementary induced potential w_E is completely neglected in the Euler equation; the insertion of w_E gives a substantial reduction of the kinetic energy expectation value. At higher densities, the two approximations give quite different results.

JTS model has been treated via an improved version of the FHNC/SOC approximation. The difference in energy between this calculation and a bare FHNC/SOC estimate is minimal, so we are confident that the JTS results are sufficiently accurate, in particular

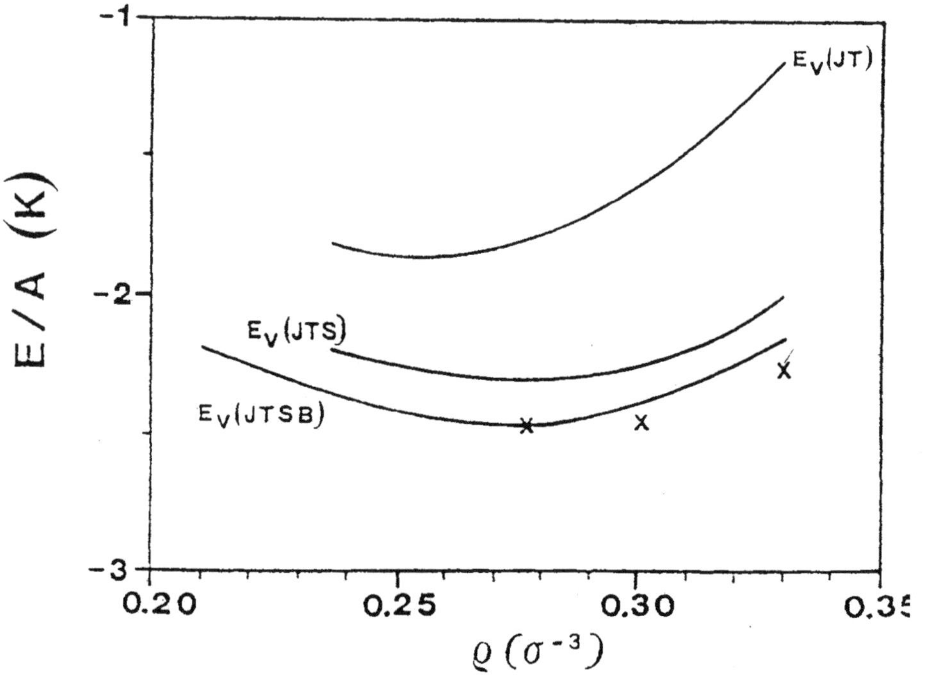

Fig. 18-*Comparison of the FHNC/α variational energies of liquid ^3He for the JT,JTS and JTSB models with the experimental data represented by crosses.*

they are expected to be more reliable than JTB. JTSB model gives energies very close to the experiments. In this case the treatment of the back-flow correlations should be more accurate than in JTB, since the energy correction due to the back-flow is ~ 20% of the analogous JTB correction . The liquid structure function $S(k)$ is given in Fig. 19 and it is compared with the experimental data. The calculated velocity of sound is $c = 185 \quad m/sec$ to be compared with the experimental value $c_{exp} = 182.9 \quad m/sec$. The above results are a strong indication that JTSB model describes accurately the g.s. properties of liquid ^3He and that HFDHE2 potential provides a realistic hamiltonian.

Let us now discuss the CBF second order correction to the g.s. energy, which has recently been calculated[52] for both the JT and JTB models. For the JTB model, if one considers only two-body back-flow contributions to the kinetic energy, the FHNC$^\lambda$ equations of section 2.2b change as shown below:

$$g_{de}(r_{ij}) \rightarrow g_{de}(r_{ij}) + \lambda g_{dd}(r_{ij}) v_\eta(r_{ij}), \tag{4.5}$$

with

$$v_\eta(r) = \frac{\hbar^2}{5m} k_F^2 \{3\eta_B(r)[1+\eta_B(r)] + r^2 \eta_B'(r)[\eta_B'(r) + (2\eta_B(r)+1)/r]\}, \tag{4.6}$$

	$\rho(\sigma^{-3})$	J	JT	JTB	JTS	$JTSB$
$FHNC/\alpha$	.237	−1.54	−1.82	−2.14	−2.19	−2.34
$FHNC/s$	.237	−1.46	−1.79	−2.12		
$FHNC/\alpha$	.277	−1.32	−1.80	−2.18	−2.31	−2.46
$FHNC/s$	.277	−1.28	−1.83	−2.22		
$FHNC/\alpha$	.301	−0.99	−1.61	−2.02	−2.26	−2.38
$FHNC/s$	.301	−1.06	−1.78	−2.19		
$FHNC/\alpha$	.330	−0.39	−1.15	−1.54	−1.99	−2.15
$FHNC/s$	.330	−0.60	−1.55	−1.97		

Table III. *Variational values of g.s. energy per particle in ^3He for different choices of the wave function. All the energies are in K.*

and

$$\hat{l}(k_F t) = \frac{\hbar^2}{2m}\left(\frac{\nabla_t^2 l(k_F t)}{2} - l'(k_F t)[r\eta_B''(r) + 4\eta_B'(r)]\right), \tag{4.7}$$

where $t(r) = r[1 + 2\eta_B(r) + 2Z_B(r)]$ and the derivatives are made respect to t ($Z_B(r)$ is the $Z(r)$ function of ref.(24)). Also the matrix elements change in the JTB model. In the TBD approximation 3.6, the quantity $\Gamma'_{dd}(k)$ is now:

$$\Gamma'_{dd}(k) = \rho \int e^{i\mathbf{k}\cdot\mathbf{t}} g'_{dd}(r) d\mathbf{r}, \tag{4.8}$$

where $\mathbf{t} = t(r)\mathbf{r}/r$. An analogous expression is valid for $\Gamma_{dd}(q)$. Moreover an extra kinetic term Δ^{TBD} has to be added to eq.(3.6) to obtain the JTB-TBD perturbative correction:

$$\Delta_{dir}^{TBD} = \frac{\hbar^2}{2m}\rho \int e^{i\mathbf{k}\cdot\mathbf{t}} g_{dd}(r) \quad [i\eta_B^{(1)}(r)\mathbf{k}\cdot\mathbf{r} - 2s^2\eta_B^2(r) - 2\eta_B^{(2)}(r)(\mathbf{s}\cdot\mathbf{r})^2] d\mathbf{r}, \tag{4.9}$$

with $\eta_B^{(1)}(r) = \eta_B''(r) + 4\eta_B'(r)/r$, $\eta_B^{(2)}(r) = \eta_B'(r)[\eta_B'(r) + 2\eta_B(r)/r]$ and $\mathbf{s} = \mathbf{p}_2 - \mathbf{h}_1$.

In Fig.20 second order perturbative evaluations of $\Delta E_0^{(2)}$, calculated by using CBF and OCBF theory, are compared for the JT model at various values of the density. The curve labelled JT gives the variational values of ref.(24). The other curves have been computed using the approximations described in the previous section for the matrix elements and adding the perturbative corrections to the variational estimates. The not negligeable differences between O3BS and 3BS results and especially those between O3BS and SEP clearly show the importance of the orthogonality spuriosities in liquid ^3He. The introduction

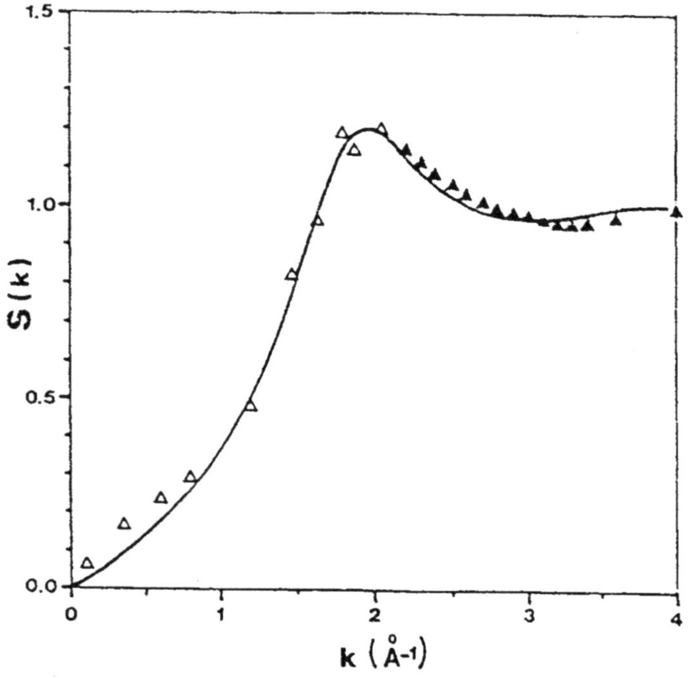

Fig. 19-*Calculated liquid structure function $S(k)$ compared with experimental data at $T = 0.41K$ (white triangles) and at $T = 0.76K$ (solid triangles).*

of the *Schmidt* and *Löwdin* orthogonalization corrections to the CBF many-body term seems to be necessary.

Similar calculations have also been performed in the JTB model. Perturbative correction are found to be consistently smaller than in JT. A comparison of the results obtained withi the TBD and O3BS approximations with the variational JTB and JTSB results is show in Fig. 21. $E_P(TBD)$ and $E_P(O3BS)$ are obtained by adding the perturbative correction to the JTB results of table III obtained in the FHNC/s scheme. An encouraging agreemen between the best variational estimates $E_V(JTSB)$ and $E_P(O3BS)$ is found at $\rho \leq \rho_0$. Th discrepancies at larger densities are partly due to the disagreement between FHNC/s an FHNC/α schemes. However, before a perturbative estimate of the g.s. energy correctio can be given at the level of accuracy required by the liquid ^3He physics, a more accurat evaluation of the many body diagrams prescribed by OCBF theory has to be performed.

5.- Nuclear matter calculations.

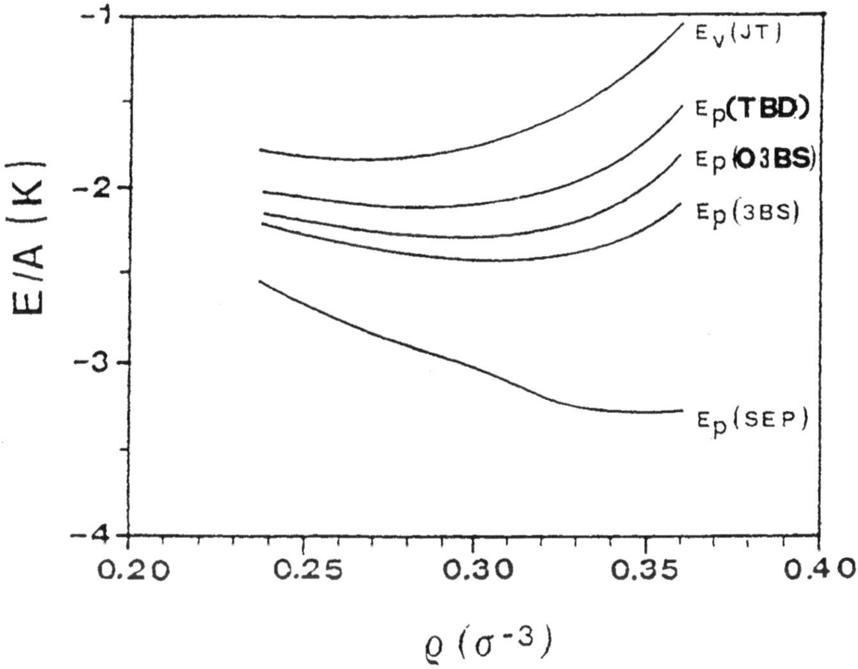

Fig. 20 *Second order perturbative corrections to the ground state energy of liquid ^3He for the JT model.*

In this section we report on a few calculations performed in nuclear matter based on CBF theory for quantities of physical interest like the binding energy, the optical potential, the momentum distribution and the response function to a longitudinal electromagnetic probe.

In these calculations it has been used the correlation operator G of eq. (1.16), where the triplet correlations have been neglected and the state dependence of $F(ij)$ has been taken of the following form

$$F(ij) = \sum_{p=1,8} f^p(r_{ij}) O^p(ij), \tag{5.1}$$

where the operators $O^p(ij)$ are given by

$$\begin{aligned} O^p(ij) =& 1; \quad \sigma_i \cdot \sigma_j; \quad \tau_i \cdot \tau_j; \quad (\sigma_i \cdot \sigma_j)(\tau_i \cdot \tau_j); \quad S(ij) \\ & S(ij)\tau_i \cdot \tau_j; \quad \mathbf{L} \cdot \mathbf{S}; \quad \mathbf{L} \cdot \mathbf{S}\tau_i \cdot \tau_j, \end{aligned} \tag{5.2}$$

with $S(ij) = 3(\sigma_i \cdot \hat{\mathbf{r}}_{ij})(\sigma_j \cdot \hat{\mathbf{r}}_{ij}) - \sigma_i \cdot \sigma_j$ being the tensor operator. The operatorial dependence of the correlation $F(ij)$ given above simulates that of the two-nucleon interaction and it has been found to give energy upperbounds which are believed to be quite close to the true eigenvalues.

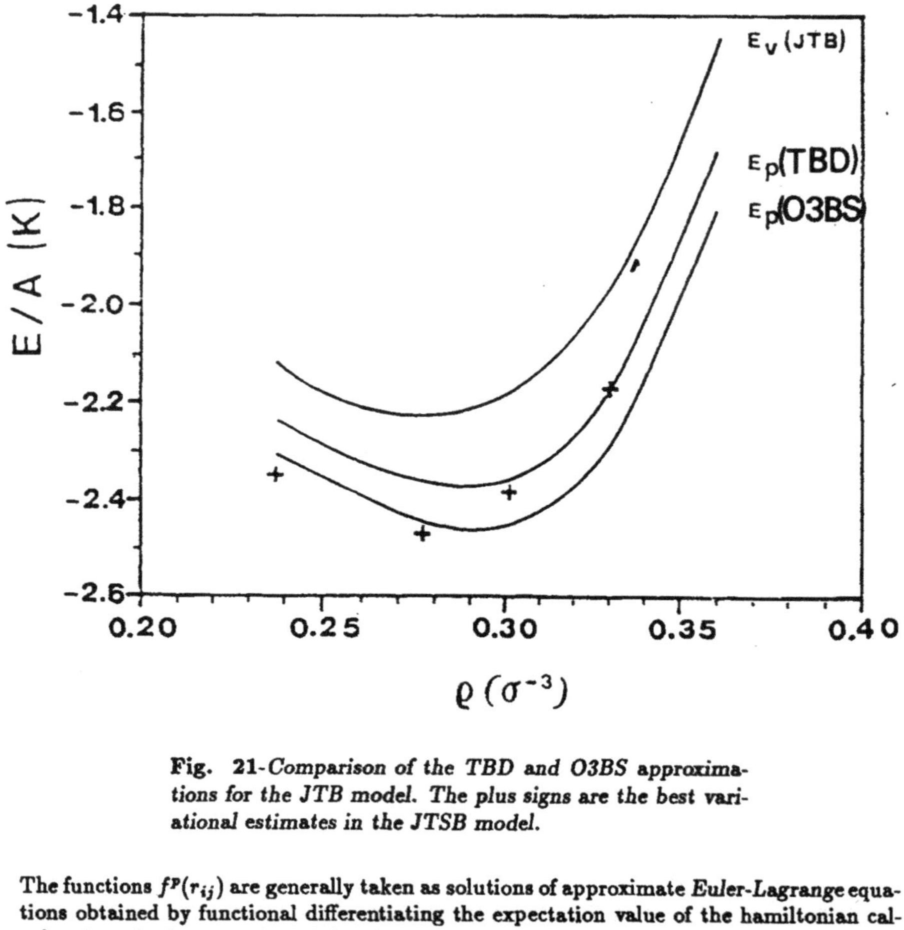

Fig. 21-*Comparison of the TBD and O3BS approximations for the JTB model. The plus signs are the best variational estimates in the JTSB model.*

The functions $f^p(r_{ij})$ are generally taken as solutions of approximate *Euler-Lagrange* equations obtained by functional differentiating the expectation value of the hamiltonian calculated at the lowest order of the cluster expansion . As a consequence of the fact that the lowest order of the cluster expansion is not sufficient to mantain a variational principle, one is forced to introduce *ad hoc* constraints which are typically taken of the form $f^p(r_{ij} > d_p) = \delta_{p1}$ and $(f^p(d_p))' = 0$ and the *healing distances* d_p are considered as variational parameters to be determined by minimizing the expectation value of the hamiltonian evaluated with a much better approximation, like for instance the FHNC/SOC discussed in section 2.2. It has been found that the most important variational parameters are the central healing distance $d_{i=1,4} = d$, the tensor healing distance $d_{i=5,6} = d_t$ and a third parameter β which is used as constant factor multiplying the non central potential terms in the solution of the approximate *Euler equations*.

In most of the nuclear matter calculation reviewed below, the *Urbana* v_{14}[67] and the *Argonne* v_{14}[68] have been adopted as a two-body potential. In the variational calculations performed to evaluate the binding energy and the variational parameters for the correlation operator G, the three-nucleon interaction *Urbana VII*-model[(69] has been used. In the calculations of the non diagonal matrix elements of the hamiltonian the three-nucleon interaction has been approximated by a density dependent two-nucleon interaction TNR^{67}. The values of the variational parameters are little dependent on the density considered for the nuclear matter and the following set of values has been used in the calculations

$$d = 1.875 r_0, \quad d_t = 3 r_0, \quad \beta = 0.8 \quad . \tag{5.3}$$

The real part of the optical potential is defined by the equation

$$e(k) = \frac{\hbar^2}{2m} k^2 + U(e, \rho). \tag{5.4}$$

Its variational estimate is obtained as discussed in section 2 and the second order perturbative corrections are given by

$$\begin{aligned} \Delta e^{(2)}(p) = & \frac{1}{2} \sum \frac{|<\mathbf{p}|H - E_p^v|\mathbf{h}_1\mathbf{p}_1\mathbf{p}_2>|^2}{e^v(p) + e^v(h_1) - e^v(p_1) - e^v(p_2)} \\ & - \frac{1}{2} \sum \frac{|<0|H - E_p^v|\mathbf{h}_1\mathbf{h}_2\mathbf{p}\mathbf{p}_1>|^2}{e^v(h_1) + e^v(h_2) - e^v(p) - e^v(p_1)}, \end{aligned} \tag{5.5}$$

$$\begin{aligned} \Delta e^{(2)}(h) = & \frac{1}{2} \sum \frac{|<0|H - E_p^v|\mathbf{h}\mathbf{h}_1\mathbf{p}_1\mathbf{p}_2>|^2}{e^v(h) + e^v(h_1) - e^v(p_1) - e^v(p_2)} \\ & - \frac{1}{2} \sum \frac{|<\mathbf{h}|H - E_p^v|\mathbf{h}_1\mathbf{h}_2\mathbf{p}_1>|^2}{e^v(h_1) + e^v(h_2) - e^v(h) - e^v(p_1)}, \end{aligned} \tag{5.6}$$

The non diagonal matrix elements appearing in the above equations have been calculated in the so called two- *plus three-body separable* approximation which consists on summing all the diagrams displayed in Fig. 15 with the exclusion of the four-body diagrams, like for instance (15.12) and of diagram (15.15) and their exchanged counterparts, by taking into account the various cancellations as discussed in section 2.3. The approximation used in nuclear matter should not be confused with the 3BS approximation discussed in secs. 3 and 4, where two- and three-body terms are built up with the dressed links $g_{dd} - 1, g'_{dd}, X_{cc}$ and X'_{cc} in place of the bare interaction and the correlation factors.Note that the matrix elements required in the above equations are characterized by the change of two orbitals between the ket and the bra as the case discussed in section 2.3. The differences in the type of excited states are easily handled by using the cluster properties described in section 2.

The perturbative corrections are rather small, as it can be seen from the differences between U_V and U_2 displayed on Fig. 22, althought U_2 is in slightly better agreement with the empirical well depths in the energy range from -60 to $+60$ MeV.

The effective mass, which is related to $U(\rho, e)$ by the equation

$$m^*(\rho, e) = 1 - \partial U(\rho, e)/\partial e, \tag{5.7}$$

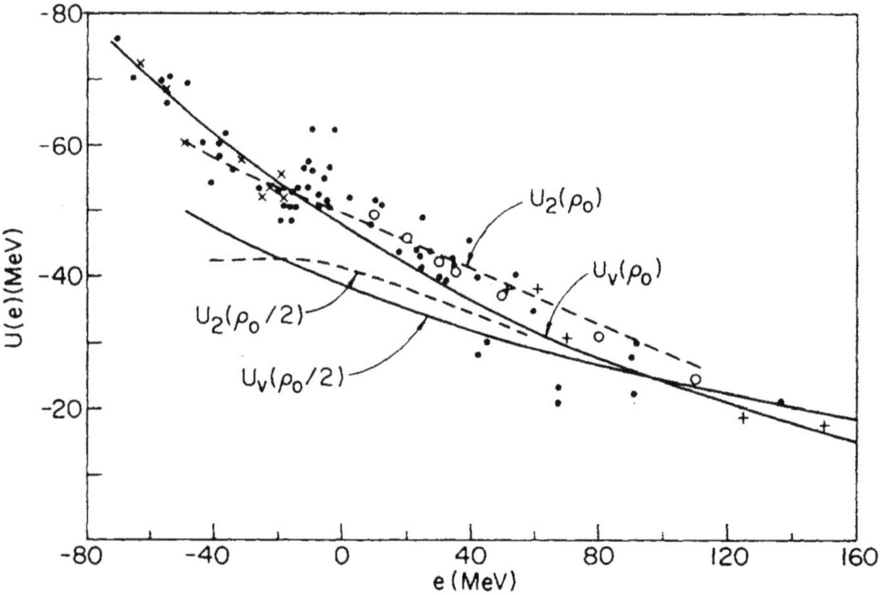

Fig. 22 - *Taken from ref. (31). The real part of the optical potential of nuclear matter calculated by using CBF theory compared with experimental data. U_v is the variational estimate and U_2 includes the second order CBF perturbative corrections due to $2-ph$ admixtures in the ground state wave function.*

determines the velocity of nucleons and the density of states in nuclear matter.

The curves labelled $m_V^*(e)$ and $m_2^*(e)$ in Fig. 23 are obtained from U_V and U_2 respectively. $m_2^*(e)$ has a peak in the region $e \sim e_F$. Such a behaviour was first suggested by *Brown et al.*[70] and later analyzed and extensively discussed by *Mahaux* and collaborators[63,71].

The imaginary part $W_0(e)$ of the optical potential, which is used to analyze elastic scattering, is related to the lifetime $\tau(p)$ of the CBF state $|\mathbf{p}>$ through the following equation

$$W_0(e) = 2\hbar\tau(p)(m^*/m). \tag{5.8}$$

The lifetime $\tau(p)$ is computed from the total transition rate of the state $|\mathbf{p}>$ to decay into two-particle one-hole states $|\mathbf{h}_1\mathbf{p}_1\mathbf{p}_2>$. This is obtained by the on shell part of the imaginary part of the *advanced* self-energy[61]

$$Im\Sigma(\mathbf{h}, E > e^v(k_F)) = \frac{\pi}{2}\sum |<0|H - E_0^v|\mathbf{h}\mathbf{h}_2\mathbf{p}_1\mathbf{p}_2>|^2 \times \delta(E + e^v(h_2) - e^v(p_1) - e^v(p_2)), \tag{5.9}$$

$$Im\Sigma(\mathbf{p}, E < e^v(k_F)) = \frac{\pi}{2}\sum |<0|H - E_0^v|\mathbf{h}_1\mathbf{h}_2\mathbf{p}\mathbf{p}_2>|^2 \times \delta(E + e^v(p_2) - e^v(h_1) - e^v(h_2)), \tag{5.10}$$

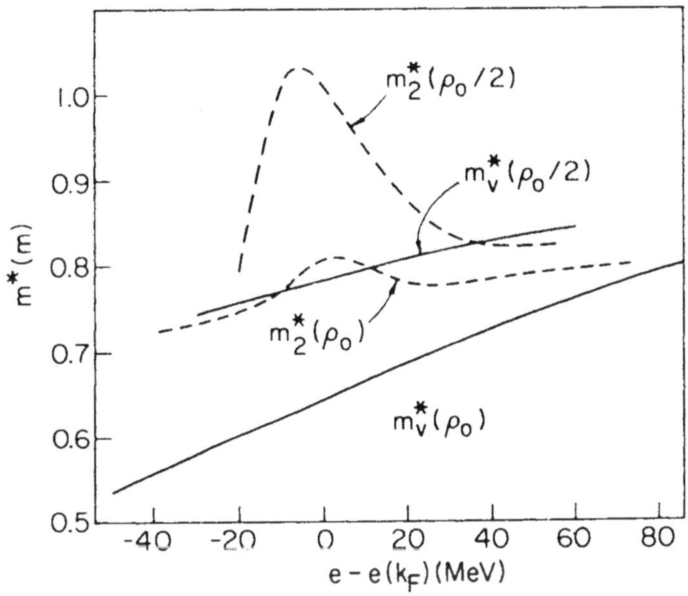

Fig. 23 -*Taken from ref. (31). The effective mass in nuclear matter at the equilibrim density ρ and half of it. The full and dashed lines give the variational and the variational plus second order CBPT calculations respectively.*

The imaginary part of the optical potential is given by

$$W_0(e) = (m^*(e)/m) Im\Sigma(E = k(e), e), \tag{5.11}$$

and it is displayed in Fig. 24.

The second order correction to the ground state binding energy is either given by eq. (2.89) or equivalently by

$$\begin{aligned} \Delta E_0 &= \frac{1}{2\pi} \sum_h \int_{e_F}^{\infty} dE \frac{Im\Sigma(\mathrm{h}, E)}{e^v(h) - E} \\ &= \frac{1}{2\pi} \sum_p \int_{-\infty}^{e_F} dE \frac{Im\Sigma(\mathrm{p}, E)}{e^v(p) - E}. \end{aligned} \tag{5.12}$$

Results of calculations done with the *Urbana* v_{14} + TNR interaction are shown in Fig. 25.

Other interesting quantities are the occupation probability $n(k)$ of states with momentum k and the longitudinal response $S_L(k, \omega)$, which can be measured by deep inelastic scattering

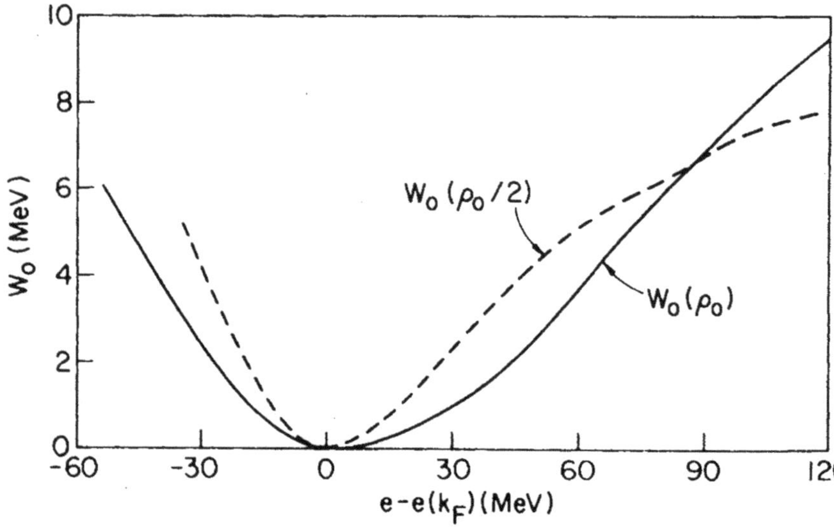

Fig. 24 - *Taken from ref. (31). The imaginary part of the optical potential of nuclear matter calculated by using CBF theory compared with experimental data.*

experiments. The deviation of both quantities from the corresponding Fermi gas estimates indicate the strength of correlations. The second order perturbative calculations of $n(k)$ are obtained from the expectation value of the operator $a^\dagger_{\mathbf{k}} a_{\mathbf{k}}$ on the state

$$|0_2> = |0> + \frac{1}{4} \sum \alpha(\mathbf{h}_1\mathbf{h}_2\mathbf{p}_1\mathbf{p}_2)|\mathbf{h}_1\mathbf{h}_2\mathbf{p}_1\mathbf{p}_2>, \tag{5.13}$$

where

$$\alpha(\mathbf{h}_1\mathbf{h}_2\mathbf{p}_1\mathbf{p}_2) = \frac{<0|H - E^v_p|\mathbf{h}_1\mathbf{h}_2\mathbf{p}_1\mathbf{p}_2>}{e^v(h_1) + e^v(h_2) - e^v(p_1) - e^v(p_2)}, \tag{5.14}$$

and keeping only the terms of order α^2 in its evaluation. Note that both $n(h)$ and $n(p)$ have infinite slopes as h or p goes to k_F. Such a singularity is well known[72] and, in the present treatment comes entirely from the perturbative correction. It results that

$$n(k \sim k_F) = n(k_F^\pm) \pm A \ln \frac{|k - k_F|}{k_F}, \tag{5.15}$$

where the plus (minus) sign is for particle (hole) state. The coeficient A in the above equation is proportional to the strength W of the quadratic term in $(e - e_F)$ of the imaginary part of the self-energy at $e \sim e_F$. In nuclei such a strength is a factor ~ 10 larger than in nuclear matter due to the coupling of single particle states with surface vibrations,

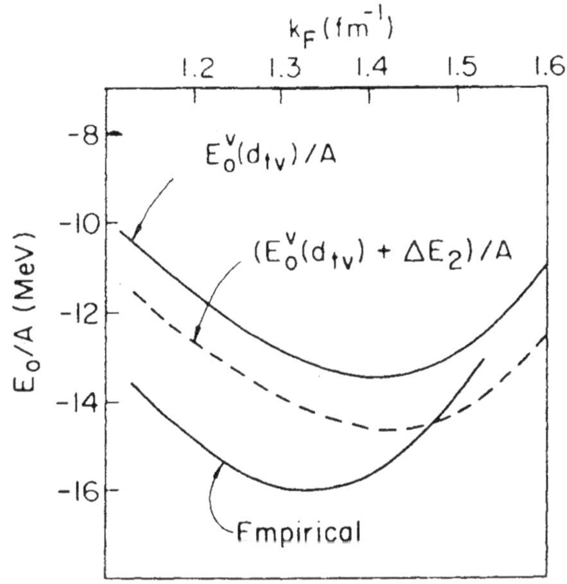

Fig. 25 -*Taken from ref. (31). Binding energy of nuclear matter. The variational estimates are obtained with the Urbana v_{14}plus model VII interaction, whereas the perturbative corrections are calculated by the approximated TNR interaction.*

which are absent in nuclear matter. Therefore one expects a smoother behaviour of $n(e)$ at $e \sim e_F$ in nuclear matter than in nuclei.

The amount of discontinuity of $n(k)$ at $k = k_F$ gives the renormalization constant Z of the Green's function, which in turn is related to the E-mass of *Mahaux* and coworkers[71]. In the CBF calculations the E-mass comes 1.43. Since the total effective mass as calculated from the real part of the optical potential , is 0.81, then the k-mass comes to be 0.57 , in fair agreement with the *Brueckner*-type calculations[71]. The calculated $Z \sim 0.7$ is also in reasonable agreement with the recent experimental data[73,74] on the difference between the charge densities of ^{206}Pb and ^{205}Te, which seem to indicate a Z of about 0.6

The discrepancies between the measured longitudinal responses and the theoretical estimates based on the Fermi gas model[75,76] cannot be completely ascribed to the presence of subnuclear degrees of freedom, but mainly to the fact that Fermi gas model gives a poor description of nuclei. The response can be calculated from the following equations

$$S_L(\mathbf{k},\omega) = \frac{1}{\pi} Im D_L(\mathbf{k},\omega), \tag{5.16}$$

$$D_L(\mathbf{k},\omega) =< 0|\rho^\dagger_{L,\mathbf{k}}(H - E_0 - \omega - i\eta)^{-1}\rho_{L,\mathbf{k}}|0 >, \tag{5.17}$$

with $\rho_{L,\mathbf{k}} = \Sigma \exp[i\mathbf{k}\cdot\mathbf{r}_i](1+\tau_{z,i})/2$ is the charge-density operator and the $D_L(\mathbf{k},\omega)$ is

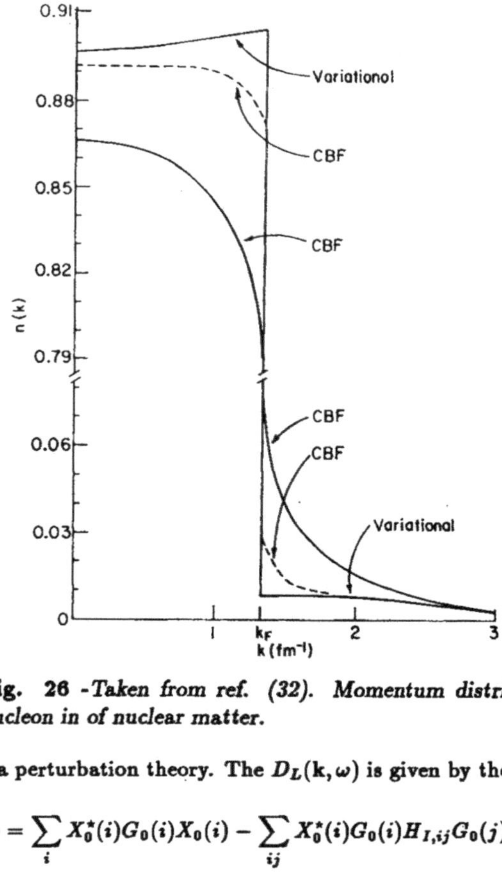

Fig. 26 *-Taken from ref. (32). Momentum distribution of a nucleon in of nuclear matter.*

calculated by a perturbation theory. The $D_L(\mathbf{k},\omega)$ is given by the following series

$$D_L(\mathbf{k},\omega) = \sum_i X_0^\star(i)G_0(i)X_0(i) - \sum_{ij} X_0^\star(i)G_0(i)H_{I,ij}G_0(j)X_0(j) + \ldots, \tag{5.18}$$

where

$$X_0^\star(i) =< 0|\rho^\dagger_{L,\mathbf{k}}|i>, \tag{5.19}$$

$$G_0(i) = \frac{1}{H_{0,ii} - E_0 - \omega - i\eta}. \tag{5.20}$$

The series is formally summed by the following Dyson-like equation

$$X(i) = X_0(i) - \sum_j H_{I,ij}G_0(j)X(j), \tag{5.21}$$

$$D_L(\mathbf{k},\omega) = \sum_i X_0^\star(i)G_0(i)X(i). \tag{5.22}$$

Improved approximations to $D_L(\mathbf{k},\omega)$ are obtained by including $1-ph$ OCS, $2-ph$ OCS etc.. in the above sums.

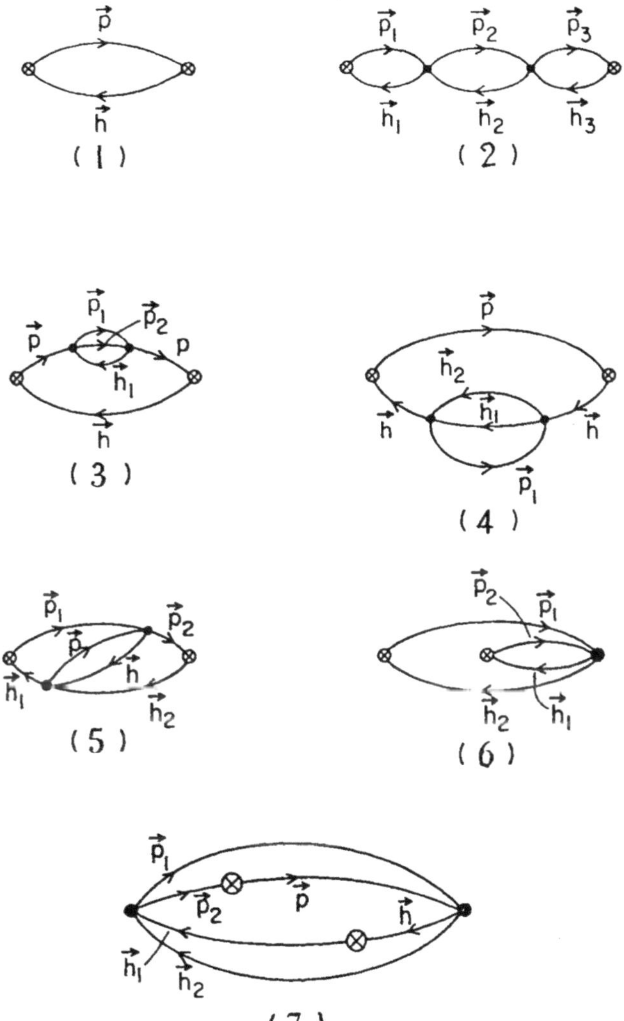

Fig. 27 *-Correlated basis function (CBF) diagrams contributing to $S(k,\omega)$. The circled crosses indicated the interactions X_0 of the nucleons with the external probe, whereas the dots denote the non diagonal matrix elements H_{ij}. The oriented lines represent particle (hole) states.*

Let us discuss the main properties of the various terms of the series, some of which are represented by diagrams in Fig. 27, as they result from the microscopic calculations on the density response[40] and, more recently, from those on the longitudinal response[54] of nuclear matter:

(i) Diagram (1) corresponds to the variational estimate of $D_L(\mathbf{k},\omega)$ given by the first term on the r.h.s of eq.(5.18). Correlation effects enter into the single particle spectrum , as well as into the matrix element $X_0(i)$ in eq.(5.19) *leading to a quenching which, in some case, is of the order of* 10% *with respect to its mean field estimate* (= 1). Figs. 28 and 29 show the differences between the uncorrelated (S_{FG}) and the correlated (S^0_{ph}) responses at $k = 262 MeV/c$ and $k = 550 MeV/c$ respectively. It should be noticed that both the uncorrelated and the correlated responses are different from zero in a finite interval of ω values, namely for $Max(e_v(k-k_F)-e_v(k_F);0) < \omega < e_v(k+k_F)+e_v(k_F)$. It is only after the inclusion of $2-ph$ OCS that the response will have nonvanishing values outside that region. In fact the inclusion of such excited states gives a width to the $1-ph$ OCS.

(ii) Diagram (2) represents a second order term having the $1-ph$ OCS as intermediate states. The solution of eq.(5.22) within the $1p-1h$ OCS includes diagrams of the type (2) with any number of $1-ph$ bubbles and it is expecially required at low momentum transfer. In this region the effects of collective excitations are important and therefore one has to use the correct combinations of $1-ph$ states to get realistic estimates of the responses. To solve the eq.(5.22) within the $1-ph$ OCS basis is equivalent to diagonalize the interaction hamiltonian in such a basis and therefore truncations of the series (5.18) are in general a bad approximation. In Figs. 28 and 29 the curves labelled with S_{ph} are obtained by solving the eq.(5.22) within the $1-ph$ OCS basis. The effective interaction H_{ij} rapidly decreases for increasing values of the momentum transfer. In fact the differences between S^0_{ph} and S_{ph} are much smaller at $k = 2.79 fm^{-1}$ then at $k = 1.33 fm^{-1}$.

(iii) Diagrams (3) and (4) represent terms having $2-ph$ OCF intermediate states, which allow for $1p$ states decaying into $2p-1h$ states and therefore are characterized by self-energy insertions on the particle (hole) lines. These processes, if calculated on shell, furnish the main contribution to the imaginary part of the optical potential of a particle (hole) of momentum $\mathbf{p}$ ($\mathbf{h}$) in nuclear matter displayed in Fig. 24 as a function of single particle energy. A reasonably good approximation is to neglect the momentum dependence of these self energy insertions and take their on shell part.

This is equivalent to modify the single particle energies by including the second order perturbative corrections U_2 (see Fig. 22) and to fold the calculated S_{ph} with the imaginary part of the optical potential W (note[31] that between W and W_0 given in Fig. 24 there is a factor m^*), namely

$$\overline{S} = \frac{1}{\pi} \int d\omega' \frac{S_{ph}(k,\omega')W(\omega)}{(\omega'-\omega)^2 + W^2(\omega)}. \tag{5.23}$$

(iv) Diagrams (5-7) are other second order diagrams that have not been included in the nuclear matter calculations of refs.(40,54). Diagrams (6) and (7) are generated when the ground state wave function is corrected for $2-ph$ admixtures[77].*Dröżdz et al.*[77] have studied the effects of $2-ph$ states on the longitudinal response function of ^{12}C and find that at $k = 300 MeV/c$ the main effects come from diagrams of the type (3,4).

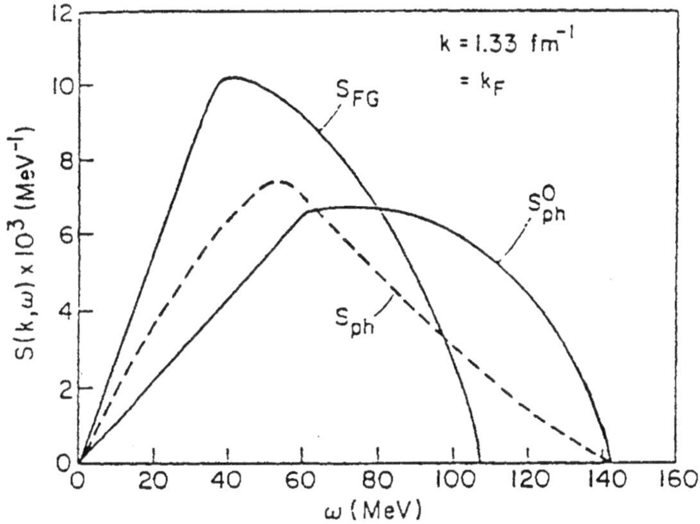

Fig. 28 - *Taken from ref. (40). Density response of nuclear matter at $k = 1.33 fm^{-1}$ in various approximations.*

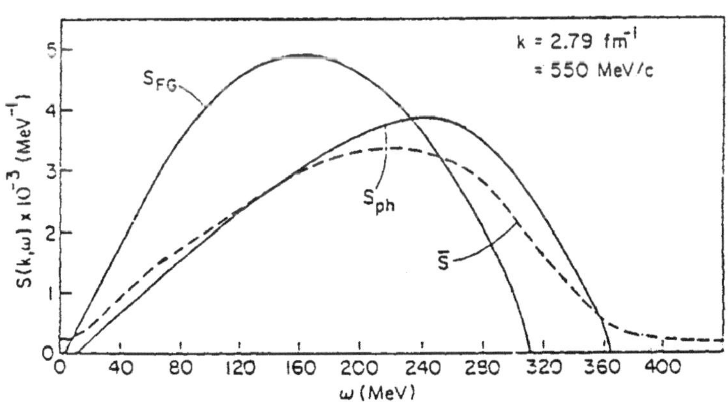

Fig. 29 - *Taken from ref. (40). Density response of nuclear matter at $k = 2.79 fm^{-1}$ in various approximations ($S_{ph} \sim S^0_{ph}$).*

The longitudinal response function is obtained by multiplying $S_L(k,\omega)$ with the squared proton factor. In Fig. 30 the experimental data are compared with the responses calculated with the covariant form factor[78] . The solid and the dashed lines refer to the density response[40] and to the longitudinal responses[54] respectively.

The results displayed in Fig. 30 show that at lower values of ω the various nuclei give similar responses which agree with the theoretical estimates both at low and high momentum transfer, and therefore they do not show evidence for a strong dependence of the nucleon form factor on the nuclear medium.

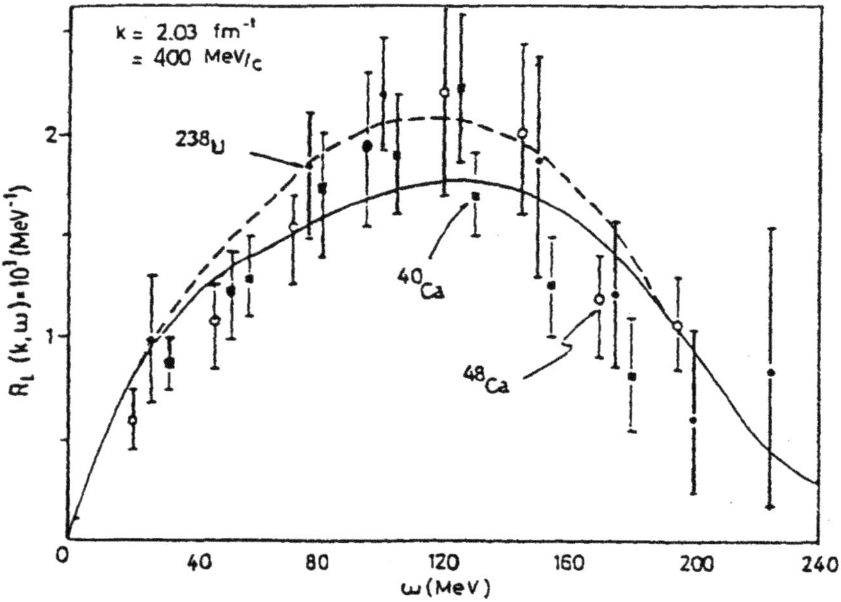

Fig. 30 -*The calculated density*[40] *and longitudinal*[54] *responses of nuclear matter at* $k = 400 MeV/c$ *are compared with experimental data on* ^{40}Ca *and* ^{48}Ca *at* $k = 410 MeV/c$ *and* ^{238}U *at* $k = 400 MeV/c$.

6. Conclusions and Perspectives

Microscopic calculations on strongly interacting Fermi systems indicate that non trivial correlation effects play an important role in their structure and that they may be underestimated in mean field theories. Such correlations on one side reflect the nature of the interparticle interaction, and, on the other side are typical phenomena of the underlying many-body problem.

Theories, like CBF, which include the main correlation effects in the basis functions appear to be good candidates to solve with the necessary accuracy the many-body Schrödinger equation.

The first step of the CBF theories is the variational theory based on correlated wave functions. This gives impressively good results for the properties of the ground state, expecially for the binding energy. A consistent perturbative scheme is then necessary in order to better describe the true eigenstates of the hamiltonian expecially for what concern the dynamical properties of the many-body system.

Suitable cluster expansions of the diagonal and not diagonal matrix elements are available which allow for expressing the various quantities of interest into proper series of cluster diagrams. Such cluster diagrams can then be summed up by means of FHNC theory and its generalization to treat state dependent correlations, non diagonal matrix elements, etc.

By using CBF theories several properties of strongly interacting Fermi systems could be quantitatively understood, mainly because the correlation effects in these theories are substantially treated in a non perturbative way, the perturbative corrections being reserved to minor details. The behaviour of the effective mass, that of the momentum distribution, the quenching of the response functions are mostly explained in terms of correlation effects.

In spite of the many successes of the CBF theories, several formal and numerical problems need still to be solved. Let us consider in the following only those amongst them which are more connected with the subject of this work.

The FHNC type of summations gives reliable results for systems at not very high density, like nuclear matter. However, for denser systems, like liquid helium the elementary diagrams give important contributions. The *scaling* or the *interpolation equation* methods have given very good results in the calculation of the expectation value of the distribution functions for the ground state. A complete understanding of the reasons for this success is lacking and, for this reason the application of the above methods to evaluate other quantities is not completely under control. Therefore a better understanding how elementary diagrams should be included in FHNC-type calculations would be very interesting.

In dealing with state dependent correlations the approximation which has been mostly used is the FHNC/SOC . Some improved approximations has revealed to give practically the same results as FHNC/SOC in the few cases of interest. However a more sistematic analysis of many body cluster diagrams containing state dependent correlations should be performed expecially in critical situations, like those one encounters in very dense systems or if strong state dependent correlations are needed.

The perturbation theory still needs a better understanding of its expansion parameter. In the case of very dense systems one does not really know how to sum the various perturbative diagrams. We have learned how important are the orthogonality corrections. Therefore we should be guided by that for summing the needed perturbative diagrams.

Finally, further numerical calculations need to be performed to study the excited states of liquid ^3He, as well as more refined calculations of the type reviewed in this work should be done in nuclear matter to better understand the role of N-N correlations in nuclear physics.

Acknowledgements. The authors are very much indebted to V. R. Pandharipande, S. Rosati and M. Viviani for fruitfull discussions. S. F. gratefully aknowledges the hospitality of the Department of Physics of the University of Pisa and A.F. that of the Department of Physics of the University of Illinois at Urbana-Champaign during the time this work has been done. This work has been partially supported by the U.S.Department of Energy, Division of Material Sciences under Grant No. DE-AC02-76ER01198.

7. References

1. M. H. Kalos, Phys. Rev. **A2** (1970) 250

   M. H. Kalos, D.Levesque and L. Verlet, Phys. Rev. **A9** (1970) 2178

2. D. M. Ceperley and M. H. Kalos, *MonteCarlo methods in Statistical Physics*, edited by K. Binder (Springer, Berlin) 1978

   R. Guardiola, this volume (and references therein)

3. M. A. Lee, K. E. Schmidt, M. H. Kalos and G. V. Chester, Phys. Rev. Lett. **46** (1981) 728

4. K. A. Brueckner, Phys. Rev. **96** (1954) 908

   K. A. Brueckner and C. A. Levinson, Phys. Rev. **97** (1955) 1344

5. H. A. Bethe, Phys. Rev. **103** (1956) 1353

6. J. Goldstone, Proc. Roy. Soc. London **A239** (1957) 267

7. B. H. Brandow, Phys. Rev. **152** (1966) 863

8. B. D. Day, *From Nuclei to Particles*, Proceedings of the International School of Physics *Enrico Fermi*, Course LXXIX, Varenna, A. Molinari editor (North-Holland, Amsterdam-New York-Oxford) (1981) 1

9. J. P. Jeukenne, A. Lejeune and C. Mahaux, Nucl. Phys. **A245** (1975) 411; Phys. Report **C25** (1976) 83

10. B. L. Friman and E. M. Nyman, Nucl. Phys. **A302** (1978) 365

11. F. Coester, Nucl. Phys. **7** (1958) 421

    F. Coester and H. Kümmel, Nucl. Phys. **17** (1960) 477

    H. Kümmel, K. H. Lührman and J G. Zabolitzky, Phys. Report **C36** (1978) 1

12. A. Lande and R. A. Smith, *Recent Progress in Many Body Theories*, Vol. I, (Plenum Press, London), A. Kallio, E. Pajanne and R. F. Bishop editors, in press (and references therein)

13. S.D.Yang and T.T.S.Kuo, Nucl.Phys. **A467** (1987) 461

14. E. Feenberg, *Theory of Quantum Liquids*, Academic Press, New York (1969)

. J. W. Clark, *Progress in Particle and Nuclear Physics* bf 2 (D.H.Wilkinson ed.) pg.89, Pergamon, Oxford, 1979)

. J. G. Zabolitzky, Phys. Rev. **A16** (1977) 1258

. E. Campani, S. Fantoni and S. Rosati, Lett. Nuovo Cimento, **15** (1976) 217

. S. Fantoni and S. Rosati, Lett. Nuovo Cimento, **10**(1974) 545 ; Nuovo Cimento **A25** (1975) 593

. E. Krotscheck and M. L. Ristig, Phys. Lett. **A48** (1974) 17; Nucl. Phys. **A242** (1975) 389

. S. Rosati, *From Nuclei to Particles*, Proceedings of the International School of Physics *Enrico Fermi*, Course LXXIX, Varenna, A. Molinari editor (North-Holland, Amsterdam-New York-Oxford) (1981) 73

. V. R. Pandharipande and R. B. Wiringa, Nucl. Phys. **A266** (1976) 269

. S. Fantoni and S. Rosati, NUovo Cimento **43A** (1978) 413

O. Benhar, C. Ciofi degli Atti, S. Fantoni and S. Rosati, Phys.Lett **70B** (1977) 1

3. V. R. Pandharipande and R. B. Wiringa, Rev. Mod. Phys. **51** (1979) 821

4. E. Manousakis, S. Fantoni, V. R. Pandharipande and Q. N. Usmani, Phys. Rev. **B28** (1983) 3770

5. M. Viviani, E. Buendia, S. Fantoni and S. Rosati, Phys.Rev. **B**, submitted

6. R. B. Panoff, *Recent Progress in Many Body Theories*, Vol. I (Plenum Press, London), A.Kallio,E.Pajanne and R.F.Bishop editors, in press

7. J. W. Clark, L. R. Mead, E. Krotscheck, K. E. Kürten and M. L. Ristig, Nucl. Phys. **A328** (1979) 45

8. S. Fantoni, Phys. Rev. **29B** (1984) 2544

9. E.Krotsheck and R. A. Smith, Phys.Rev. **B27** (1983) 4222

0. J. W. Clark and E. Krotscheck, Lecture Notes in Phys. **198** (1983) 127

1. S. Fantoni, B. L. Friman and V. R. Pandharipande, Nucl. Phys. **A386** (1982) 1;**A399** (1983) 1

2. S. Fantoni and V. R. Pandharipande, Nucl. Phys. **A427** (1984) 473

3. S. Fantoni and V. R. Pandharipande, Phys. Rev. **C**, in press

4. P. O. Löwdin, Jour. Chem. Phys. **18** (1950) 365

5. S. Fantoni and S. Rosati, Nuovo Cimento **20A** (1974) 179

6. L. Lantto and P. J. Siemens, Phys. Lett. **68B** (1977) 308 ;

J. C. Owen, Phys. Lett. **98B** (1980) 303; Phys. Rev. **B23** (1981) 2169

7. S. Rosati and M. Viviani, in this volume

38. S. Fantoni, Nuovo Cimento **A44** (1978) 191;

S. Fantoni and V. R. Pandharipande, Nucl. Phys. **A427** (1984) 473

39. E. Krotscheck and J. W. Clark, Nucl. Phys. **A328** (1979) 73; **A333** (1980) 77

40. S. Fantoni and V. R. Pandharipande, Nucl. Phys. **A473** (1987) 234

41. E. Krotscheck, Annals of Phys. **155** (1984) 1

42. M. L. Ristig, *From Nuclei to Particles*, Proceedings of the International School of Physi *Enrico Fermi*, Course LXXIX, Varenna, A. Molinari editor (North-Holland, Amsterdai New York-Oxford) (1981)

43. R.Abe, Progr.Theor.Phys. **21** (1959) 421

44. J.G.Zabolitzky, Adv.Nucl.Phys. **12** (1981) 1; Phys.Rev. **A16** (1977) 1258

45. S.Fantoni and S.Rosati, Phys.Lett. **84B** (1979) 23

46. F.Lado, J.Chem.Phys. **47** (1967) 5369

47. J.G.Zabolitzky, Phys.Rev. **B22** (1980) 2353

48. Q.N.Usmani,B.Friedman and V.R.Pandharipande, Phys.Rev. **B25** (1982) 4502;

Q.N.Usmani,S.Fantoni and V.R.Pandharipande, Phys.Rev. **B26** (1983) 6123

49. A.Fabrocini and S.Rosati, Nuovo Cimento **D1** (1982) 567; **D1** (1982) 165;

M.Viviani,E.Buendia,A.Fabrocini and S.Rosati, Nuovo Cimento **D8** (1986) 561

50. R.B.Wiringa, Nucl/Phys. **A338** (1980) 57;

R.B.Wiringa,V.Fiks and A.Fabrocini, Phys.Rev. **C**, submitted

51. I.E.Lagaris, Anales de Fisica **81** (1985) 39

52. A.Fabrocini and V.R.Pandharipande, private communication

53. B.Friedman and V.R.Pandharipande, Phys.Lett. **100B** (1981) 205

54. A.Fabrocini and S.Fantoni, in preparation

55. C.W.Woo, Phys.Rev. **151** (1966) 138

56. P.M.Morse and H.Feshbach, *Methods of Theoretical Physics* (McGraw-Hill, NewYorl 1953)

57. R.Schiavilla and V.R.Pandharipande, Phys.Rev. **C**, in press

58. J.Goldstone, Proc.Roy.Soc. (London) **A239** (1957) 267

59. A.Fabrocini,S.Fantoni,S.Rosati and A.Polls, Phys.Rev. **B33** (1986) 6057

60. S.Fantoni *et al.* ,*Recent Progress in Many Body Theories*, Vol. I (Plenum Press, London A.Kallio,E.Pajanne and R.F.Bishop editors, in press

61. A.A.Abrikosov,L.P.Gorkov and I.E.Dzydoshinsky, *Methods of Quantum Theory in Stati: tical Physics*, Prentice-Hall, Englewood Cliffs, 1963

A.L.Fetter and J.D.Walecka, *Quantum Theory of Many Particles Systems*, (McGraw-Hill, NewYork, 1971)

R.Sartor and C.Mahaux, Phys.Rev. **C21** (1980) 1546

A.Fabrocini *et al.*, Nuovo Cimento **56A** (1980) 33

R.A.Aziz *et al.*, J.Chem.Phys. **70** (1981) 349

R.Ahlrichs *et al.*, Chem.Phys. **19** (1876) 119

I.E.Lagaris and V.R.Pandharipande, Nucl.Phys. **A359** (1981) 349

R.B.Wiringa,R.A.Smith and T.L.Ainsworth, Phys.Rev. **C29** (1984) 1207

R.Schiavilla,V.R.Pandharipande and R.B.Wiringa, Nucl.Phys. **A449** (1986) 219

G.E.Brown.G.h. Gunn and P.Gould, Nucl.Phys. **46** (1963) 598

J.P.Jeukenne,A.Lejeunne and C.Mahaux, Phys.Rep. **25C** (1076) 83

A.B.Migdal, Sov.Phys.JETP **49** (1957) 333

J.M.Cavedon *et al.*, Phys.Rev.Lett. **49** (1983) 978

B.Frois and C.N.Papanicolas, Ann.Rev.Nucl.Part.Sci. **37** (1987)

W.Czyz, Phys.Rev. **131** (1963) 2141

E.Moniz *et al.*, Phys.Rev.Lett. **26** (1971) 44

S.Drodz,G.P.Co',J.Wambach and J.Speth, Phys.Lett. **185B** (1987) 287

C.Ciofi degli Atti, Progr.in Part.and Nucl.Phys. **1** (1980) 163

MONTECARLO TECHNIQUES IN THE MANY BODY PROBLEM

by
R.Guardiola(*)
Departament of Nuclear Physics, University of Granada (Spain)

(*)Departamento de Física Nuclear, Universidad de Granada
18071 Granada (España)

MONTECARLO TECHNIQUES IN THE MANY BODY PROBLEM

by R. Guardiola

MONTECARLO TECHNIQUES IN THE MANY BODY PROBLEM

R. Guardiola
Departamento de Física Nuclear
Universidad de Granada
18071 GRANADA (Spain)

ABSTRACT

These lecture notes have been prepared with the objective of being an introductory course to Montecarlo methods in quantum problems, and particularly in the quantum many body problem. In this sense, it has been assumed that the reader has no previous knowledge on Montecarlo methods in general, and even on mathematical statistics. A particular emphasis has been put in the practical character of the notes, and many examples have been included (even with a working Fortran code) to stimulate the reader to practice by himself most of the algorithms introduced.

The course starts with a brief introduction to the statistical concepts of interest for the remaining chapters, and with a quite detailed description of the ways of generating random numbers. The powerful Metropolis algorithm is presented with some theoretical background in the third chapter. Physics starts in chapter four, which is devoted to Variational Montecarlo. Cahpters six and seven are devoted to Diffusion Montecarlo and to Green function Montecarlo, respectively. These two theories share many ideas, and should not be read independently.

Along the lecture notes, some physical systems have been briefly described, with a particular emphasis on helium atoms systems. This description is very incomplete, and the reader should consult the specialized works or reviews referred in the notes to complement the description.

The notes include (quite unsually) several Fortran programs. Most of them do not require big computers or long time runs, and even with a personal computer can be satisfactorily run. These codes are far from being optimized, the major emphasis being in having a direct correspondence with the algorithms previously described. I suggest the reader to copy and run these programs, and realize that Montecarlo really works. This is a good stimulus to go forward with the subject.

1. BASIC CONCEPTS OF STATISTICS

1.1 Random variables and probability distribution functions

Consider a physical event or an action, such as throwing a dice or shooting on a target, in such a manner that we may repeat it as many times as needed. The result of this action may be expressed in terms of a numerical value, by assigning a numerical value to the outcome of the action. So we associate a random variable x which takes a concrete value at each trial.

The exhaustive set of all possible results is the domain of the random variable, e.g. $(x_1, x_2, \ldots, x_n)$. After many empirical trials we may observe that some of the outcomes appear more frequently than some others. To each of the outcomes we associate a probability, and the ordered set of all these probabilities (ordered in correspondence with the set of possible results), $(p_1, p_2, \ldots, p_n)$ is the probability distribution of the random variable. The exhaustive character of the domain implies that the sum of the probabilities is 1.

Summarizing, a random variable is defined by means of its domain, the set of possible values, and its probability distribution, the set of corresponding probabilities.

A simple example is the random variable defined as the sum of points after throwing two dices. As we know very well we have:

Domain:	2	3	4	5	6	7	8	9	10	11	12
Prob.Distr:	$\frac{1}{36}$	$\frac{2}{36}$	$\frac{3}{36}$	$\frac{4}{36}$	$\frac{5}{36}$	$\frac{6}{36}$	$\frac{5}{36}$	$\frac{4}{36}$	$\frac{3}{36}$	$\frac{2}{36}$	$\frac{1}{36}$.

The random variable can be univariate, as in the previous example, or multivariate. We may also consider random variables which take values in the continuum, in which case the density distribution function is defined.

TABLE 1.1. Concepts in discrete and continuous random variables

	Density Distr. Function	Cumulative Distr. Function	Domain
Discrete random v.	p_i=Prob($X=x_i$)	P_i=Prob($X\leq x_i$)	$x_1, x_2, \ldots, x_n$
Continuous random v.	$f(x)dx$=Prob($x\leq X\leq x+dx$)	$F(x)$=Prob($X\leq x$)	(a,b)

These concepts are summarized in Table 1. There are some simple properties that probability distribution functions must satisfy:

i) Must be positive definite,

$0 \leq p_i \leq 1$, or

$f(x) \geqslant 0$, but without an upper bound.

ii) Bounds on the cumulative distribution functions,

$0 \leqslant P_i \leqslant 1$, or

$0 \leqslant F(x) \leqslant 1$

iii) Both P and F are non-decreasing functions,

$P_i \geqslant P_j$, if $i \geqslant j$, or

$F(x_1) \geqslant F(x_2)$, if $x_1 \geqslant x_2$

iv) Normalization,

$P_n = 1$, or

$F(b) = 1$, where b is the upper limit of the continuous domain.

1.2 Examples of probability distributions

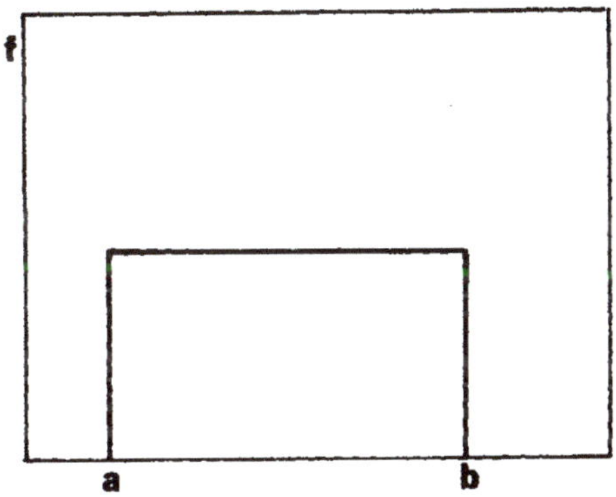

Figure 1.1. The uniform distribution

Uniform distribution U(a,b)

$f(x)=0, \; x < a$

$$f(x)=1/(b-a), \; a \leqslant x \leqslant b \quad (1.1)$$

$f(x)=0, \; x > b$

The standard form of this distribution corresponds to a=0 and b=1. This distribution is basic in the random number generation algorithms.

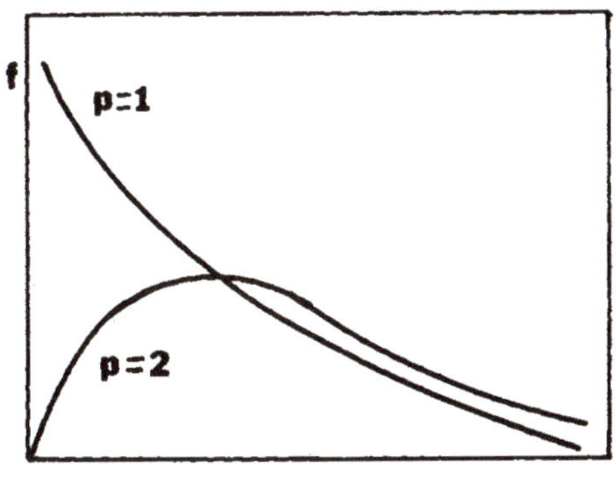

Figure 1.2 . The gamma distribution

Gamma distribution

$$f(x)=x^{p-1}\exp(-x)/\Gamma(p) \quad (1.2)$$

$0 \leqslant x < \infty$, and

$0 < p < \infty$.

This distribution is related to the Green function of a free particle in multi-dimensional spaces.

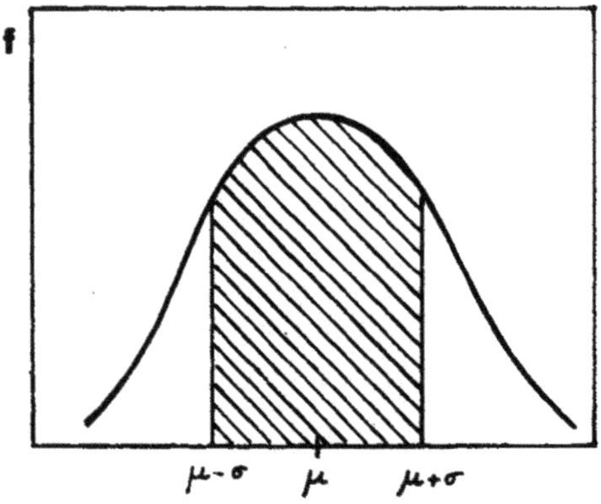

Figure 1.3. The normal distribution

Normal distribution

$$f(x)=\frac{1}{\sqrt{2\pi}}\,\frac{1}{\sigma}\,\exp\left\{-\frac{(x-\mu)^2}{2\sigma^2}\right\} \qquad (1.3)$$

μ and σ are parameters of the distribution, and

$$-\infty < x < \infty \, .$$

The normal distribution has been written so that its mean is μ, and its variance is σ^2. These concepts will be introduced below. The standard form corresponds to a mean of 0 and to a standard deviation $\sigma=1$.

The value of σ is usually interpreted as a confidence limit in the following form. The probability of having x in the interval $(\mu-\sigma,\mu+\sigma)$ is 0.68, and doubling the interval increases this probability to 0.95. In this manner, when we say 'we are within one standard deviation', this means a mathematical probability of 68 p.c.

1.3 Expectation values

Let g(x) be an arbitrary function of the random variable x , this random variable corresponding to a probability distribution function f(x) defined in the interval (a,b). Arbitrary means that g(x) must not satitify any positivity condition, i.e., it may have any shape, provided the integrals to be defined below are well defined.

The expectation value of g is defined as

$$\langle g \rangle = \int_a^b g(x)\, f(x)\, dx \qquad (1.4)$$

Some particular forms of g(x) have an special interest:

i) MEAN value of the distribution, when g(x)=x,

$$\mu = \int_a^b x\, f(x)\, dx \qquad (1.5)$$

ii) CENTRAL MOMENTS, i.e., moments of the distribution with respect to the mean value,

$$g(x) = (x-\mu)^n$$

$$\mu_n = \int_a^b (x-\mu)^n\, f(x)\, dx \qquad (1.6)$$

The second moment μ_2 is called the variance of the probability distribution, and its square root is called the standard deviation which usually is represented by the greek letter σ .

$$\sigma = \sqrt{\mu_2} \qquad (1.6a)$$

We may now recall the example of eq. (3) referring to the normal distribution. The parameters μ and σ introduced there correspond to the mean and standard deviation of this distribution, respectively.

iii) CHARACTERISTIC FUNCTION, when $g(x)=\exp(itx)$,

$$\phi(t)= \int_a^b \exp(itx)\ f(x)\ dx \qquad (1.7)$$

The n-th derivative of the characteristic function is proportional to the n-th (non-central) moment

$$\left\{\frac{d^n}{dt^n}\ \phi(t)\right\}_{t=0} = i^n \langle x^n \rangle \qquad (1.8)$$

There is an important theorem regarding the moments of a given distribution: under quite mild conditions on the distribution function, the set of moments uniquely determines the probability distribution.

1.4 The Central Limit theorem

Let x be a random variable with probability distribution function f(x). Consider a new random variable z

$$z = (x_1+x_2+ \dots +x_n) / n \qquad (1.9)$$

i.e., the average of n random values of the variable x. The question is to determine the probability distribution function of this new variable. It can be computed using the following formula

$$g(z)=\int dx_1\ dx_2 \dots dx_n\ f(x_1)f(x_2)\ \dots f(x_n)\ \delta\left(\frac{x_1+x_2+\dots+x_n}{n} - z\right) \qquad (1.10)$$

where δ represents Dirac's delta function. Using the integral representation of the delta function, shifting the integration variables with the mean value and reordering the integrals, eq. (10) is converted into

$$g(z)= \frac{1}{2\pi}\int_{-\infty}^{\infty} dt\ e^{-i(z-\mu)t} \left\{\int f(x)\ e^{it(x-\mu)/n}\ dx\right\}^n \qquad (1.11)$$

This gives an explicit way of evaluating the new distribution function, which consists in determining the characteristic function of the original distribution with a scaled and shifted variable, then rising it to the power n and finally computing the inverse Fourier transform. In general eq. (11) is not simple to evaluate. However, in the case of large n we may approximate this equation by expanding the second exponential of (11) and keeping only the dominant terms,

$$\int f(x)\ \exp(it(x-\mu)/n)\ dx \simeq \int f(x)dx\ \left\{1 + it(x-\mu)/n - t^2(x-\mu)^2/2n^2\ \dots\right\}$$

The second term gives a zero contribution, and the third term is related to the variance of the distribution, i.e.,

$$\int f(x)\ \exp(it(x-\ )/n)\ dx \simeq 1 - t^2\sigma^2/2n^2 \dots$$

Substituting in eq. (11) and carrying out the remaining integral there results the following value for the new distribution function,

$$g(z) = \frac{1}{\sqrt{2\pi}} \frac{1}{(\sigma/\sqrt{n})} \exp - \frac{(z-\mu)^2}{2(\sigma/\sqrt{n})^2} \qquad (1.12)$$

i.e. we obtain the normal distribution, cf. eq. (3), with a mean which is the mean of the original distribution, and a variance equal to the variance of the original distribution divided by n.

The above statement is known as central limit theorem. According with this theorem, the average of n samples is an unbiased estimator of the mean of the distribution. Moreover, its variance decreases with 1/n, and its standard deviation, i.e., the error in the determination of the mean, decreases with the square root of n. This is the idea which governs Montecarlo method. Note that in order to halve the error we must multiply the number of samples times 4.

The central limit theorem can be generalized to other types of expectation values. In general, if we have the expectation value

$$m = \langle h \rangle = \int f(x)\, h(x)\, dx \qquad (1.13)$$

and define

$$s^2 = \int f(x)\, (h(x)-m)^2\, dx, \qquad (1.14)$$

and consider the random variable

$$z = (h(x_1)+h(x_2)+...+h(x_n))/n \qquad (1.15)$$

then it turns out that z is distributed with a normal distribution of mean m and variance s^2/n.

1.5 Summary

This section describes some simple statistical concepts basic for the Montecarlo method. The most relevant concept of this section is Central Limit theorem, which provides an algorithm to determine empirically average values of arbitrary functions with respect to a given probability distribution function f(x). In other words, it permits the evaluation of definite integrals. The estimator is the average value of the function, eq. (15). To carry out in practice this algorithm it is necessary to devise a method of sampling a random variable x accordingly with a prescribed probability distribution function f(x). This will be the subject of next chapter.

Along this chapter we have made reference only to univariate distributions. All these ideas may be extended to multivariate probability distributions $f(x_1,x_2,...x_n)$. When a multivariate distribution can be written as a product of univariate distributions, i.e.,

$$f(x_1, x_2,...,x_n) = f_1(x_1)\, f_2(x_2)\, ...\, f_n(x_n)$$

that variables are called uncorrelated.

This section does not include references. All concepts discussed here may be found in any elementary textbook of statistics.

2. ON THE GENERATION OF RANDOM NUMBERS WITH PRESCRIBED DISTRIBUTION FUNCTIONS

A task of primary interest in Montecarlo work is the generation of random numbers which follow a given probability distribution function. In this chapter we are going to consider the basic methods, which unfortunately will not solve the general problem. The methods considered here are the change of variables, the sum or product of random numbers and the acceptance rejection method. The powerful Metropolis algorithm will be studied latter on, as well as the composition method.

To complement this section the reader may consult Chapter 3 of Hammersley and Handscombe book /1/, or the article of Zelen and Severo /2/.

2.1 The uniform distribution U(0,1)

Usually it is a routine of your computer, callable from many programming languages (Fortran, Pascal, Basic,...) which is based in a congruential relation

$$N_i = (a\, N_{i-1} + b)\ (\text{modulus } M)$$
$$x_i = N_i/M \tag{2.1}$$

In the above equations N_i is called a seed, and is an integer number, and x_i is the random number in the interval (0,1). Note that 1 can never be reached, but 0 is included in the generation interval. The constants a, b and M are chosen so as to satisfy the following three requirements:

i. The resulting distribution is uniform, and the random numbers generated are uncorrelated.
ii. The period of eq. (1) is as long as possible, and
iii. The algorithm is fast.

A quite complete analysis of congruential methods may be found in Ref. /3/. Currently, the number M is determined by the number of bits of the integer word of the computer, i.e., is the larger unsigned integer. In this form the modulus operation is automatically carried out by discarding the overflow digits in the previous integer operation. Analogously, the division N/M is never carried out, being just a shift of bits and the corresponding adjustment of the floating exponent.

Randomness cannot be easily studied. One should analyze the posssible correlations between N_i and N_{i+s}, s arbitrary, as well as the case in which random numbers are considered by pairs, triplets and so on. The routines are empirically tested. Binder /4/ describes several problems of the common computer routines, and also some simple ways to improve the quality of the sequence.

The generation of uniformly distributed random numbers is very interesting because it is the basic algorithm to generate random numbers corresponding to other probability distributions. With this routine all Montecarlo work can be carried out.

From $x \in U(0,1)$ one can obtain $z \in U(a,b)$ by the transformation

$$z = a + (b-a)\, x \tag{2.2}$$

2.2 Change of variables

Let x correspond to a distribution f(x). Define a new variable $y=h(x)$. The question is to determine the distribution function g(y) of the new variable y.

We must have

$$f(x)\, dx = g(y)\, dy$$

so that

$$g(y) = f(\, h^{-1}(y))\ \ dh^{-1}(y)/dy \tag{2.3}$$

where h^{-1} is the inverse function of h. When f(x) is the uniform distribution we should only consider the derivative term,

$$g(y) = dh^{-1}(y)/dy \tag{2.4}$$

Example 1: Let $y=x^2$. The inverse function is the square root, and the distribution of y is

$$g(y) = d\, y^{1/2}/dy = 1/2y^{1/2} \tag{2.5}$$

The new variable is also defined in the interval (0,1).

We are currently interested in the inverse problem, i.e. g(y) of eq. (4) is known and we seek for the change of variables h(y). Integrating eq. (4) we get

$$h^{-1}(y) = \int^{y} g(x)\, dx = G(y) \tag{2.6}$$

so that the function which defines the change of variables is the inverse of the cumulative distribution function.

For this procedure to be useful we then require i) to be able to compute the cumulative distribution function, and ii) to be able of inverting it. These are two strong requirements which can be fulfilled only in few cases.

Example 2: Consider the example 1. The cumulative distribution is

$$G(y) = \int_0^{y} dx\ /2\ x\ = y^{1/2}$$

and the change of variables is $y=x^2$.
Sampling from this distribution is simple: draw $x \in U(0,1)$ and

square it.

2.3 The normal distribution

In the case of the normal distribution its cumulative distribution function is the complementary error function, which cannot be exactly inverted. This problem may be circunvected by considering the bivariate normal distribution

$$f(x_1,x_2) = \frac{1}{2\pi} \exp \left\{ -\frac{1}{2} (x_1^2 + x_2^2) \right\} \tag{2.7}$$

and change to polar coordinates

$$\begin{aligned} x_1 &= r \cos\varphi \\ x_2 &= r \sin\varphi \end{aligned} \tag{2.8}$$

so that the distribution function in terms of the new variables is

$$g(r,\varphi) = r \exp(-r^2/2) / 2\pi \tag{2.9}$$

i.e., $r \exp(-r^2/2)$ corresponding to r and uniform $U(0,2\pi)$ corresponding to φ. Now the cumulative functions can be computed and inverted, and the sampling algorithm is

$$\begin{aligned} x_1 &= (2 \log z_1)^{1/2} \cos(2\pi z_2) \\ x_2 &= (-2 \log z_1)^{1/2} \sin(2\pi z_2) \end{aligned} \tag{2.10}$$

where z_1 and z_2 are U(0,1) random variables. In this form one obtains pairs of normally distributed random numbers. Currently both numbers can be taken, but this may amplify possible correlations already present in the uniform distribution algorithm. This may be avoided by considering x_1 or x_2 with probability 1/2 each. The final sampling algorithm is as follows:

i. Draw z_1 and z_2 form U(0,1)
ii. Compute x's with eqs. (10)
iii. Draw z_3 from U(0,1). If it is larger than 1/2 then consider x_1, otherwise consider x_2.

2.4 Sum and product of random variables

Consider x and y corresponding to probability distribution functions f(x) and g(y), respectively. The distribution corresponding to the new variable

$$z = x+y \tag{2.11}$$

is given by

$$h(z) = \int f(x)\, dx \quad f(y)\, dy \;\; \delta(x+y-z) \tag{2.12}$$

i.e., the convolution

$$h(z=x+y) = \int f(x)\, g(z-x)\, dx \tag{2.13}$$

In a similar way we may obtain the distribution of the product of two random variables, $z=xy$, just by changing the argument of the delta function in eq. (12) to $(xy-z)$. The result is

$$h(z=xy) = \int f(x)\, g(z/x)\, dx \,/\, x \tag{2.14}$$

Example 3: Let both $f(x)$ and $g(x)$ be the uniform distribution $U(0,1)$. Then xy is defined in the range $(0,1)$ also. We may evaluate eq. (14) by means of the step function, i.e.

$$h(z) = \int_0^1 dx\, \theta(z/x)\, \theta(1-z/x) \,/\, x.$$

The product of the step functions is different form zero only when $x>z$, so that the integral is efectively extended from z to 1. The distribution of the product is then $-\log(z)$.

Similar steps may be followed in the case of the sum $z=x+y$. Now the range is $(0,2)$ and the distribution is

$$\begin{array}{ll} z & \text{when } z<1 \\ 2-z & \text{when } z>1. \end{array}$$

Example 4: See Ref. /5/

Consider now the distribution $h(z,n)$, where

$$z = -\log(x_1 x_2 \dots x_n) \quad , \; x_i \in U(0,1) \tag{2.16}$$

The new variable is defined in the range $(0,\infty)$. The form of the distribution may be obtained by induction. In the case $n=1$, i.e., $z=-\log(x)$ we can use the change of variables method, sect. 2.2, to obtain

$$h(z,1) = \exp(-z) \tag{2.17}$$

When $n=2$, z corresponds to the sum of two random variables which are distributed exponentially, eq. (17). Using eq. (13) there results

$$h(z,2) = z \exp(-z) \tag{2.18}$$

For arbitrary n we guess

$$h(z,n) = z^{n-1} \exp(-z) \,/\, (n-1)! \tag{2.19}$$

and this can be confirmed by computing the convolution of $h(z,n-1)$ with $h(z,1)$. Equation (19) is the Gamma distribution, and eq. (16) gives the sampling algorithm. This example is quite interesting, because this distribution appears in the determination of the free Green function of a many body system.

2.5 Acceptance-Rejection method

This method was invented by von Neumann /6/. Let $f(x)$ be defined in a finite interval, and bounded by the value M. Random numbers corresponding to this distribution can be obtained in the following form:

i. Draw x from U(a,b)
ii. Draw p from U(0,M)
iii. If $p \leq f(x)$ then accept x as the new random number. Otherwise, reject both x and p and repeat the algorithm from the beginning.

These three steps may be condensed in a single sentence: draw x from U(a,b) and accept it with probability f(x)/M.

This method is illustrated in figure 1, which shows a case of acceptance and a case of rejection. The proof of the algorithm is quite simple, just write the previous statement mathematically,

$$\int_a^b \frac{1}{(b-a)}\, dx \int_0^M \frac{1}{M}\, dp\ \delta(x-z)\ \theta(f(x)-p) = \frac{1}{M(b-a)}\, f(z) \qquad (2.20)$$

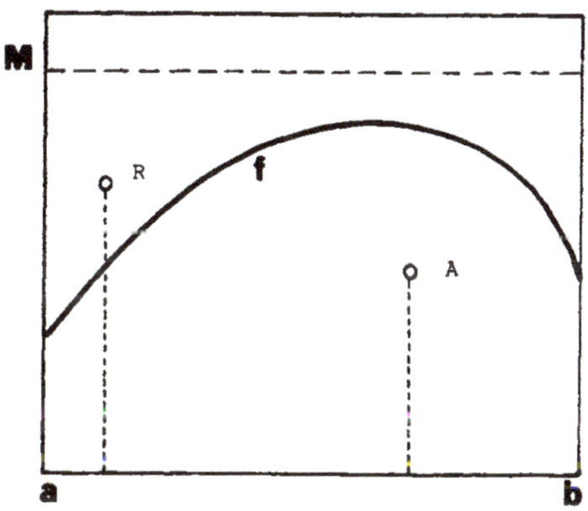

Figure 2.1: The acceptance-Rejection method. The figure shows one point accepted (A), and one rejected (R).

The quantity 1/(b-a) is the uniform distribution of x, U(a,b). The delta function says the result of sampling z is x. The step function means that only when f(x) is greater than p the value of x will be accepted. The integral is trivially carried out and the result is in the rhs of eq. (20). We obtain the desired distribution, but divided by M(b-a), which is always greater than 1. The lack of normalization of the resulting distribution reflects the loss of points due to rejections. For N trials, only N/M(b-a) will be accepted. To have high efficiency M should be as small as possible, namely the maximum value of f(x) in the interval (a,b).

The acceptance-rejection method may be improved so as to have maximum efficiency, i.e., so as to obtain on the average N points from N trials. The improved algorithm is the following:

i. Draw x from U(a,b)

ii. Compute f(x)

iii. Draw an integer n with mean value f(x)/M, where M=1/(a-b). The way of doing it is

$$n = \mathrm{Int}(z + f(x)/M)$$

where z is from U(0,1).

iv. Count n times the value x, i.e., if n=0 then reject it, if n=1 count it once, and so on.

The trials rejected are compensated by the trials counted more than once, and on the average we will end up with N random numbers after N trials. Note that now M is not an upper bound to the distribution, but 1/(b-a). This improved algorithm is illustrated in figure 2.

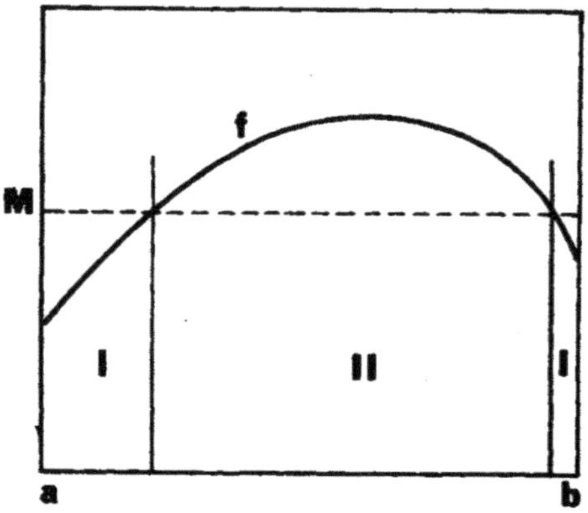

Figure 2.2: Improved acceptance-rejection method.
Points in regions labelled I are not counted, or are counted only once. Points in region labelled II are counted once or twice

2.6 Empirical determination of probability distributions

The various algorithms above described may be easily used and tested by means of even a home computer. The method is quite simple. Just generate a long set of random numbers using the appropriate sampling formula and accumulate the results in the form of an histogram, i.e. count an plot the number of times the generated random numbers are within x_i and x_{i+1}. These values of x define the intervals used to define the histogram. To compare with the theoretical form of the distribution the obtained numbers must be scaled. The number of times a random number is in the interval (x_i, x_{i+1}) is given by

$$N_s \int_{x_i}^{x_{i+1}} f(x)\,dx \approx N_s(x_{i+1}-x_i)\,f(x_i)$$

where N_s is the total number of samples.

Table 2.1. Examples of probability distributions, the sampling algorithms and results. The first column refers to figure (2.4)

Figure	Distribution	Range	Mean	Variance	Sampling	Empirical Mean	Empirical Variance
A	1/2 x	(0,1)	.333	.0888	$x = z^2$	.344	.0911
B	exp(-x)	(0,∞)	1	1	x=-log(z)	1.01	1.057
C	2x	(0,1)	.666	.0555	$x=z^{1/2}$	.671	.0526
D	Normal	(-∞,∞)	0	1	See text	-.03	.9767

```
C
C       EMPIRICAL DETERMINATION OF P.D.F.
C
        DIMENSION IBIN(20)
        XMIN=0.0                          ! LOWER LIMIT FOR HISTOGRAM
        XMAX=1.0                          ! UPPER LIMIT
        STEP=0.05                         ! SIZE OF BIN
        NBINS=(XMAX-XMIN)/STEP            ! NUMBER OF BINS
        NSAMPLES=1000
        AVERAGE=0.0
        SQAVERAGE=0.0
        DO 10 I=1,NSAMPLES
        X=RAN(K)**2                       ! SAMPLING FORMULA
                                          ! CHECK CALLING OF RANDOM ROUTINE
        AVERAGE=AVERAGE+X                 ! COMPUTE MEAN
        SQAVERAGE=SQAVERAGE+X**2          ! COMPUTE VARIANCE
        IF (X .GT. XMAX) GOTO 10          ! DISCARD IF OUTSIDE RANGE
        IX=(X-XMIN)/STEP+1
        IBIN(IX)=IBIN(IX)+1               ! ACTUALIZE HISTOGRAM
10      CONTINUE                          ! END OF SAMPLING
        XMEAN=AVERAGE/NSAMPLES
        VARIANCE=SQAVERAGE/NSAMPLES-XMEAN**2
C
C       PLOT AND PRINT RESULTS
C
        DO 20 I=1,NBINS
20      PRINT*,I,IBIN(I)
        PRINT*,'    MEAN=',XMEAN
        PRINT*,'VARIANCE=',VARIANCE
        END
```

Figure 2.3 . A FORTRAN program to determine empirically the shape of a probability distribution function.

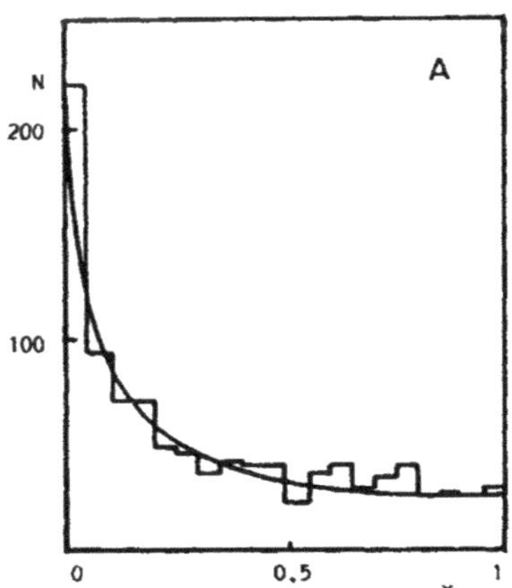

Figure 2.4.a; The distribution 1/$\sqrt{x}$

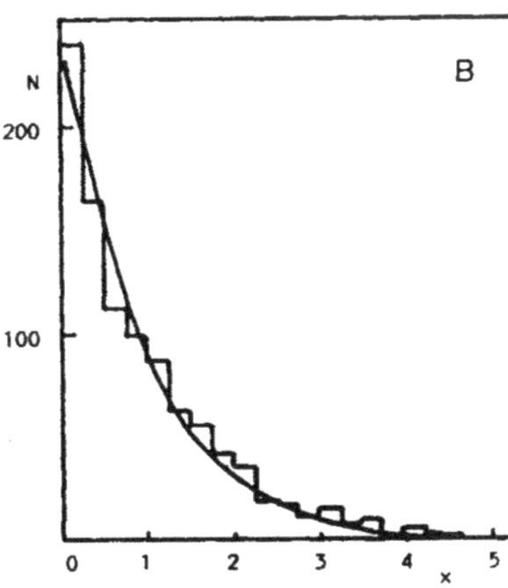

Figure 2.4.b: The exponential distribution exp(-x)

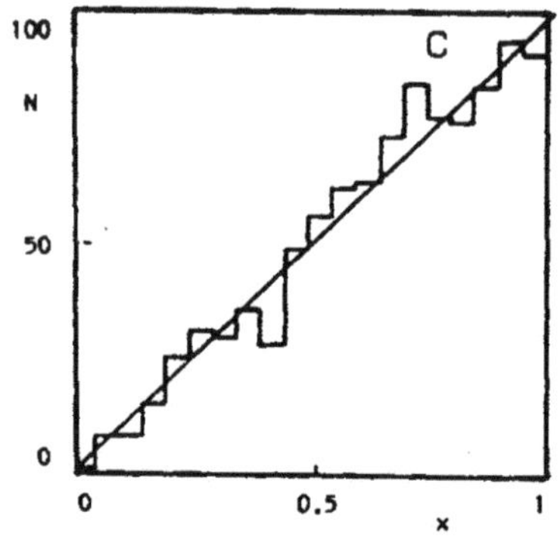

Figure 2.4.c: The triangular distribution 2x

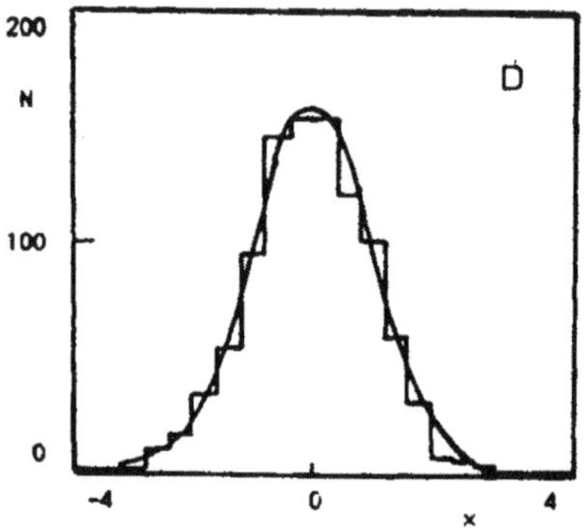

Figure 2.4.d: The normal distribution

This process is exemplified in the FORTRAN program presented in Figure 2.3, which do not requires further explanations because of its simplicity. Using this program the distributions listed in Table 2.1 have been sampled, using a total of 1000 points. The global parameters corresponding to these distributions are compared in Table 1 with their exact value, and Figures 4.a to 4.d show the corresponding histograms.

2.7 Summary of methods

1. The uniform distribution is an algorithm to be found in the system library of the computer or in a general purpose Fortran library.

2. The inverse method can be used when the cumulative distribution function can be both computed and inverted.

3. The acceptance-rejection method is useful in bounded, finite range distributions.

4. A clever use of sums, products and quotients of uniform random numbers may produce some simple sampling algorithms in special cases.

5. There remain still a large amount of sampling methods which cannot be classified. For example, the algorithm $x=\max(x_1,x_2)$, both x_1 and x_2 uniformly distributed random numbers, generates the triangular distribution 2x. If x_1, x_2, and x_3 are U(0,1), and they are ordered, then the distribution of $x=x_2$, the middle value, is $6x(1-x)$.

6. This section has not covered two very important methods which are the basis of the Variational and Green Functions Montecarlo theories. They are the Metropolis algorithm and the composition method, which will fully described below.

3. MONTECARLO QUADRATURE AND METROPOLIS ALGORITHM

3.1 Crude Montecarlo quadrature

We are interested in the computation of the integral

$$I = \int_a^b h(x)\, dx \tag{3.1}$$

where h(x) is an arbitrary (positive or negative) function, provided the integral is defined.

Let $(x_1, x_2, \ldots, x_N)$ be a set of random numbers uniformly distributed in the interval (a,b). Because of central limit theorem we can write

$$I = \lim_{N \to \infty} \frac{1}{N} \sum_{i=1}^{N} h(x_i) \tag{3.2}$$

The variance may also be determined

$$\sigma^2 = \lim_{N \to \infty} \frac{1}{N} \sum_{i=1}^{N} h^2(x_i) \quad - I^2 \tag{3.3}$$

If the sampling is limited to N random numbers one usually writes the result of the integral as

$$I \pm (\sigma^2 / N)^{1/2} \tag{3.4}$$

where the error corresponds to the 68% of confidence.

The error of the quadrature depends on two quantities, the variance of the function h(x) to be integrated and the number of samples. Note that the variance depends <u>only</u> on the function h(x),

$$\sigma^2 = \int_a^b (h(x)-I)^2 dx \tag{3.5}$$

and has no relation with the number of samples or the sampling method. As a consequence, to diminish the computational error in this crude method the only way is to increase the number of samples. Specifically, to halve the error we must work four times more. The variance measures the quadratic deviation of our function h(x) from its mean value I. Rapidly varying functions will have a large variance and poor precision.

To improve the efficiency (precision versus work) of crude Montecarlo several techniques have been developed, like importance sampling, stratified sampling, control variates, anatitetic variates and so on. Ref. /1/ gives a quite complete description and analysis of all these techniques. We will concentrate here on importance sampling because of its direct relation with the techniques in the many body problem.

3.2 Importance sampling

The idea of importance sampling is to modify the function to be sampled, as well as the sampling algorithm, so as to significantly decrease the variance of the resulting function. Consider a probability distribution function f(x) defined in the same interval (a,b) of the quadrature. Equation (1) may be rewriten as

$$I = \int_a^b (h(x)/f(x)) \; f(x)\, dx \tag{3.6}$$

and the Montecarlo estimate of the integral can be obtained by drawing random numbers from the probability distribution f(x) and computing the average

$$I = \lim \frac{1}{N} \sum (h(x_i)/f(x_i)) \tag{3.7}$$

The new value of the variance is

$$\sigma^2 = \int_a^b (h(x)/f(x))^2 \; f(x)\, dx \; - I^2 \tag{3.8}$$

With a judicious choice of f(x) me may achieve a small variance, so that the extra work required by the new formula (3.7) may be compensated by the need of a smaller number of samples.

Example 3.1: The integral of the function h(x)= (exp(x)-1)/(e-1) , in the (0,1) interval, is 0.418, and its variance is 0.082. If the importance sampling function f(x)=2x is used, the new variance is lowered to 0.0027. Of course, the value of the integral remains the same.

3.3 Markov chains

The main problem in the importance sampling technique is the determination and use of the importance sampling function. To lower significantly the variance, the importance sampling function should mimic the function h(x) to be integrated. This loose criterium does not help too much in chosing the required distribution function. On the other hand, there remains still the problem of carrying out the required sampling with this function. Both problems are formidable in multivariate integrals. In many cases, the importance sampling function is dictated by the mere structure of the physical problem at hand, as in the case of the variational many body problem. In this case, the importance sampling function is the trial wave function squared. However, there remains still the problem of sampling from a normally very complicated distribution function. To solve this problem, Metropolis, Rosenbluth, Rosenbluth, Teller and Teller /7/ have devised a stochastic algorithm, i.e., a random walk in the multivariate coordinate space which generates asymptotically the desired set of random numbers. This algorithm is known with the name of Metropolis, or the acrostic $M(RT)^2$.

This random walk is of the type known as Markov chain or Markov process. Consider a system which can be in a discrete set of states S_1, S_2... S_N. The evolution of this system is governed by a clock, so that at every jiffy the system jumps from its current state to any other state of the set, itself included. This jump is characterized by a transition probability matrix p_{ij}, which represents the probability of jumping from the state i to any other state j. Note that p_{ii} is also included, and that these matrix elements must fulfill the following conditions:

$$0 \leqslant p_{ij} \leqslant 1 \tag{3.9}$$

$$\sum_j p_{ij} = 1 \tag{3.10}$$

In addition, it is required that the transition from i to j is independent on the previous occupied states during the evolution.

Equations (9) and (10) are obvious. Equation (9) says that the probability of jumping from i to j is a positive number less or equal than 1. Equation (10) is the sum rule corresponding to these probabilities. Note that the sum is over the second index, actually there are no restrictions on the corresponding sum extended to the first index. A matrix which satisfies both (9) and (10) is called a stochastic matrix, and the process related to it is called a Markov chain.

The generalization to the case of the continuum is called a Markov process. We should use now x to label the states, and define a transition density p(x,x') with the properties

$$p(x, x') \geqslant 0$$
$$\int p(x, x')\, dx' = 1 \tag{3.11}$$

The following question emerges naturally. After a very long walk, which is the probability P_k of having passed on some state S_k? . More precisely, assume we carry out N jumps, and let N_k be the number of stops at S_k. The searched probability is $P_k = \lim N_k/N$.

For this question to have sense we must add two new conditions to the stochastic matrix

i. The random walk must be endless, i.e., there is none state with $p_{ii}=1$. If that state would exist, then our walker would be trapped at it. This kind of state is called an absorbing wall. Terminating random walks have a deep interest in many physical problems, like shielding of radiation, but we are not interested at them here.

ii. The chain is irreducible. Probably the opposite concept is easier to understand: if we can classify the states in two subsets so that there cannot be a transition from one subset to the other, then the chain is termed reducible. A closely related word to this concept is ergodicity.

3.4 Solution of the direct problem

The algebraic determination of the probabilities P_k is quite simple. The probability of arriving to a state S_i is the product of the probability of being previously in another state S_k times the transition probability, i.e.,

$$P_i = \sum_k P_k \ p_{ki} \qquad (3.12)$$

This is a set of homogeneous linear equations, which as expected are not independent, as it can be checked by summing up over the free index i. It should be supplemented by the normalization condition

$$\sum P_i = 1 \qquad (3.13)$$

and the system can now be solved.

There is however an alternative way of facing up this problem, just by walking along the random walk process. Instead of describing the method we present in Figure 1 the corresponding Fortran program in a simple case: start in a given state (I=1) and do NMOVES jumps. Sentence labelled 300 decides the destination of a given jump by comparing with a U(0,1) random number.

The results corresponding to several values of the total number of moves are shown in Table 1, as well as the exact results. Certainly the method works.

```
C       GIVEN A RANDOM WALK DETERMINE THE PROBABILITY DISTRIBUTION
C
        DIMENSION TRANSPROB(4,4),PROB(4)
C
C       ARRAY STORED BY COLUMNS
        DATA TRANSPROB / 0.2, 0.1, 0.3, 0.1, 0.2, 0.1, 0.4,
        1    0.5, 0.2, 0.1, 0.2, 0.2, 0.4, 0.7, 0.1, 0.2/
C
        I=1                              ! CURRENT STATE
        PROB(I)=1                        ! COUNT STOPS AT I
        TYPE*,' INPUT NUMBER OF MOVES AND SEED FOR RANDOM GENERATOR'
        ACCEPT*, NMOVES, ISEED
        DO 100 L=2, NMOVES
        Z=RAN(ISEED)                     ! THROW A DICE
        J=1                              ! DESTINATION STATE
300     IF(Z.LT.TRANSPROB(I,J)) GOTO 200! ACCEPT DESTINATION
        Z=Z-TRANSPROB(I,J)
        J=J+1                            ! TRY ANOTHER STATE
        GOTO 300
200     PROB(J)=PROB(J)+1                ! COUNT STOPS AND
        I=J                              ! RENAME CURRENT STATE
100     CONTINUE
C       PRINT RESULTS AFTER RENORMALIZATION OF PROBABILITIES
        DO 400 I=1,4
400     PRINT*, I, PROB(I)/NMOVES
        END
```

Figure 3.1. A random walk solution of a system of linear equations, equations (12) and (13).

Table 3.1: A random walk solution of a system of linear equations.

	EXACT	100 Moves	1000 Moves	10000 Moves
P(1)	0.14859	0.10	0.149	0.1494
P(2)	0.31325	0.36	0.310	0.3088
P(3)	0.16867	0.16	0.172	0.1710
P(4)	0.36984	0.38	0.369	0.3708

We have discovered a surprising form of solving systems of linear equations. Many other simple and surprising examples of random walks may be found in the book of Martin Gardner, Ref. /8/. This algorithm has been generalized /1/ to a class of linear equations of the form

$$x_i = a_i + \sum_j H_{ij} x_j \tag{3.14}$$

where the euclidean norm of the matrix H is less than one, i.e.,

$$\max_i \left\{ \sum_j H_{ij} \right\} < 1 \tag{3.15}$$

but otherwise is arbitrary. The formal solution of (14) is

$$x = \frac{1}{1-H} a \tag{3.16}$$

and condition (15) guarantees that the series expansion of (16),

$$x = (1 + H + H^2 + H^3 + \dots) a \tag{3.17}$$

converges. The generalized algorithm consists in constructing this series by means of a random walk.

In the case of the continuum the equivalent to (14) is a nonhomogeneous linear integral equation. Techniques for the solution of this kind of integral equations may be found in Ref/ /9/.

3.5 The inverse problem: Metropolis algorithm

In the previous section we have solved a direct problem: given a random walk determine a probability distribution. Our problem, namely sampling random numbers with a prescribed probability distribution, is just the inverse of the previous problem. Now we know the value of P_i and we want to determine a random walk, i.e., the stochastic matrix p_{ij}. The problem is not determined, and the solution is not unique.

Metropolis and coworkers /7/ have determined a solution. Let q_{ij} be a stochastic and symmetric matrix, otherwise arbitrary. Then our random walk matrix is given by

$$p_{ij} = q_{ij} \qquad \text{if } P_i > P_j \ , \ i \neq j \tag{3.18}$$

$$p_{ij} = q_{ij} \, P_j/P_i \qquad \text{if } P_i \leq P_j \ , \ i \neq j \tag{3.19}$$

$$p_{ii} = q_{ii} + \sum_k{}' \, q_{ik}(1-P_k/P_i) \tag{3.20}$$

The value of p_{ii}, eq. (20) has been fixed so as to fulfill condition (10) of the stochastic matrix. The prime on the summation symbol means that only states which satisfy $P_k < P_i$ are considered.

The meaning of above equations is the following. Assume we are sitting at S_i. Then choose a trial state S_j with probability q_{ij}. If the probability of the trial state is greater than our current probability, then the move is accepted. This is eq. (18). Otherwise, the trial state is accepted with probability P_j/P_i, which is now smaller than one. In the case the new state is not accepted, then the new state (i.e., the new random number) is the old state S_i, which is then counted twice.

The proof of Metropolis form is quite simple. Actually we should only check that this p_{ij} satisfies the set of equations (12), because the stochastic conditions (9) and (10) are satisfied by construction. An intermediate step is to show that this transition matrix satisfies the microreversibility or detailed balance condition

$$P_i \, p_{ij} = P_j \, p_{ji} \tag{3.21}$$

There is no summation on repeated indexes in eq. (21). Once this condition is checked, eq. (12) turns out just by summing up in (21) on index i. We leave the proof of (21) as an exercise to the reader.

Metropolis algorithm can be generalized to the case of a continuous probability distribution p(x). The moves corresponding to the auxiliary matrix q_{ij} can now be carried out by means of a uniformly distributed displacement within a segment line of length D centered at the current position x.

The algorithm is as follows. Assume we are at point x, then

i. Try a move to x'=x + D (2z-1) , where z is from U(0,1).

ii. If p(x')>p(x) then go to step iv.

iii. If we arrive here is because p(x') is smaller than p(x). Then sample a new random number z from U(0,1) and do the following test:
IF p(x')/p(x)>z then go to iv,
ELSE
put x'=x, i.e., remain at the old position.

iv. Put x' in the list of random numbers, rename x' to x and go back to step i.

Distance D is arbitrary, the only condition on the auxiliary matrix q being its symmetry. When D is very small, the numbers of acceptances in steps ii and iii is large, but a lot of moves is necessary to cover the space and the successive random numbers are strongly correlated. On the contrary, when D is large the number of rejections is large, so that we may spend a long time at every position. It is usually admitted that the appropriate value of D should give a ratio of acceptances between 50% and 70%.

The same algorithm can be applied to the case of multivariate distributions. The most common problem in physics involves a probability distribution which depends on the coordinates of N particles, i.e., 3N variables. The random walk is ussually carried out in N steps, moving in turn each particle (not necesarily in the same order), and accepting or rejecting the individual motions with the usual test. After all particles have been tried, the resulting point is the new multidimensional random number. Note that some of the new coordinates may have the same value than the old coordinates.

A usual procedure to avoid correlations among the succesive positions is the so called thermalization. Metropolis algorithm is repeated a number of times before accepting the new point for sampling. This will increase the computing time. In any case, we should always have in mind that Metropolis algorithm is asymptotic.

4. VARIATIONAL MONTECARLO

4.1 General description

The usual form of a many body hamiltonian is

$$H = -\frac{h^2}{2m} \sum_i \nabla_i^2 + \sum_{i<j} V(i,j) \tag{4.1}$$

involving one body kinetic energy terms and a two body potential energy interaction. The two body interaction may depend on the relative distance between the particles as well as on spin or isospin degrees of freedom.

One of the features which makes the many body problem so complex is the fact that in almost all interesting many body systems the two body potential shows a strong repulsion at short distances, in such a form that plain perturbative methods cannot be straightforwardly applied.

A simple alternative to the exact solution of eq. (1) is to compute a variational upper bound to the ground state energy using a trial function which takes into account appropriately the repulsive core of the two body interaction. The simplest form is known with the name of Jastrow /10/, and consists in two pieces, a model wave function made up of single particle orbitals and a correlation factor which is the product of all pairs of two body correlations.

Some of the physically interesting forms are:

i. A finite system of bosons (drops)

$$\Psi_T = \prod_i \varphi(r_i - R) \prod_{i<j} f(r_{ij}) \tag{4.2}$$

The first term clusters all particles around their center of mass

R, and the second is the correlation factor.

ii. An infinite system of bosons (a liquid or a gas)

$$\Psi_T = \prod_{i<j} f(r_{ij}) \tag{4.3}$$

In this kind of problem the product is extended to a finite number of particles which stay in a simulation box, and afterwards periodically extended to cover all space.

iii. A bosonic crystal

$$\Psi_T = \prod_i \varphi(r_i - A_i) \prod_{i<j} f(r_{ij}) \tag{4.4}$$

where A_i are the lattice sites of the crystalline structure, and the single particle wave functions tie each particle to a lattice site. Again this wave function is limited to a finite number of particles and periodically extended to all space.

Systems of fermions may be described in a similar way, the main change being the presence of a Slater determinant to account for the antisymmetric character of the wave function.

More general forms of trial functions may include three body correlations, momentum dependence of the correlation factor and spin or isospin dependence. I will mainly consider the simplest forms only.

The upper bound to the ground state enegy is given by the multidimensional integral

$$E_V = \frac{\int \Psi_T^*(R)\ H\ \Psi_T(R)\ dR}{\int \Psi_T^*(R)\ \ \Psi_T(R)} \tag{4.5}$$

where R represents the set of the 3N coordinates. When particles are spinless and the hamiltonian is a local operator, eq. (5) may be writen in a form appropriate for Montecarlo calculations. Define a <u>local energy</u> $E_L(R)$ as

$$E_L(R) = \frac{1}{\Psi_T(R)}\ H\ \Psi_T(R) \tag{4.6}$$

and introduce a multivariate probability distribution

$$p(R) = |\Psi_T(R)|^2 \Big/ \int |\Psi_T(R)|^2\ dR \tag{4.7}$$

Note that p(R) is always positive and normalized to 1. Then the variational energy is given by

$$E_V = \int p(R)\ E_L(R)\ dR \tag{4.8}$$

This integral has been writen in the standard form (3.6) of Montecarlo quadrature with p(r) the importance sampling function, and $E_L(R)$ the function to be sampled. The Montecarlo evaluation of the variational energy is a simple task: draw a set of random variables $R_1, R_2, \ldots R_N$ using Metropolis algorithm and obtain an estimate of the energy with the quantity

$$E_V = \frac{1}{N} \sum_i E_L(R_i) \tag{4.9}$$

with an statistical error

$$E_V = \left\{ \frac{1}{N} \sum_i E_L^2(R_i) - E_V^2 \right\}^{1/2} \Big/ N^{1/2} \tag{4.10}$$

Note that sampling E_L^2 is not the same that computing the expectation value of H^2, unless H is a local operator.

The form of the local energy is not unique. The contribution of the potential energy term of the hamiltonian is, of course, the potential energy. On the other hand, the kinetic energy involves the computation of derivatives and the straightforward way of carrying out these derivatives gives rise to three body terms. A clever use of integration by parts technique can produce other forms, which may be simpler than the direct form of eq. (6). In any case all these forms give the same value for the variational energy, but the value of the variance and statistical error may be different. The direct way, i.e., acting with Laplace operator on the rhs wave function, has zero variance in the limit case in which the exact wave function is used as a trial wave function. This may be a good recipe to organize the calculations, but as we have mentioned above, this originates in general three body terms so that computations may be expensive.

4.2 A simple example: a nucleon bound to a gaussian potential

Consider a nucleon in the gaussian potential

$$V(r) = V_0 \exp(-(r/b)^2) \tag{4.11}$$

with V_0=-45 MeV, b=2.09 fm and a mass given by $\hbar^2/m$=41.50 MeV.fm^2.

The energy of the ground state may be obtained by numerical integration of the radial Schrodinger equation, with the result of -10.813 MeV.

A simple and quite accurate trial wave function is the 1s harmonic oscillator state

$$\Psi_T = \frac{a^{3/2}}{\pi^{3/4}} \exp\left\{-\frac{1}{2} a^2 r^2\right\} \tag{4.12}$$

the local energy being

$$E_L(r) = \frac{\hbar^2}{2m} (3a^2 - a^4 r^2) + V(r) \tag{4.13}$$

The variational energy may be easily evaluated, with the result

$$E(a) = \frac{3\hbar^2}{4m} a^2 + V_0 \left\{ \frac{(ab)^2}{1+(ab)^2} \right\}^{3/2} \tag{4.14}$$

One can compute the expectation value of the energy in two ways.

First, by using Montecarlo quadrature algorithm with direct sampling of the wave function squared, which is a normal distribution. Secondly, we can use Metropolis algorithm to carry out the generation of random numbers.

```
C       VARIATIONAL MONTECARLO METHOD WITH GAUSSIAN POTENTIAL
C       TRIAL FUNCTION IS H.O. WITH PARAMETER A PROP. TO EXP(-0.5*(A*R)**2)
        COMMON/RAND/NR
        V(X,Y,Z)=-45.*EXP(-(X*X+Y*Y+Z*Z)/4.3681)! POTENTIAL
        D=20.75                               ! HBAR**/2M
        TYPE*,' ENTER INTEGER FOR RANDOMNESS'  ! TO CHANGE THE
        ACCEPT*,NR                            ! RANDOM SEQ.
        TYPE*,' ENTER NUM. SAMPLES AND VARIATIONAL PARAMETER A'
        ACCEPT*,NITER,A
        ENERG=0.
        ENERG2=0.
        A2=A*A
        T=1./A2                               ! PARAMETER OF GAUSS DISTR.
        DO 100 ITER=1,NITER
        CALL GAUSS(X,Y,Z,T)                   ! GET RANDOM VECTOR
        VP=V(X,Y,Z)+D*A2*(3-A2*(X*X+Y*Y+Z*Z))  ! LOCAL ENERGY
        ENERG=ENERG+VP
        ENERG2=ENERG2+VP*VP
100     CONTINUE
        EAV=ENERG/NITER
        EAV2=ENERG2/NITER
        ERR=SQRT((EAV2-EAV*EAV)/(NITER-1))
        PRINT*,' ENERGY OF G.S.=',EAV,' ERROR=',ERR,' NUMBER SAMPLES',NITER
        END
C       ==================================================================
        SUBROUTINE GAUSS(X,Y,Z,ALFA)    ! GENERATES THREE POINTS OF A GAUSS
        COMMON/RAND/NR                  ! DISTRIBUTION WITH PARAM. ALFA
        LOGICAL ODD                     ! EXP(-X**2/ALFA**2), NOT STANDARD
        DATA ODD/.TRUE./                ! FORM
        IF (ODD) THEN
                FI=6.283185*RAN(NR)     ! GENERATE X AND
                R=SQRT(-ALFA*ALOG(RAN(NR)))
                XOLD=R*COS(FI)          ! SAVE ONE FOR LATER USE
                X=R*SIN(FI)
        ELSE
                X=XOLD                  ! RECOVER OLD VALUE
        END IF
        FI=6.283185*RAN(I1,I2)          ! GENERATE Y AND Z
        R=SQRT(-ALFA*ALOG(RAN(I1,I2)))
        Y=R*COS(FI)
        Z=R*SIN(FI)
        ODD= .NOT. ODD
        RETURN
        END
```

Figure 4.1. Variational Montecarlo with direct sampling of the wave function squared, corresponding to the example of section 4.2.

```
C       VARIATIONAL MONTECARLO METHOD WITH GAUSSIAN POTENTIAL
C       TRIAL FUNCTION IS H.O. WITH PARAMETER A PROP. TO EXP(-0.5*(A*R)**2)
C       METROPOLIS SAMPLING
        V(R)=-45.*EXP(-R/4.3681)          ! POTENTIAL, R IS X**2+Y**2+Z**2
        FI(R,A2)=EXP(-A2*R)               ! WAVE FUNCTION SQUARED
        D=20.75                           ! HBAR**2/2M
        TYPE*,' ENTER INTEGER FOR RANDOMNESS'
        ACCEPT*,NR
        TYPE*,' ENTER NUM. MOVES and SCALE OF MOVES'
        ACCEPT*,NITER,SCAL
        TYPE*,' ENTER VARIATIONAL PARAMETER'
        ACCEPT*,A
        A2=A*A
        C1=3.*D*A2                        ! CONSTANTS TO EVALUATE
        C2=D*A2*A2                        ! LOCAL ENERGY
        EAV=0.                            ! TO COMPUTE AVERAGES
        EAV2=0.
        NACEP=0                           ! COUNT NUMBER OF ACCEPTS.
        X=RAN(NR)-0.5                     ! STARTING POINT
        Y=RAN(NR)-0.5
        Z=RAN(NR)-0.5
        R=X*X+Y*Y+Z*Z
        FIOLD=FI(R,A2)                    ! OLD W.F. SQUARE
        ENERG=0.
        ENERG2=0.
        DO 100 ITER=1,NITER               ! RANDOM SAMPLING LOOP
        XP=X+SCAL*(RAN(NR)-0.5)           ! ATTEMPT A MOVE
        YP=Y+SCAL*(RAN(NR)-0.5)           ! MAXIMUM OF 0.5*SCAL
        ZP=Z+SCAL*(RAN(NR)-0.5)           ! IN EACH AXIS
        RP=XP*XP+YP*YP+ZP*ZP
        FINEW=FI(RP,A2)                   ! NEW W.F. SQUARE
C       ------------------------------------------------------------
        IF(FINEW/FIOLD .LT. RAN(NR)) GOTO 5000   ! MOVE REJECTED     !
        X=XP                                     ! MOVE ACCEPTED     !
        Y=YP    !-------------------!            ! UPDATE POSITIONS  !
        Z=ZP    ! METROPOLIS BLOCK  !                                !
        R=RP    !-------------------!                                !
        FIOLD=FINEW                              ! AND WAVE FUNCTION!
        NACEP=NACEP+1                            ! COUNT ACCEPTS     !
5000    CONTINUE                                                     !
C       ------------------------------------------------------------
        EN=V(R)+C1-C2*R                   ! COMPUTE LOCAL ENERGY
        ENERG=ENERG+EN                    ! AND AVERAGES
        ENERG2=ENERG2+EN**2
100     CONTINUE
        E=ENERG/NITER
        E2=ENERG2/NITER
        ERR=SQRT((E2-E*E)/NITER)
        PRINT*,' ENERGY OF G.S.=',E,' ERROR=',ERR
        PRINT*,' NUMBER OF SAMPLES=',NITER, 'NUM. ACCEPTED=',NACEP
        END
```

Figure 4.2. Variational Montecarlo with Metropolis Random Walk method to generate random points. The program corresponds to the example of section 4.2.

Figure 4.1 contains a simple Fortran program corresponding to the problem at hand with a direct sampling from the squared wave function. The bulk of the program is the set of statements in DO-loop labelled 100, its structure being quite obvious. Draw a point (its three cartesian coordinates), with subroutine GAUSS, compute the local energy and accumulate to compute at the end the local energy average and the squared local energy average. The method of drawing normal distributed random points was already explined in Section 2.3. With the algorithm presented there, random numbers are generated by pairs, and the routine takes advantage of this fact by generating alternatively four points, and saving one for next time, and two points, recovering the previously saved value. This is controlled by the logical variable ODD.

Figure 4.2 is the Fortran program for the same problem but using now Metropolis method to generate the sequence of random positions. The structure of the program is very much the same than the structure of the previous program, with small differences. First of all, Metropolis algorithm consists in a move from the old position (X,Y,Z) to a new position (XP,YP,ZP), moving symmetrically around the starting position. The maximum length of the step at each move is controlled by the variable SCAL, which should be adjusted so as to have a reasonable ratio of acceptances. The block between the two COMMENT lines with equal signs is the acceptance/rejection test.

Table 4.1. Variational determination of the ground state of a gaussian potential with various methods. All energies are in MeV and the variational parameter a is in 1/fm. The first row of the table indicates the number of samples.

a	Analytic $\langle E_L \rangle$	$\langle E_L^2 \rangle$	Gaussian sampling 1000	10000	Metropolis sampling 1000	10000
0.4	-6.894	120.3	-6.58±.26	-6.89±.08	-8.19±.30	-7.07±.09
0.5	-9.190	135.6	-9.07±.22	-9.11±.07	-9.61±.23	-9.12±.07
0.6	-10.301	131.2	-10.49±.16	-10.31±.05	-10.86±.16	-10.45±.05
0.7	-10.069	121.5	-9.94±.13	-10.10±.04	-9.66±.10	-9.98±.04
0.8	-8.525	130.2	-8.70±.24	-8.53±.08	-9.13±.24	-8.45±.08

Some results corresponding to this case example are presented in Table 4.1. The columns labelled ANALYTIC show the exact values, and include both the energy expectation value and the expectation value of the hamiltonian squared. The next two columns are the results corresponding to a direct generation of random numbers from the gaussian and the last two columns are the results corresponding to Metropolis method. The calculations have been carried out with the programs listed above, for 1000 and 10000 samples, and the scale factor was 3 fm, which gives around a 50% of acceptances.

The columns labelled GAUSSIAN are in a very good agreement with the analytic results. In all cases the obtained values are within one standard deviation of the exact values. This is not the case of

Metropolis method with 1000 points, whereas the calculation with 10000 is again in very good agreement with the exact values. The conclusion is that 1000 steps are not enough to randomly cover all configuration space and that the random walk must be longer. In this case the value of the statistical error is meaningless. In real problems the excat value is not known a priori, and a large amount of computing is necessary to ascertain the validity of Montecarlo estimates.

4.3 Extended systems and periodic boundary conditions

Certainly one cannot attempt the direct description of a system with an infinite number of particles. The usual way of dealing with this problem, /7/, /11-12/, inherited form Classical Statistical Mechanics, is assuming that the system is partitioned in an infinite set of simulation boxes of side L which are repetitions of a fundamental cell. This picture is known with the name of establishing periodic boundary conditions.

The number of particles in each cell and the length of the simulation cube L are chosen so as to have the desired density

$$\rho = \text{number of particles} \,/\, L^3 \qquad (4.15)$$

The value of the number of particles is taken so as to allow for a crystalline order in the case this is the phase preferred by the system. At high density one can expect a close packing of the atoms, and the expected structure will be fcc (face centered cubic) or hcp (hexagonal close packed). In a fcc structure the number of atoms per unit cell is four, 6 atoms at the corners each shared by six cells plus six atoms at the faces each shared by two cells. If the simulation cube contains n^3 cells, the total number of particles will be $4n^3$. The 'magic numbers' generated in this form are 32, 108, 256, 500, 864, ...and most of variational Montecarlo work in bosons is carried out with one of these numbers.

The magic numbers corresponding to fermion systems have a different origin, being related to the quantization of single particle states momentum, and the occupancy of levels up to a given value of the Fermi momentum. The allowed values of the momenta are determined in terms of three integer numbers (n_1, n_2, n_3), positive or negative, and levels are filled in increasing order of the distance to the origin. The first point is (0,0,0) which corresponds to 1 state. At a distance of 1 we may have (1,0,0), (0,1,0), and (0,0,1), and the corresponding negative values, with a total of 6 new states. Next, at distance $\sqrt{2}$ we have (1,1,0), (1,0,1) and (0,1,1) and the negative values, with a total of 12 states, and so on. The cumulative numbers are 1,7,19,27,33,57,... to be multiplied times the degeneracy (2 or 4) of the system.

To use consistently periodic boundary conditions is necessary to follow certain rules. Each time a particle exits of the fundamental cell (i.e., the attempted and succesful Metropolis move forces the particle to go outside the simulation cube) another image of this particle will enter through the opposite wall. To test the acceptance of a possible move, one should test the change in the wave function due to move of this particle with consideration of all boxes. To make the computation feasible, it is usually assumed that the correlation factor is unity at distances greater than L/2. Then, we draw a sphere of radius L/2 around the particle under consideration, and count the contribution to the wave function of all particles inside this sphere, even if some of them are not in the fundamental cell. If this sphere contains a given particle, then none of the other periodic images of this particle will be inside the sphere. In conclusion, to compute the change of the wave function due to the motion of one particle, we must determine only its nearest neighbours generated from the particles in the fundamental cell by adding and substracting the length L to any of its cartesian coordinates. This mechanism is depicted in Figure 4.3

Once all particles have been tested by means of Metropolis algorithm, we have to compute the local energy corresponding to this new configuration. The local energy has two terms, one related to the interaction potential and the other involving derivatives of the correlation factor. Because of the healing conditions of the correlation factor at L/2, only particles inside this sphere must be considered to evaluate the kinetic energy. Warning: you must ensure that the correlation factor is smooth around the distance L/2, both function and derivative, to compute correctly the kinetic energy. A simple way to overcome this question is to take a length L large enough so that the correlation is negligible at distances longer than L/2. The price of this solution is that a large number of particles is necessary at high densities.

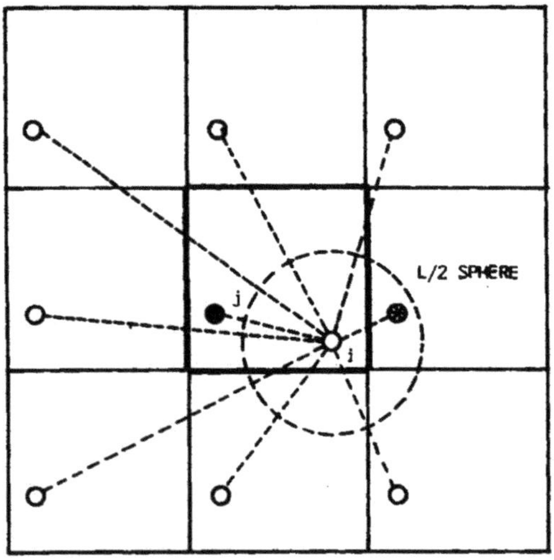

Figure 4.3. Determination of the nearest neighbour of particle i related to particle j (black circle) and all its images (open circles). The circle with an asterisk is the nearest neighbour. Note that there is only one particle of family j in the L/2 sphere drawn around particle i. This figure is adapted from Wood and Parker, /11/.

All these problems are not present when computing potential energy. In that case we should consider the interaction of particle i with particle j and all images of particle j, and this for all pairs of particles of the fundamental cell. If the interaction is short ranged we can discard all particles at a distance larger than an appropriate cutoff value. However, in the case of long range interactions, like e.g. Coulomb potential, all particles must be considered. The way of doing it is to compute Ewald's like summations.

The effect of these cutoffs related to boundary conditions can be corrected in the case of simple interactions. This will be explicitly discussed in the next section for the case of a Lennard-Jones interaction.

4.4. Study of liquid Helium 4

Liquid helium 4 was the first quantum system studied with variational Montecarlo method, already in 1965 by McMillan /13/. The interaction between helium atoms is quite well described in terms of a 6-12 Lennard-Jones potential

$$V(r) = 4\varepsilon\left(\left\{\frac{\sigma}{r}\right\}^{12} - \left\{\frac{\sigma}{r}\right\}^{6}\right) \tag{4.16}$$

where σ and ε are parameters characteristic of helium adjusted by De Boer and Michels to the values $\sigma=2.556$ A and $\varepsilon=10.22$ K. The other constant describing the system is the helium mass, its value being $\hbar^2/m=12.1209$ KA^2. A simple form of the two body correlation factor is

$$f(r_{ij}) = \exp\left(-\frac{1}{2}\left\{\frac{b}{r}\right\}^{n}\right) \tag{4.17}$$

This correlation factor satisfies the Schrodinger equation for the relative motion of two particles at short distances, i.e. in the region of repulsion, when n=5

$$\left[-\frac{h^2}{m}\nabla^2 + V(r)\right] f(r) = 0 \tag{4.18}$$

actually, only the dominant term of this equation r^{-12} is satisfied. The corresponding value of b is

$$b = (\,16\varepsilon\, m/25\,\hbar^2\,)^{1/12}\,\sigma \tag{4.19}$$

corresponding to 1.13 sigmas. With this simple form of the correlation the only free parameters are the correlation range b, the power n and the density of the system. This form of the correlation factor was suggested by McMillan /13/, and subsequently used by Schiff and Verlet in a molecular dynamics study of the quantum problem. Attempts to improve this two body correlation have resulted in a very small gain in energy (see the review article of Ceperley and Kalos /15/ for a complete analysis). We should however mention that an important correction to the simple form (17) is the introduction of three body correlations /16/, particularly at high densities.

Table 4.2 shows some of the helium calculations using Lennard-Jones interaction, with variational Montecarlo and two body correlations only and including three body correlations, and also with the exact Green functions Montecarlo method, to be studied in chap. 7. From this table we observe that McMillan's form explains almost all binding energy. The minimum is obtained at a density 0.9 times the experimental value, and at n=5 and b=1.16 sigmas, very close to the suspected value. Three body correlations are also important giving a value for the energy very close to the exact GFMC method. Note finally that GFMC energies are still a five per cent higher (less bound) than the experimental value. To obtain the experimental value is then necessary to improve the two body interaction. Nowadays it is beleived that Aziz interaction /17/ is the best interaction describing helium liquids.

Table 4.2: Liquid helium 4 with Lennard Jones potential

	Two body correlation	Three body correlation	GFMC	Experiment
E(K)	-5.9 ± 0.1	-6.53 ± 0.02	-6.84 ± 0.02	-7.14
Density (σ^{-3})	0.333	0.365	0.365	0.3648

In the rest of this section I will describe the simplest variational calculation using the correlation (17) with n=5. The squared wave function is given by

$$\Psi^2 = \exp\left\{ - \sum_{i<j} \left[\frac{c}{x_{ij}}\right]^5 \right\} \qquad (4.20)$$

where $c=b/\sigma$, and $x_{ij}=r_{ij}/\sigma$ are the dimensionless interatomic distances.The local energy can be reduced by means of an integration by parts to the two body form /14/

$$E = \sum_{i<j} \left\{ \frac{5\hbar^2 c^5}{m\sigma^2} \frac{1}{x_{ij}^7} + 4\varepsilon \left[\frac{1}{x_{ij}^{12}} - \frac{1}{x_{ij}^6} \right] \right\} \qquad (4.21)$$

The best way to describe all computing steps is to follow the Fortran code shown in Figure (4.4). Note that this code has not been optimized for the sake of simplicity. However, it works and produces correct results.

The positions of the particles are stored in array X(I,J), the first index corresponding to the component (I=1,2,3) and the second to the particle. DO loop 10 places all particles at random inside the simulation cube and the purpose of DO loop 100 is to carry out a series of statistically independent averages to evaluate the mean value of the energy and its statistical error. Each of these calculation blocks is carried out in DO loop 200, which is a long series of sequential moves of all particles. Particles are moved inside a cube of side SCAL/2, the value of SCAL being fixed in the program to one fourth of the simulation cube side L. In the range of b values of interest this results in around a 50% of acceptances. Of course, the value of SCAL may be changed to improve the stability or just to understand its actual role in the calculation.

Moves are carried out id DO loop 1000. From the old position we attempt a new position in the usual way. Afterwards, the new coordinates are displaced, if necessary, to replace particles in the fundamental cell. Once a new position is determined, we must compute the quotient of the old wave function by the new wave function by means of eq. (20). However, given that only one particle has been moved, only the terms in (20) involving this particle will contribute to the change of the wave function and it is not necessary to compute all pairs of particles in eq. (20). This part of the calculation is made in DO loop 320, running over all particles with the exception of the particle we are attempting to move. Each of the contributions to the exponential part of eq. (20) requires the determination of the nearest neighbour (this is done in subroutine CLOSE) and also the test on the cutoff of the correlation factor, i.e., if the distance to the nearest neighbour is greater than L/2 it is assumed to be L/2, so that the correlation is constant outside the interesting region.

The rest of the code is as expected: Metropolis test is carried out, and once all particles have been tested the local energy is computed and accumulated for the statistical average.

The determination of the nearest neighbour is the task of subroutine CLOSE, and given that the algorithm is quite straightforward I beleive it does not require further explanations. This routine also determines the cartesian coordinates of the nearest neighbour with respect to the particle under consideration. Conventionally these coordinates have been taken as positive always (even if it is not the case) and their value is required by the routine ENER which computes the local energy.

In the calculation of the local energy, the kinetic part has a cutoff at L/2, as required by consistency with Metropolis algorithm. However, the potential energy has been computed with a larger cutoff (L), and because of that more than one of the images can contribute to it. This is taken into account in Do loops 20. Here is where our criterium of considering all cartesian coordinates positive enters: the only images of interest are obtained with no displacement or with a -L displacement from the nearest neighbour.

Because of the cutoff in the interaction, the final energy must be corrected to have the physical value. A simple way of evaluating this correction is the following /14/. If the cutoff distance is large enough, we may consider that the two body distribution function is already 1 at distances longer than the cutoff. The contribution to the energy per particle coming from large distances can be approximated by

$$\Delta E = \frac{1}{2} \int_L d^3r\, g(r)\, V(r) = \frac{1}{2} \int_L d^3r\, V(r) \qquad (4.22)$$

See the lectures of Sergio Rosati to find the definition of the two

```
C       MACMILLAN'S VARIATIONAL MONTECARLO IN LIQUID HELIUM-4
C       WITH LENNARD-JONES POTENTIAL
        DIMENSION X(3,108),XOLD(3),XNEW(3)
        REAL L,LSQU,LSQDIV4
        COMMON X,NP,L,LSQU,HL,LSQDIV4   ! Arguments for subroutines
        TYPE*,' INTEGER FOR RANDOMNESS, NUMBER PARTICLES AND CORRELATION'
        ACCEPT*,NR,NP,C                 ! C=B/SIGMA, Dimensionless
        DENS=0.3648*0.9                 ! SCHIFF-VERLET in SIGMA units
        L=(NP/DENS)**0.3333333          ! Simulation cube
        LSQU=L*L
        LSQDIV4=LSQU/4.
        HL=L/2.
        SCAL=L/4.                       ! Scale of Metropolis moves
        C5=C*C*C*C*C
        TYPE*,' ENTER NUMBER BLOCKS and NUMBER OF MOVES'
        ACCEPT*,NBLK,NMOV
        DO 10 I=1,NP                    ! Starting position
        DO 10 J=1,3                     ! at random in simulation box
        X(J,I)=L*RAN(NR)
10      CONTINUE
C       METROPOLIS MOVES
        ETOT=0.                         ! Block average
        ETOT2=0.
        DO 100 IBLK=0,NBLK
        EAV=0.
        DO 200 IMOV=1,NMOV
        DO 300 IPAR=1,NP                ! Move particles sequentially
        DO 1000 I=1,3
        XOLD(I)=X(I,IPAR)                        ! Old position
        XNEW(I)=XOLD(I)+SCAL*(RAN(NR)-0.5)       ! Attempted move
        IF(XNEW(I).LT.0.) XNEW(I)=XNEW(I)+L      ! Move particle to the
        IF(XNEW(I).GT.L)  XNEW(I)=XNEW(I)-L      ! fundamental cell
1000    CONTINUE
        SUM=0.                                   ! Old and New W.F.
        DO 320 JPAR=1,NP
        IF (JPAR.EQ.IPAR) GOTO 320               ! Skip particle IPAR
        CALL CLOSE (XOLD,JPAR,R2)                ! Old neighbour: R2=squared
        IF (R2.GT.LSQDIV4) R2=LSQDIV4            ! Cuttoff at L/2
        CALL CLOSE (XNEW,JPAR,S2)                ! New neighbour: S2=squared
        IF (S2.GT.LSQDIV4) S2=LSQDIV4            ! and cuttoff
        SUM=SUM+1./(R2*R2*SQRT(R2)) - 1./(S2*S2*SQRT(S2))
320     CONTINUE
C       QUOTIENT OF PROB: (W.F.NEW/W.F.OLD)**2
        EXPR=C5*SUM                              ! Exponent
        IF(EXPR.GT.0.) GOTO 330                  ! Accepted
        PROBQ=EXP(C5*SUM)                        ! Probability quotient
        IF(PROBQ.LT.RAN(NR)) GOTO 300            ! Test Acceptance-Rejection
330     DO 1010 I=1,3                            ! Accepted move
1010    X(I,IPAR)=XNEW(I)                        ! Update position
300     CONTINUE
        CALL ENER(E,C)                           ! Sample energy
        EAV=EAV+E
200     CONTINUE
```

```
      EAV=EAV/NMOV
      TYPE*,IBLK,EAV                          ! Print iterations
      IF (IBLK.EQ.0) GOTO 100                 ! Neglect first block
      ETOT=ETOT+EAV
      ETOT2=ETOT2+EAV**2
100   CONTINUE
      ETOT=ETOT/NBLK/NP                       ! Block Average
      ETOT2=ETOT2/NBLK/NP**2
      ERR=SQRT((ETOT2-ETOT**2)/(NBLK-1))
      PRINT*,'  E =',ETOT, 'ERR=',ERR
      END
C     ===============================================================
      SUBROUTINE CLOSE (XX,JPAR,R2)
C     DETERMINES THE IMAGE OF PARTICLE JPAR COLSEST TO IPAR
C     THE RESULT R2 IS THE SQUARE OF THE DISTANCE
      DIMENSION X(3,108),XX(3),XCLOSE(3)
      REAL L,LSQU,LSQDIV4
      COMMON X,NP,L,LSQU,HL,LSQDIV4
      COMMON/COORD/XCLOSE               ! Components of (Ri-Rj)
      R2=0.
      DO 1000 I=1,3
      XCLOSE(I)=ABS(XX(I)-X(I,JPAR))    ! Always positive
      IF (XCLOSE(I).GT.HL) XCLOSE(I)=L-XCLOSE(I)
1000  R2=R2+XCLOSE(I)**2                ! Squared distance
      RETURN
      END
C     ===========================================================
      SUBROUTINE ENER(E,C)
      DIMENSION X(3,108),XCLOSE(3)
      REAL L,LSQU,LSQDIV4
      COMMON X,NP,L,LSQU,HL,LSQDIV4
      COMMON/COORD/ XCLOSE
      DATA A,B/9.27647,40.88/           ! A is 5*(HBAR**2/M)/SIGMA**2
      E=0.                              ! B is 4*Epsilon
      CKIN=A*C*C*C*C*C
      DO 10 IPAR=1,NP-1                 ! Count only PAIRS
      DO 10 JPAR=IPAR+1,NP
      CALL CLOSE (X(1,IPAR),JPAR,R2)
      IF(R2.GT.LSQDIV4) GOTO 100        ! Cutoff of correlation
      E=E+CKIN/(R2*R2*R2*SQRT(R2))      ! Kinetic energy, only
C                                       ! nearest neighbour
100   DO 20 K1=-1,0                     ! Check of images, X axis
      Z1=(XCLOSE(1)+K1*L)**2
      DO 20 K2=-1,0
      Z2=(XCLOSE(2)+K2*L)**2                    ! - Y axis, and
      DO 20 K3=-1,0
      Z3=(XCLOSE(3)+K3*L)**2                    ! - Z axis
      R2=Z1+Z2+Z3
      R6=R2**3                          ! This is R**6
      IF (R2.GT.LSQU) GOTO 20           ! Cutoff of Potential at L
      E=E+B*(1./R6-1.)/R6
20    CONTINUE
10    CONTINUE
      RETURN
      END
```

Figure 4.4. Variational calculation of liquid He-4 with Lennard-Jones interaction.

body distribution function and eq. (22).

The integration in (22) is extended to all space outside the sphere of radius L. The most important contribution to eq. (22) comes from the term r^{-6} in Lennard-Jones interaction, and the value of the energy correction which results is

$$E = -\frac{2\pi}{3} 4\varepsilon\rho\sigma^6/L^3 \tag{4.23}$$

In the case of 108 particles at a density 0.9 times the equilibrium density the value of this correction is -0.08K.

4.5 Fermion systems

There are not essential differences in the case of fermion systems with respect to the case of boson systems. There are however some technical problems whose main effect is the increasing of computing time. The first problem is the mere calculation of determinants at every Metropolis move. The second problem is the need of taking into account spin and isospin degrees of freedom in the case the interaction or the correlation or both depend on such operators.

The trial wave function is now the product of the Jastrow correlation and a Slater determinant. In the case of spin independent interactions there are as many Slater determinants as fermion species, e.g. one corresponding to spin up and the other corresponding to spin down. Note that one can include also various kinds of two body correlations, namely, spin up-spin up, up-down and down-down correlations. Each determinant is constructed with single particle orbitals φ_α, so that the fermion part is $\det(D_{i\alpha})$, with $D_{i\alpha} = \varphi_\alpha(r_i)$. Greek indexes will run over orbitals, and roman indexes over particles. Once the position of the particles is determined, this determinant is a number and the acceptance criterium in a Metropolis move is the squared quotient of two determinants, apart from the correlation factor.

Here is where the problem comes in, because the number of floating operations involved in the computation of a determinant is proportional to N^2 and consequently moving N particles involves N^3 operations. Ceperley, Chester and Kalos /19/ have found an algorithm to reduce the number of operations to N^2. The clue is to use and update the matrix of cofactors $C_{i\alpha}$. Each element of this matrix is obtained by zeroing in the original matrix row i and column α, but leaving the element (i,α) equal to 1. Matrix C is related to the inverse traspose of matrix D,

$$C_{i\alpha} = \det(D)\,\bar{D}_{i\alpha} \tag{4.24}$$

Once C is known, the value of the determinant may be computed as

$$\det(D) = \sum_\alpha C_{i\alpha}\, D_{i\alpha} \tag{4.25}$$

where the sum runs only over α, the particle label i being arbitrary.

This arbitrarieness is the clue of the simplicity of the subsequent algorithm.

Let us attemp a move of particle i. Then:

i. Matrix D does not change, with the exception of row i, its new value being

$$D'_{i\alpha} = \varphi_\alpha(r'_i) \quad , \alpha = 1,\dots N \qquad (4.26)$$

ii. All elements of matrix C will change, with the exception of row i which remains with its old value.

The new determinant is

$$\det(D') = \sum_\alpha C_{i\alpha} \, \varphi_\alpha(r'_i) \qquad (4.27)$$

where i corresponding to the particle moved. The ratio of the new to the old determinant is (put eqs. (24) and (27) together)

$$q = \frac{\det(D')}{\det(D)} = \sum_\alpha \bar{D}_{i\alpha} \, \varphi_\alpha(r'_i) \qquad (4.28)$$

This part of the calculation is simple, and involves only N floating operations. If the move is accepted, all of the elements of D must be updated. The way of doing it is the following:

$$\begin{aligned} &\bar{D}_{i\alpha} \longrightarrow \bar{D}_{i\alpha}/q \quad , \alpha = 1,\dots N \\ &\bar{D}_{j\alpha} \longrightarrow \bar{D}_{j\alpha} - \frac{\bar{D}_{i\alpha}}{q} \sum_\beta \bar{D}_{j\beta} \, \varphi_\beta(r'_i) \quad , \alpha = 1\dots N, \text{ and } j \neq i \end{aligned} \qquad (4.29)$$

which involves N^2 operations. Ceperley and coworkers /19/ discuss also the way of computing the energy from matrix D.

The question of spin dependence is quite more involved. The way of dealing with this problem is to carry out a double random walk to integrate over the spatial coordinates and to sum up over spin and isospin variables.

Assume an ordering convention so that a given single particle state $|\phi_\alpha\rangle$ is the direct product of a coordinate space state $|\varphi_\alpha\rangle$ and a spin state $|\chi_\alpha\rangle$,

$$|\phi_\alpha\rangle = |\varphi_\alpha\rangle \, |\chi_\alpha\rangle \qquad (4.30)$$

The numerical value, i.e., the wave function, is obtained by projecting on the bra $\langle r\nu|$, where ν is a discrete index representing the two spin orientations or the four spin/isospin orientations. Then

$$\phi_\alpha(r,\nu) = \langle r\nu | \varphi_\alpha \chi_\alpha \rangle \qquad (4.31)$$

and its value is $\varphi_\alpha(r)$ if χ_α is state ν, and 0 otherwise.

This is extended to a many fermion state, so that, e.g.,

$$\langle\Psi|\Psi\rangle = \sum_{\nu_1 \ldots \nu_N} \int dr_1 \ldots dr_N \, \langle\Psi|r_1\nu_1, \ldots\rangle\langle r_1\nu_1, \ldots|\Psi\rangle \qquad (4.32)$$

Note that $|\Psi\rangle$ is antisymmetric under permutations of single particle labels, but the intermediate states are not necessarily restricted. Of course, non antisymmetric intermediate states will give a null contribution to (32).

The numerical value of the wave function is a Slater determinant with many zeros. For example, assume a spin 1/2 four particle system with two spatial orbitals φ_1 and φ_2. The single particle states may be named in the following form

$$\begin{aligned} |1\rangle &= |\varphi_1, +\rangle \\ |2\rangle &= |\varphi_1, -\rangle \\ |3\rangle &= |\varphi_2, +\rangle \\ |4\rangle &= |\varphi_2, -\rangle \end{aligned} \qquad (4.33)$$

We may construct $2^4=16$ Slater determinants corresponding to the possible values of ν_i. Only that cases in which two particles have spin up and two particles have spin down are not null. This is the case of

$$\langle r_1+, r_2+, r_3-, r_4-|\Psi\rangle = \begin{vmatrix} \varphi_1(r_1) & \varphi_1(r_2) & 0 & 0 \\ 0 & 0 & \varphi_1(r_3) & \varphi_1(r_4) \\ \varphi_2(r_1) & \varphi_2(r_2) & 0 & 0 \\ 0 & 0 & \varphi_2(r_3) & \varphi_2(r_4) \end{vmatrix} \qquad (4.34)$$

A Metropolis move is carried out in two steps:

i. Try a displacement of particle i,

$$r_i \longrightarrow r_i + d(\xi - 0.5)$$

ii. Try a swap of spin of particle i with the spin of any of the particles, itself included.

The first part results in changing the value of the coordinate r_i. The second part acts like a column exchange in the determinant. For example, interchanging the spins of particles 1 and 3 in determinant (34) gives a new determinant

$$\langle r_1-, r_2+, r_3+, r_4- | \Psi \rangle = \begin{vmatrix} 0 & \varphi_1(r_2) & \varphi_1(r_3) & 0 \\ \varphi_1(r_1) & 0 & 0 & \varphi_1(r_4) \\ 0 & \varphi_2(r_2) & \varphi_2(r_3) & 0 \\ \varphi_2(r_1) & 0 & 0 & \varphi_2(r_4) \end{vmatrix} \tag{4.35}$$

Note that in all cases the determinant degenerates in two blocks after an appropriate reordering of rows and columns.

Finally, the energy is sampled with the hamiltonian matrix

$$\langle \psi | H | \psi \rangle = \int dr_1 dr_2 \dots dr_N \sum_{\nu_1 \cdots \nu_N} \sum_{\mu_1 \cdots \mu_N} \langle \Psi | r_1 \nu_1 \ \dots \ r_N \nu_N \rangle \langle r_1 \nu_1 \ \dots \ r_N \nu_N | H | r_1 \mu_1 \ \dots r_N \mu_N \rangle \langle r_1 \mu_1 \ \dots r_N \mu_N | \Psi \rangle \tag{4.36}$$

To complement this brief section on fermion problems we suggest the reader to consult Refs. /20/ for the case of few particles, and Ref. /21/ for a case with more particles.

4.6 Variational Montecarlo: summary

We have seen that Metropolis algorithm can be used to sample from almost everything. In other words, from the variational point of view almost any kind of trial wave function describing a many body system can be used. Note however that programming codes must be carefully writen to make calculations feasible.

Many systems of bosons and fermions have been studied in this form. Generally the calculations have been limited to infinite systems with spin independent interactions. A good description of many of the physically interesting systems studied with VMC can be found in the review article of Ceperley and Kalos /15/. We should say, however, that this algorithm (as any other many body theory) has many troubles when dealing with spin or isospin dependent interactions.

A final question, and not trivial, is that we are interested in the minimum value of the energy. This means that each VMC calculation must be repeated several times for different values of the parameters describing the trial function. This is a quite slow method implying a huge amount of computing time consumption. The problem is still a little more complex because the statistical fluctuations of the results may give rise to an incorrect placement of the minimum. A procedure appropriate to simple forms of the interaction was discovered by McMillan /13,14/. His method, based in scaling properties of the various terms of the local energy, seems not to be known in other many body theories. Another nice and general purpose method, known as the method of reweighting configurations, is described in Ref. 19/.

5. MONTECARLO INTEGRATION OF THE MANY BODY SCHRODINGER EQUATION

A completely different approach to the many body problem in the framework of Montecarlo is the direct integration of Schrodinger equation. Two methods have been developed within this spirit, which are known as DIFFUSION MONTECARLO (DFMC) and GREEN FUNCTION MONTECARLO (GFMC). The first method integrates the imaginary time dependent Schrodinger equation, whereas the second integrates directly the time independent Schrodinger equation. Actually, GFMC is a time-integrated version of DFMC.

These two methods have many features in common. Both represent the many body wave function by means of a large set of multidimensional points R_i, where as usual R represents the set of 3N coordinates of a N particle system. This set is such that its elements are distributed accordingly with the ground state wave function of the system. Each point R is called a psip in DFMC or a walker in GFMC. It is very important to stress that the quantity interpreted as a probability distribution function is the wave function of the system, and not its square, as in VMC. Of course, this requires the wave function to be positive, as in the case of a boson ground state of a given hamiltonian.

This very particular way of representing the wave function is depicted in Figure (5.1) for a one dimensional case. The set of random points, which are vertical bars along the X-axis, is more concentrated in the regions where the wave function is important, and their density is low in the regions where the wave function is small. Along the iterations, going from a wave function to another wave function means to change the random numbers of the set (called a generation) to a new set. Each psip or walker generates a number of descendants (none, one or more than one) which will accomodate to the new probability distribution function.

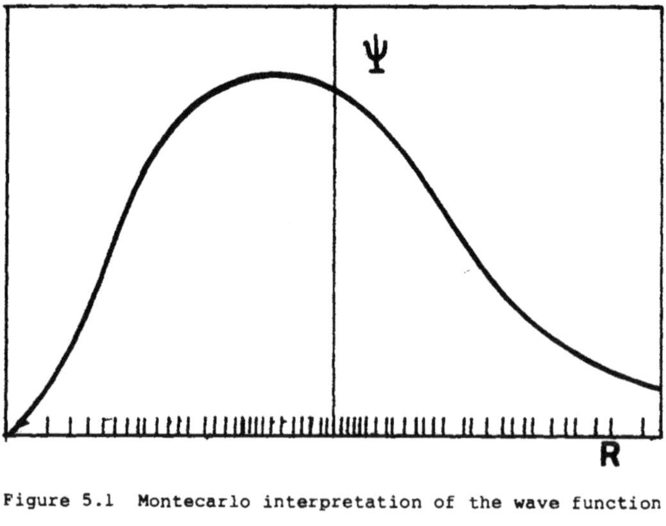

Figure 5.1 Montecarlo interpretation of the wave function

Both methods are iterative, and the ground state is approached exponentially in both methods. Again the technique of importance sampling may be introduced, improving the quality of the results as well as reducing computer time. The way of taking into account importance sampling is quite similar in both theories.

In their simplest form, DFMC and GFMC are only appropriate to describe the ground state of boson systems. In the case of fermions the ground state wave function is not everywhere positive, so the wave function cannot be interpreted as a probability distribution function. Importance sampling combined with the so called fixed node approximation permits the description of fermion systems.

To end this short introductory chapter let me mention that both techniques are intrinsically exact, with only statistical errors which may be controlled by increasing the statistics or by improving the importance sampling function. In the case of DFMC there is an additional error coming from the step by step integration in time.

I will discuss first DFMC method in a rather detailed form, in chapter 6. Finally chapter 7 will be devoted to GFMC.

6. DIFFUSION MONTECARLO

6.1 Introduction

The possibility of integrating the time dependent Schrodinger equation by means of stochastic methods was suggested by Metropolis and Ulam many years ago /22/, and subsequently applied by Donskar and Kac /23/. Undoubtedly, computers at that time were small and slow, and the idea remained dormant until 1975, when Anderson /24/ applied it to an atomic system. Since that time the method has been deeply developed /25-33/ and many applications have appeared in the mean time, mainly in the field of quantum chemistry and electron systems. Just for didactical interest I suggest the introductory reader to study the works of Reynolds et al /30/ and Vrbik and Rothstein /33/.

The starting point is the Schrodinger equation in imaginary time,

$$-\frac{\partial\Psi(R,t)}{\partial t} = (H - E)\,\Psi(R,t) \tag{6.1}$$

where R represents the set of all coordinates and E is an energy shift, at this moment arbitrary. In eq. (1) time is measured in units of inverse energy.

The formal solution of eq. (1) is

$$\Psi(t) = \exp\left\{-t(H-E)\right\}\Psi(0) \tag{6.2}$$

We may expand the starting function in the basis of eigenfunctions of H,

$$H\,\phi_n = E_n\,\phi_n \qquad E_0 < E_1 < \cdots \tag{6.3}$$

which have been ordered by their energy. Then eq. (2) is converted into

$$\begin{aligned}\Psi(0) &= \sum C_n\,\phi_n \\ \Psi(t) &= \sum C_n\,\phi_n\,\exp\{-t(E_n - E)\}\end{aligned} \tag{6.4}$$

The amplitudes change with time, increasing or decreasing accordingly with the sign of $(E_n - E)$. Independently of the value of the energy shift E, the most important amplitude at large time is that one corresponding to the state with energy E_0, i.e., the ground state, provided the starting wave function has a non null projection of the g.s., i.e., provided C_0 is not null.

This behaviour may be interpreted in terms of a diffusion problem. With the usual form of the hamiltonian, eq. (1) is

$$-\frac{\partial\Psi(R,t)}{\partial t} = -D\nabla^2\Psi(R,t) + V(R)\,\Psi(R,t) - E\,\Psi(R,t) \tag{6.5}$$

where $D=\hbar^2/2m$ is called the diffusion constant, and ∇^2 is a shorthand representation of the Laplace operator on all coordinates. This term represents an isotropic diffusion. The other term, (V(R)-E) is a rate term, with creation or dissapearance of the magnitude diffused. These two concepts will be clarified below.

Equation (5) may be improved in the sense of importance sampling introduced at the beginning of these lectures. Consider a trial function $\Psi_T(R)$, independent of time, and construct an equation analogous to (5) for the quantity

$$f(R,t) = \Psi_T(R)\,\Psi(R,t) \tag{6.6}$$

This equation turns out to be

$$-\frac{\partial f(R,t)}{\partial t} = -D\nabla^2 f(R,t) + D\vec{\nabla}(\vec{F}\,f(R,t)) + (E_L(R)-E)\,f(R,t) \tag{6.7}$$

where F is called the <u>drift</u> force, and given by

$$\vec{F} = \frac{2}{\Psi_T(R)}\vec{\nabla}\,\Psi_T(R)\,, \tag{6.8}$$

and is directly related to the trial wave function. E_L is the local energy, the same quantity introduced in VMC,

$$E_L(R) = \frac{1}{\Psi_T(R)}\,H\,\Psi_T(R) \tag{6.9}$$

The objective of DFMC is the integration in time, step by step, of eq. (5) or eq. (7), up to a value of the time large enough so that the only important amplitude is the corresponding to the ground state wave function. This integration is carried out by means of the integral equation equivalent to (5) or (7), by using a Montecarlo description of the instantaneous wave function through a set of random points, as we have described at the end of the previous chapter.

6.2 Time dependent Green functions

The formal solution of the generic equation

$$- \partial_t \Psi = A \Psi$$

is, as we have already mentioned

$$\Psi(t) = \exp(-At) \quad \Psi(0)$$

In coordinate representation, i.e., projecting on bra $\langle R'|$, we have

$$\langle R' | \Psi(t) \rangle = \int \langle R' | \exp(-At) | R \rangle \langle R | \Psi(0) \rangle \, dR \qquad (6.10)$$

The matrix element of the exponential of A is the time dependent Green function

$$G(R',R,t) = \langle R' | \exp(-At) | R \rangle \qquad (6.11)$$

and the integral equation is usually writen in the form

$$\Psi(R',t) = \int G(R',R,t) \; \Psi(R,0) \, dR \qquad (6.12)$$

The point is that in general G(R',R,t) is not known, but reliable approximations can be derived for small values of t. Then instead of solving directly eq. (12) one proceeds step-by-step in time, using repeteadly the formula

$$\Psi(R',t+\Delta t) = \int G(R',R,\Delta t) \; \Psi(R,t) \, dR \qquad (6.13)$$

The next point is the calculation of the Green function for small time intervals. Let us introduce three operators related to eq. (7)

$$\begin{aligned} A_1 &= -D \nabla^2 \\ A_2 &= D \left\{ (\vec{\nabla}\vec{F}) + \vec{F}\vec{\nabla} \right\} \\ A_3 &= E_L(R) - E \end{aligned} \qquad (6.14)$$

so that A in eq. (11) is the sum $A_1+A_2+A_3$. Analogously we may define the individual Green functions

$$G_i(t) = \exp(-A_i \, t) \qquad (6.15)$$

The total Green function G may be approximated by

$$\exp(-At) = \exp(-A_3 t/2)\ \exp(-A_2 t/2)\ \exp(-A_1)$$
$$\exp(-A_2 t/2)\ \exp(-A_3 t/2) + O(t^3) \tag{6.16}$$

This may be easily checked by expanding both sides of eq. (16) in powers of t. Warning, the various operators A_i do not commute among themselves. In coordinate representation eq. (16) is converted into

$$G(R',R,t) = \int dR_1\, dR_2\, dR_3\, dR_4\; G_3(R',R_1,t/2)\; G_2(R_1,R_2,t/2)$$
$$G_1(R_2,R_3,t)\; G_2(R_3,R_4,t/2)\; G_3(R_4,R,t/2) \tag{6.16.a}$$

Eq. (16) is not the unique $O(t^3)$ approximation, but it is the most convenient form for the subsequent reduction of the total Green function.

The computation of the three individual Green functions may be carried out by integrating the differential equation they satisfy,

$$-\partial_t\, G_i(R',R,t) = A_i\, G_i(R',R,t) \tag{6.17}$$

with the boundary condition

$$G_i\ (R',R,0) = \delta(R'-R) \tag{6.18}$$

The following results are obtained

$$G_1(R',R,t) = (4\pi Dt)^{-3N/2} \exp\left\{-(R-R')^2/4Dt\right\}, \tag{6.19}$$

where N is the number of particles,

$$G_2(R',R,t) = \delta\,(R' - \mathcal{R}\,(t)\,) \tag{6.20}$$

where $\mathcal{R}(t)$ is the solution of the equation of motion

$$\frac{d\mathcal{R}}{dt} = D\ F\ (\mathcal{R}(t)\ ) \tag{6.21}$$

$$\mathcal{R}(0) = R \tag{6.22}$$

Note that in eq. (21) F ($\mathcal{R}$(t)) means the drift force, eq. (8), computed at the value $\mathcal{R}$(t). From eq. (21) we see that the appropriate name for F is velocity drift. Equation (6.21) is not simple, and the reader may consult /33/ for its derivation.

Finally

$$G_3(R',R,t) = \exp\left\{(E-E_L(R)\ )\ t\right\}\ \delta(R-R') \tag{6.23}$$

Note that G_1 and G_2 are normalized to 1, i.e.,

$$\int G_{1,2}\ (R',R,t)\ dR' = 1 \tag{6.24}$$

whereas G_3 is not normalized.

The particular form of the individual Green functions has been the reason to choose the particular ordering of operators in eq. (16), because now all integrals appearing in eq. (16.a) may be carried out with the help of the delta functions. Substitute all values (19-23) into eq. (16) to obtain the approximate complete Green function /33/

$$G(R',R,t) = (4\pi Dt)^{-3N/2} \exp\left\{\left[E - (E_L(R)+E_L(R'))/2\right] t\right\} \exp\left\{-\left[R' - R - Dt(F(R)+F(R'))/2\right]^2 /4Dt\right\} \det\left\{I - D\,t\,H(R')\,/\,2\right\} + O(t^3) \qquad (6.25)$$

where I is the 3Nx3N unit matrix, and H is the 3Nx3N matrix of the spatial derivatives of F

$$H_{pq} = \frac{\partial^2}{\partial R_p\,\partial R_q}\left\{2\ \ln \Psi_T(R)\right\}.$$

A simpler, less accurate form of the complete Green function is /30/

$$G(R',R,t) = (4\pi Dt)^{-3N/2} \exp\left\{\left[E - (E_L(R)+E_L(R'))/2\right] t\right\} \exp\left\{-\left[R' - R - Dt\ F(R)\right]^2 /4Dt\right\} + O(t^2) \qquad (6.26)$$

In the next section, when describing the Montecarlo simulation of this problem, we will refer to eq. (26) There is still a simpler form corresponding to the case where importance sampling is not taken into account. This is equivalent to consider the trial wave function to be constant everywhere. Then the drift force is null, and the local energy is just the potential energy, so that

$$G(R',R,t) = (4\pi Dt)^{-3N/2} \exp\left\{\left[E - (V(R)+V(R'))/2\right] t\right\} \exp\left\{-\left[R' - R\right]^2 /4Dt\right\} + O(t^3) \qquad (6.27)$$

Up to this moment the energy shift appearing in all previous formulae is an arbitrary quantity. Of course, it is in no way related to the expectation value of the hamiltonian in the trial importance sampling function. We already know that when E is equal to the exact ground state energy, then the solution of the differential equations (5) and (7) will stabilize at large times approaching the ground state wave function, namely

$$\Psi(R,t) \longrightarrow C\ \phi_0(R)\ , \text{ non importance sampling}$$

$$f(R,t) \longrightarrow C\ \ \Psi_T(R)\ \ \phi_0(R)\ , \text{ with importance sampling,}$$

with a coefficient C_0 unknown. This behaviour in the large time limit can be used to compute the energy, using the mixed matrix element

$$E_0 = \frac{\int \Psi_T(R)\ H\ \phi_0(R)}{\int \Psi_T(R)\ \ \phi_0(R)} \qquad (6.28)$$

In the case of non importance sampling just delete the trial wave function from the above equation. Equation (28) is obvious, because ϕ_0 is the ground state of the hamiltonian. In practical applications it is convenient to act with the hamiltonian operator on the left state. In this form we obtain

$$E_0 = \frac{\int E_L(R) \quad \Psi_T(R) \, \phi_0(R)}{\int \Psi_T(R) \quad \phi_0(R)} \tag{6.29}$$

or, in other words,

$$E_0 = \frac{\int E_L(R) \quad f(R, t \to \infty)}{\int f \; (R, t \to \infty)} \tag{6.30}$$

This last equation is of deep interest in the DFMC method. After many steps in time we will end up with a set of random points R distributed with the probability distribution function f(R,t). The we may appeal to central limit theorem to get the following estimate for the energy

$$E_0 = \frac{1}{M} \sum_i E_L \; (R_i \;) \tag{6.31}$$

where $\{ R_i, \; i=1, \, 2 \, \ldots \, M \}$ is the set of random points representing $f(R,t)$.

A similar result may be derived in the case of non importance sampling. The only difference is that instead of E_L in eq. (31) there appears V(R).

6.3 Montecarlo solution

After all that mathematical preliminaries we may jump to the practical solution of our problem, namely to eq. (13). The general approach consists in generating a set of random points R'_i, i=1, ... M', corresponding to the lhs wave function, from the set of random points R_i, i=1, ... M, by means of eq. (13) Note that each point R, i.e. each psip, corresponds to 3N coordinates of the N particles. From each psip R we generate new psips R' by using $G(R',R,\Delta t)$ as a conditional probability distribution. Given that G is not normalized, this means that from and old psip we may end up with none, one or more than one new psips R'. In other words, after each iteration the population of the resulting configuration may grow or may decline. On the average growing or declining is controlled by the value of the energy shift E, this change on the population being the reflect of the change in normalization discussed in eq. (4). After many iterations, the population will be stable only if the energy shift equals the ground state energy. Using eq. (31) one can guess the appropriate value of E.

The Montecarlo algorithm goes through the following steps (this is adapted from Ref. /30/)

0. Preliminaries

Guess a trial function ψ_T, obtain the formula for the local energy, assume a time step and also a value for the energy shift E.

1. Sample the initial distribution, e.g. form $|\psi_T|^2$. This initial distribution is, to a large extent, arbitrary. Usually one starts with a distribution with a very low population, and with an energy shift larger than the ground state energy. In the first iterations the population will grow quite rapidly, and this growing will go towards the correct ground state distribution. Remember that the various components of the starting function do grow more rapidly the smaller their energy is.

Let's call old set this initial distribution, its points being referred as R_i.

2. From each point R_i generate a new point R' using the gaussian part of eq. (26). The new value R' is given by

$$R' = R + D\,t\,F(R) + \chi \tag{6.32}$$

where χ is drawn from the 3N-variate gaussian distribution

$$(4\pi Dt)^{-3N/2} \exp(-\chi^2/4Dt)$$

Note that the gaussian part of eq. (26) is normalized to 1 when integrating with respect to R'. For that reason we obtain at this step one new point from one old point. What we are doing in this step is a displacement from R. This displacement has two parts: a spherically symmetric part χ coming form the Laplace operator term of the hamiltonian, which corresponds to a isotropic diffusion from point R, and a drift DtF(R) directly related to the importance sampling function. The drift force is the log derivative of the trial wave function, and this drift will move the psips towards the region where the importance sampling function is larger.

3. Compute the other term of the Green function, namely the exponential involving the local energies and the energy shift. This part of the Green function is not normalized, and is the responsible of the change in population. This is called the rate term.

4. Branching of psips. The new point has a multiplicity such that on the average equals the value of the rate term. To carry out this step we draw an integer from a distribution which value is the rate term, namely

$$\text{Multiplicity} = \text{Integer}\left\{ \exp\left\{ \left[E-(E_L(R)+E_L(R')\right] t \right\} + \xi \right\} \tag{6.33}$$

where ξ is from U(0,1).

5. If in the previous step the multiplicity is null, then discard the new point R' and go again to step 2 unless the old set is exhausted. Otherwise copy this value R' into the new set as many times as the value of the multiplicity, and add the same number of times the value of $E_L(R')$ to compute the average energy of the iteration, eq. (31). Again go back to step 2.

6. We arrive here when all psips corresponding to the old configuration have evolved. Now we have advanced one time step and have obtained a new configuration. The next task is to advance another time step, so rename the new configuration to the old configuration and go back to step 2. This block should be repeated as many times as needed to reach the asymptotic limit. We may actualize the energy shift E by using eq. (31) to stabilize the population once a critical population has been reached.

When time is long enough we should still advance many time steps to evaluate the expectation value of the energy and control the statistical error. So, the last step is

7. Rename the new set to the old set, test some exiting criterium and go to step 2.

There are some additional comments on this algorithm. Along the iterations, the succesive averages of the energy are strongly correlated if the time step is small, as it should be. To avoid these correlations it is convenient to carry out a block average. The final result will depend on the value of the time step. This is the only systematic error of the algorithm. This error can be corrected by carrying out several runs with different time steps, and extrapolating the result to a null time step value. Finally, this algorithm may be easily modified to deal with the other forms of the time dependent Green function given in eqs. (25) and (27).

6.4 A nucleon in a gaussian potential revisited

As a case example let's consider again the case of a nucleon in a gaussian potential in the same conditions than in section (4.2). Many of the equations necessary have been already deduced in that section, and we will repeat them for convenience:

$$V = - V_0 \exp(-(r/b)^2) \quad , V_0 = -45 \text{ Mev}, b= 2.09 \text{ fm}$$

$$\Psi_T = \exp (- \alpha^2 r^2/2) \ , \ \alpha = 0.625 \text{ fm}^{-1} \text{ for variational minimum}$$

$$E_L = V(r) + D (3\alpha^2 - \alpha^4 r^2)$$

$$D = \hbar^2/2m = 20.75 \text{ MeV fm}^2$$

and the drift force is given by

$$\vec{F}(r) = -2\alpha^2 \vec{r}$$

and points always towards the origin, where the maximum of the importance sampling function is.

In this case it is quite simple to use the more complex and precise form of the Green function, eq. (25), because of the simplicity of the trial function. The normalized gaussian part of the time dependen Green function is given by

```
C       ==============================================================
C       DIFFUSION MONTECARLO WITH IMPORTANCE SAMPLING
C       AND WITH BLOCK AVERAGE
C       ==============================================================
        DIMENSION PSIP(4,5000,2)        ! PSIP(1,...), (2,...) (3,...) is point
C                                       ! PSIP(4,...) is Local Energy
        COMMON/RAND/NRAN                ! For Gauss subroutine
        V(X,Y,Z)=-45.*EXP(-(X*X+Y*Y+Z*Z)/4.3681)         ! Potential
C
C       ====      BLOCK 0, Starting parameters ======
C
        TYPE*,' INTEGER FOR RANDOMNESS'
        ACCEPT*,NRAN                    ! Integer for randomness
        TYPE*,' ENTER NUMBER OF BLOCKS, ITERATIONS PER BLOCK AND TIME STEP'
        ACCEPT*,NBLK,NITER,T            ! Blocks, Iterations and Time step
        TYPE*,' ENTER TRIAL ENERGY and TRIAL PARAMETER'
        ACCEPT*,E0,HOALF                ! Importance sampling parameter
        E=E0                            ! Starting energy
        C1=20.75*3*HOALF*HOALF          ! Constants for local energy
        C2=C1*HOALF*HOALF/3.
        D=20.75                         ! Diffusion constant
        STRC=(1.-D*T*HOALF*HOALF)/(1+D*T*HOALF*HOALF)   ! Scaling of Gaussian
        TGAUSS=4*D*T/(1+D*T*HOALF*HOALF)**2             ! G(x,y) parameter
        TYPE*,' ENTER CRITICAL POPULATION AND STARTING POPULATION'
        ACCEPT*,ICRIT,NPOP              ! Critical and starting population
C
C       ==========  BLOCK 1 : Starting configuration ==========
C
        IN=1                            ! IN and IO control the old
        IO=2                            ! and the new PSIP
        EAV=0
        TGAUS=1./HOALF/HOALF            ! Distribution of Trial Fnct Squared
        DO 10 I=1,NPOP                  ! Generate starting PSIP
        CALL GAUSS(X,Y,Z,TGAUS)         ! from variational trial state
        PSIP(1,I,IN)=X
        PSIP(2,I,IN)=Y                  ! Coordinates and ...
        PSIP(3,I,IN)=Z
        PSIP(4,I,IN)=V(X,Y,Z)+C1-C2*(X*X+Y*Y+Z*Z)       ! ... Local energy
10      CONTINUE
C       ======================================================
        EB=0.
        EB2=0.                          ! Block average
        DO 2000 IBLK=0,NBLK
        NUMBER=0                        ! Counts mean population in block
        ENERG=0.                        ! Time iter. average
        DO 100 ITER=1,NITER
        EAV=0                           ! Set average
        JPOP=0                          ! Population of new set
        DO 1000 I=1,NPOP
        X=PSIP(1,I,IN)*STRC             ! ========================
        Y=PSIP(2,I,IN)*STRC             !
        Z=PSIP(3,I,IN)*STRC             !       Step 2
        EL=PSIP(4,I,IN)                 !
        CALL GAUSS(XP,YP,ZP,TGAUSS)     ! Sample new PSIP
        XP=X+XP                         !
        YP=Y+YP                         !
        ZP=Z+ZP                         ! ========================
```

```
         ELP=V(XP,YP,ZP)+C1-(XP*XP+YP*YP+ZP*ZP)*C2       ! Local Energy  (3)
         NSONS=INT(EXP(T*(E-0.5*(EL+ELP)))+RAN(NRAN))    ! and Branching (4)
         IF (NSONS.EQ.0) GOTO 1000          ! Descendant rejected
         DO 1010 J=1,NSONS                  ! ========================
         JPOP=JPOP+1                        !
         PSIP(1,JPOP,IO)=XP                 ! Step 5
         PSIP(2,JPOP,IO)=YP                 !
         PSIP(3,JPOP,IO)=ZP                 ! Copy NSONS times sampled point
         PSIP(4,JPOP,IO)=ELP                ! and
         EAV=EAV+ELP                        ! actualize averages
         NUMBER=NUMBER +1                   !
 1010    CONTINUE                           ! ========================
 1000    CONTINUE
         NPOP=JPOP                          ! ========================
         II=IN                              ! Renaming of OLD and NEW psips
         IN=IO                              ! Step (7)
         IO=II                              !
         E=EAV/JPOP                         ! Update trial energy
         IF(IBLK.EQ.0) GOTO 110             ! Discard averaging in first BLOCK
         ENERG=ENERG+E
 110     IF (JPOP.LT.ICRIT) E=E0            ! If population is LOW, increase it
 100     CONTINUE
         IF(IBLK.EQ.0) GOTO 2000
         EB=EB+ENERG/NITER
         EB2=EB2+(ENERG/NITER)**2
         PRINT*,IBLK,NUMBER/NITER,ENERG/NITER    ! Print iterations
 2000    CONTINUE
         EB=EB/NBLK                         ! Block average
         EB2=EB2/NBLK
         ERR=SQRT((EB2-EB**2)/(NBLK-1))     ! and statistical error
         PRINT*
         PRINT*,' ENERGY =',EB,' ERROR=',ERR
         STOP
         END
 c       Insert here subroutine GAUSS, in Fig. (4.1)
```

Figure 6.1. Diffusion Montecarlo with importance sampling.

$$\exp\left[-\left[(1+Dt\alpha^2)\vec{r}\,' - (1-Dt\alpha^2)\vec{r}\right]^2/4Dt\right](1+Dt\alpha^2)^3/(4\pi Dt)^{3/2}$$

This is a normalized gaussian or the form

$$\exp\left[-(\vec{r}\,' - \vec{X})^2 / A\right]$$

where $\vec{X}$ is a scaled coordinate

$$\vec{X} = (1-Dt\,\alpha^2)\,\vec{r}\,/(1+Dt\,\alpha^2)$$

and the gaussian parameter A is

$$A = 4Dt/(1+Dt\,\alpha^2)^2.$$

When $\alpha=0$ all that formulae are the corresponding to the non importance sampling algorithm. Note that the drift is always in the direction of $\vec{r}$, so that step 2 of previous section is simplified. We first scale the actual position r to a new, closer to the origin, position X, and afterwards just add the isotropic gaussian move.

The algorithm described in section (6.3) has been coded in the Fortran program listed if Fig. (6.1). This code follows step by step the mentioned algorithm and with the comments added can be easily followed.

The results corresponding to this simple example are presented in Figure (6.2). We plot the expectation value of the energy computed in the current configuration at every time step. The open circles correspond to the evolution in the absence of importance sampling, and the black circles correspond to the importance sampling case using the optimal harmonic oscillator wave function. The critical population was of 700 points. This means that the energy shift was adjusted to stabilize the population once the number of psips reached this critical value. The time step used was 0.005 MeV^{-1}. The first 60 steps were discarded in obtaining the energy average and the horizontal line shown in the figure is the average of 200 steps more. Its value is -10.9 MeV with a statistical error of 0.1 Mev, to be compared with the exact energy, -10.813 Mev. This is a rather simple and modest calculation, and the results are very gratifying.

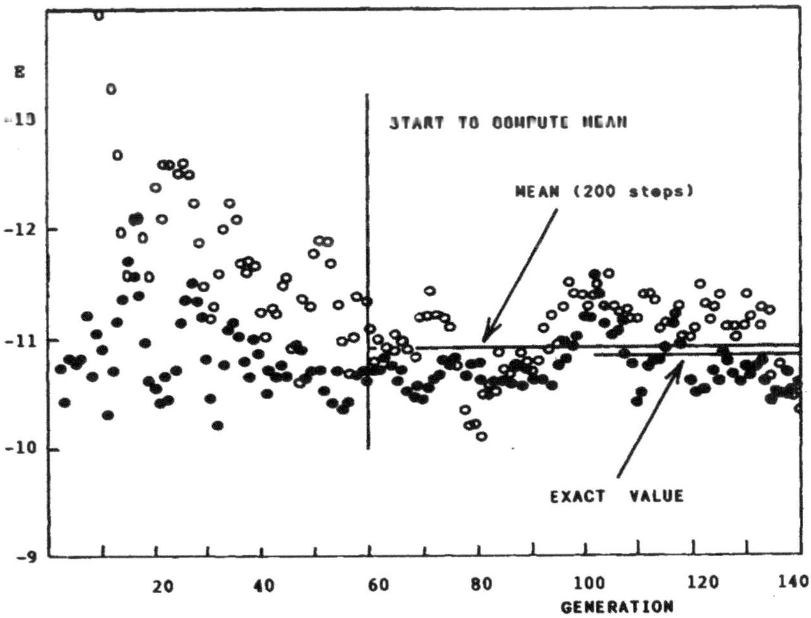

Figure 6.2. Diffusion Montecarlo whithout (open circles) and with (black circles) importance sampling. Energy in MeV.

The calculation with the importance sampling function starts with an already good value for the energy, and the dispersion in the initial iterations is large because the population is small. On the contrary, the values computed without importance sampling show an exponential approach to the correct values in the initial steps. In both cases we observe correlations between the successive steps which suggest the convenience of a block average calculation to get statistical independence. This block averaging is included in the Fortran code of Figure 1.

6.5 Complements

The simple scheme presented in section (6.3) and (6.4) has to be complemented to deal with two common situations, namely the case of extended systems and the case of fermions.

In infinite systems one should use periodic boundary conditions, like in VMC. Whenever a psip exits outside the simulation cell, one should consider its lattice image which entered in the fundamental cell through the opposite wall. Furthermore, the presence of all images must be taken into account when computing the local energy, using an appropriate cutoff or carrying out Ewald's summations.

The problem is not so simple when the system is made up of fermions. Note that the hamiltonian ignores the nature of the particles, and if the previous algorithm is followed as it stands, we will en up with a boson ground state, even if it is not physical. The fermionic character must be imposed externally. The other question is that a fermion wave function is not positive everywhere, and in the regions where it is negative it cannot be interpreted as a probability distribution. This problem was not present in VMC, because the quantity interpreted there as a probability distribution was the wave function squared.

The usual procedure to deal with fermions is the so called fixed node approximation. By means of a trial function Ψ_T one defines a set of nodal surfaces which are the surfaces where Ψ_T is null. These surfaces divide the space in regions where Ψ_T does not change sign. Then one assumes that the exact wave funtion Ψ has the same nodal structure than the trial function, so that $f = \Psi \cdot \Psi_T$ is always positive. These nodal surfaces are assumed to be perfectly absorbing walls: when a psip tries to move to the other side of a nodal surface, it is killed, i.e., is not included in the new generation.

The fixed node algorithm is not exact, but it may be proved /28,30/ that the average of the local energy computed with this approximation is an upper boung to the fermionic ground state energy.

The fixed node approximation can be improved up to some extent, by means of the so called nodal relaxation mechanism. Once fixed node approximation has stabilized, one allows for jumps across the nodal boundaries, but taking track of a signature (+ or -) of the psips. When a psip crosses a nodal surface, its current signature is changed.

After some iterations, every region will have a positive population and a negative population: to compute averages in that region one must add the values corresponding to positive signature psips, and substract the values corresponding to negative signature psips. Because of that, the statistical uncertainty will increase. From the physical point of view what is happening is that a boson population is being superimposed on the fermion population. If many time steps were carried out with node relaxation, we will end up with the boson ground state /27/.

6.6 An application: diatomic molecules

Many of DFMC calculations have been devoted to molecules, the simplest case being diatomic molecules like H_2, LiH, and so on. In all these calculations (as well as in almost all molecular structure calculations) the Born-Oppenheimer approximation is used. This approximation consists in assuming a fixed position for the atomic nuclei, which are static pieces.

When dealing with diatomic molecules one can compute several quantities:

i. The total energy, its meaning being obvious.

ii. The dissociation energy, which is the total energy minus the energy of the non interacting atoms, i.e., when the atomic constituens are at an infinite distance. Chemical measurements determine directly the dissociation energy, corresponding to the equilibrium distance of the nuclei.

iii. Atomic interaction potential, which is the dissociation energy as a function of the distance of the nuclei.

iv. Correlation energy. Is the difference between the total energy and the self-consistent field energy. This quantity has no experimental equivalent, but it is a convenient measure of the goodness of the theoretical approaches.

Calculation in Quantum Chemistry (see Ref. 34 for a full description) have various levels of complexity. The simplest way of carrying out molecular calculations is known as MO-LCAO, and this means that Molecular Orbitals are expressed as a Linear Combination of Atomic Orbitals. The simplest form of these atomic orbitals are known with the name of Slater orbitals. They are hidrogen-like orbitals with an effective charge. The self consistent field (SCF) approach is the expansion of molecural orbitals in a large basis, usually of Slater type. SCF calculations do not include the electron-electron correlation, but they explain the bulk of the dissociation energy. The physically interesting part is the remaining energy, which is called the correlation energy. It may be computed by means of a Configuration Interaction (CI) expansion, or using an explicit dependence on the e-e relative coordinates (Hylleraas form).

In table (6.1) I have collected some recent results corresponding to the LiH molecule, obtained by means of a variety of methods. We cannot enter here in a detailed discussion of these results, but as a general comment we can say that the numbers of Table 1 are somehow relaxing: quite different theories give the correct result within small deviations. Needless to say, any of these calculations are quite costly in terms of computing time.

Table 6.1. Correlation energy of LiH molecule. The SCF energy is 7.987 hartrees (1 hartree = 2 Ry = 27.2 eV). The VMC energy was computed with the importance samplig function of DFMC. Is a poor function, but the nodes are good.

Corr. Energy Milihartrees	Method	Reference
-76.5	CI with Slater orbitals	35
-78.2	CI with elliptic orbitals	36
-80 ± 2	DFMC, fixed node	30
-54 ± 8	VMC	30
-81.5	Coupled Cluster SUB(2)	37
-83.2 0.1	Exact (mixed Theory and Exp.)	38

6.7 Jellium

Jellium is a simplified many body system consisting in N electrons moving in a fixed, uniformly charged background of positive charge which provides electric neutrality. It was proposed 50 years ago by Wigner /39/ as a simplified model of conducting metals. This system may be described in terms of a dimensionless parameter r_s which is the mean interparticle distance measured in units of Bohr radius. High values of r_s correspond to a low density system, and conversely small values fo r_s represent a high density system. A nice feature of Jellium is its expected crystaline configuration (bcc) at low density.

Table 6.2. Correlation energy of jellium in miliRydbergs as a function of r_s, computed by means of various many body theories.

Method $r_s \to 0$	1	2	4	10	20	Ref.
DFMC	-120	-90.2	(-63.6)	-37.2	-23.0	27
RPA 62.2 log r_s	-157	-123	-93.6	-61.3	-42.8	41
CC(4) 62.2 log r_s	-122	-90.4	-63.4	-37.0	-23.6	42
FHNC	-138	-98	-67	-37		43
VMC	-122	-87.4	-62.4	-36.3	-22.5	44

This system has been studied by means of almost all many body theories. A set of results of the correlation energy per particle is displayed in Table 6.2. The Random Phase approximation (RPA) and

Coupled Cluster (CC) theories predict the correct behaviour at high density, but as expected RPA fails at low density. A bosonized Fermi Hypernetted Chain (FHNC) theory predicts a high densith behaviour (57 log r_s) which is not bad /40/. However, FHNC is particularly valid at low densities (large r_s) as Table 2 shows.

A nice feature regarding Jellium is its complex phase structure, as predicted by DFMC /27/. At high density the system is paramagnetic (unpolarized fluid), including the region where physical metals should be placed (around r_s=20), and cristalizes at around r_s=100, after passing through a ferromagnetic (polarized) phase. See Ref. /27/ for further details.

7. GREEN FUNCTION MONTECARLO

7.1 Introduction

The basic ideas of GFMC method were developed almost 25 years ago by Malvin Kalos in a nice and simple work /45/, which actually was devoted to the inverse problem: determine the depth of a potential well of prescribed form so as to bind three and four nucleon nuclei at the appropriate energy. The first many body calculation with GFMC for a boson fluid, helium four, appeared several years later, /46/ in 1970, and was limited to a quite small number of particles in the simulation cell. Liquid helium four was subsequently studied with an already complete theroy by Kalos, Levesque and Verlet /47/ coincidindg with the revitalization of many body theories of the last ten years. The role played by VMC and GFMC since 1970 has been decisive to clarify and settle down many of the newly born or newly rediscovered theories, particularly Fermi Hypernetted Chain and Coupled Cluster (exp S) theories.

Further developement of GFMC was devoted to many fermion studies /48/ and to few component systems, both fermions /49/ and bosons /50/. At this moment GFMC are a tour de force to large scale (vector) computers. Up to my knowledge, the last GFMC calculation corresponds to liquid helium three /51/.

There exists some nice reviews of GFMC /52-54/, and the reader should consult them to complement or clarify the description of this theory presented here.

GFMC has many features in common with DFMC, and they also share many of the problems and troubles. Actually, GFMC is a time integrated version of DFMC, but DFMC is not a necessary step to describe GFMC. In fact, GFMC is older than DFMC. Again the objective is to solve the many body Schrodinger equation in a stochastic way, by iterating the equivalent integral equation with Montecarlo sampling techniques. The wave function is represented by a set of random points called walkers, and succesive iterations produce generations closer to the ground state wave function.

In this chapter I will skip some of the technical points of GFMC method, and I will pay mainly attention to the conceptual content of this method. As in previous chapters, some examples will be described and documented with the hope of increasing the clarity of the basic ideas.

7.2 Green Functions

Consider the many body problem ($\hbar^2/2m=1$)

$$\left[-\nabla_R^2 + V(R) + V_0\right] \Psi(R) = E\,\Psi(R) \tag{7.1}$$

where V_0 is an energy shift so as to have all eigenvalues positive. The value of V_0 is not known a priori, unless the interaction is bounded from below, $V(R) > -C$, with C positive. Then $V_0=C$ can serve as an appropriate shift. In the following V_0 will be included in the potential, so it will disappear from the formulae. The role of the energy shift here is very different of the energy shift in DFMC. Here it is required to have a positive Green function. Below we will introduce a trial energy which will play the same role than the energy shift in DFMC.

The 3N-dimensional Green function satisfies the differential equation

$$\left(-\nabla_R^2 + V(R)\right) G(R,R') = \delta(R-R') \tag{7.2}$$

subjected to the same boundary conditions than eq. (1). So, in finite, localized systems, the wave function drops to zero when any coordinate goes to infinity. In extended systems one should impose periodic boundary conditions, as usual.

From G(R,R') one can determine the wave function by solving the homogeneous integral equation

$$\Psi(R) = E \int G(R,R')\, \Psi(R')\, dR' \tag{7.3}$$

Formally equation (2) is the coordinate representation of the operatorial equation

$$H\,G = 1 \tag{7.4}$$

its solution being obviously G=1/H. Note that because of the positivity of H, forced throug the introduction of the energy shift, the solution of eq. (4) is well defined.

Using the complete set of eigenstates of H, $\{\phi_n\}$, eq. (6.3), we may derive the following two forms for the Green function,

$$G = \sum_n |\phi_n\rangle \frac{1}{E_n} \langle\phi_n| \tag{7.5}$$

and, in coordinate representation

$$G(R,R') = \langle R | G | R' \rangle = \sum_n \phi_n(R) \frac{1}{E_n} \phi_n^*(R') \qquad (7.6)$$

Equation (7.3) may be solved by iteration. Start form an arbitrary state

$$\psi^{(0)} = \sum c_n \phi_n \qquad (7.7)$$

and apply repeteadly the operator $E_T G$ (E_T is a trial energy). Then we obtain after p iterations

$$\psi^{(p)} = \sum_n c_n \left(\frac{E_T}{E_n} \right)^p \phi_n \qquad (7.8)$$

After a large number of iterations the amplitude of ϕ_0 has grown more rapidly than the remaining amplitudes (provided C_0 is not null) and asymptotically only the state ϕ_0, i.e., the exact ground state, will survive. Note that the norm is not maintained along the iterations, and it will grow or decrease when (E_T/E_0) is greater or smaller than 1, respectively. This fact may be used to determine E_0 just by adjusting the trial energy which stabilizes the norm.

The underlying idea of GFMC is the iteration of eq. (3) by means of Montecarlo method. From a generation of points corresponding to iteration number p construct a new generation, corresponding to the nex iteration, by generating new walkers R from the old walkers R' by means of the conditional probability distribution G(R,R').

7.3 Sampling from G(R,R'): Born series

Unfortunately, the Green function is not known in general. So, a pending question of the previous section is how can we sample random points from a distribution function which is not known. The problem is solved by means of a Born series expansion of the complete Green function from an approximate Green function G_0.

Consider the simple problem of a bounded, always attractive potential. We may define a normalized potential v(R) by the equation

$$V(R) = -U\, v(R)$$

with v(R) varying between 0 and 1. Shift both potential and energy by U and use dimensionless coordinates

$$\vec{Z} = \alpha \vec{R}\ , \quad \alpha = (2mU/\hbar^2)^{1/2} \qquad (7.9)$$

Then the Schrodinger equation simplifies to

$$\left[-\nabla^2 + (1 - v(Z)) \right] \psi(Z) = \lambda \psi(Z) \qquad (7.10)$$

with

$$\lambda = (U+E)/U \qquad (7.11)$$

Note that $E>0$, and $E<-U$, so that λ is between 0 and 1.

Define now a complete Green function

$$\left[-\nabla^2 + (1-v(Z))\right] G(Z,Z') = \delta(Z-Z') \tag{7.12}$$

and a free Green function

$$\left[-\nabla^2 + 1\right] G_0(Z,Z') = \delta(Z-Z') \tag{7.13}$$

The integral equation to be solved now is

$$\Psi(Z) = \lambda \int G(Z,Z')\, \Psi(Z')\, dZ' \tag{7.14}$$

We will consider eqs. (10), (12) and (13) extended to all space. This condition will be relaxed below.

Some interesting relations may be obtained from these equations. Integrate eq. (10) to obtain the eigenvalue

$$\lambda = \int (1-v(Z)\ \Psi(Z)\ dZ \Big/ \int \Psi(Z)\ dZ\ . \tag{7.15}$$

Analogously integrate eq. (13) to get the normalization condition of the free Green function

$$\int G_0(Z,Z')\ dZ = 1 \tag{7.16}$$

and also integrate eq. (12)

$$\int (1-v(Z))\ G(Z,Z')\ dZ = 1\ . \tag{7.17}$$

To obtain eqs. (15) to (17) we have integrated by parts the laplacian terms of the required equations. When the boundary is not infinity there will have appeared contributions from the normal derivatives at the boundary surface.

It may be proved that both G and G_0 are positive. Moreover $G(Z,Z')$ is symmetric, and in this particular case G_0 is also symmetric. The symmetry of G_0 will be lost below because of some special boundary conditions.

Next step is to relate G with G_0. Formally

$$G = 1/(T + 1 - v)$$

and

$$G_0 = 1/(T+1),$$

with $T=-\nabla^2$. Use the operatorial identity (this identity is valid when the inverses of A and B are well defined. Again be careful in the case of special boundary conditions)

$$\frac{1}{A} - \frac{1}{B} = \frac{1}{B}\,(B-A)\,\frac{1}{A}$$

with $A=T+1-v$ and $B=T+1$, to obtain

$$G = G_0 + G_0\ v\ G \tag{7.18}$$

This is the integral equation relating G and G_0. In coordinate representation eq. (7.18) is

$$G(Z,Z') = G_0(Z,Z') + \int dZ'' \; G_0(Z,Z'') \; v(Z'') \; G(Z'',Z') \qquad (7.19)$$

Solving (18) or (19) iteratively we get the Born series expansion of G,

$$G(Z,Z') = \sum_p G^{(p)}(Z,Z'),$$

with

$$G^{(0)}(Z,Z') = G_0(Z,Z'), \qquad (7.20)$$

and

$$G^{(p)}(Z,Z') = \int dZ_p \; G_0(Z,Z_p) \; v(Z_p) \; G^{(p-1)}(Z_p,Z') \qquad (7.21)$$

The Montecarlo integration algorithm is devised by considering simultaneously the two iterative processes implicit in eq. (14) and in Born series eqs. (20) and (21). The skeleton of the algorithm is the following:

Step 1: Draw an initial state configuration $\{Z_1, Z_2, \dots Z_N\}$ and assume a value for λ. It is convenient to start with a small population and a value of λ greater than the value corresponding to the ground state, so as to have the population increasing in the correct way.

Step 2: Loop over all walkers of the previous configuration. For each point Z_i of the old set carry out the following steps

a: Draw a new Z from $G_0(Z,Z_i)$. The fisrt time we go through this step it will correspond to the first term of the Born series. Given that G_0 is normalized to 1 we get one new point at this step.

b: Consider now eq. (14). So, with probability λ accept this value Z for the new configuration. Moreover, accumulate (1-v(Z)) to have an estimate of the eigenvalue, eq. (15). Even if this Z is not accepted for the wave function of the next generation, go on to the next step. This Z will contribute to the total Green function.

c: Rename Z_i=Z.

d: Compute $v(Z_i)$. We are going to take into account the remaining terms of the Born series, eq. (21). Then sample ξ from U(0,1). If $v(Z_i)$ is less than ξ then go to step e (Born series finished), otherwise go back to step a to consider next term of Born series.

e: Test if i is the last term of the current configuration. If not, repeat from step a for the next walker.

Step 3: At this point one iteration of the integral equation (14) has been completed. We may compute the new estimate for λ and go back to Step 2 a sufficiently large number of times to reach the asymptotic stability and to accumulate statistics.

There still remains to describe how to sample from $G_0(Z,Z')$. The solution of eq. (13) is well known /45/,

$$G_0(Z,Z') = \frac{1}{(2\pi)^{3N/2}} K_{3N/2-1}\left(\left|\vec{Z}-\vec{Z}'\right|\right) / \left|\vec{Z}-\vec{Z}'\right|^{3N/2-1} \qquad (7.22)$$

corresponding to N particles (3N coordinates). K is the imaginary argument Bessel function. Sampling Z when Z' is known corresponds to determine a displacement $\vec{d}$ so that $\vec{Z}=\vec{d}+\vec{Z}'$. The modulus of this displacement follows the above probability distribution, and is isotropically oriented. Kalos /45/ describes the way of carrying out this sampling.

The algorithm presented in this section is quite general, but has some limitations. It can only be applied to bounded attractive potentials, and to a problem extended to all space. It can be directly applied to finite systems with simple interactions, as we will do in the next section.

7.4 A nucleon in a gaussian potential revisited

Consider again the simple example described in sections (4.2) and (6.4), this time with GFMC method as described above.

Carrying out the change of variables (7.9) we have

α = 1.4726 fm ,

U = -45 MeV

$v(Z) = \exp(-Z^2/9.4730)$

and the exact eigenvalue is

$\lambda_{exact} = 0.7597$

The free Green function may be sampled from the general form eq. (22) for N=1. However, it is convenient here to use the explicit form of the distribution (the Bessel K function 3/2 has a simple form), which in spherical coordinates is

$G_0(Z) = Z \exp(-Z) / 4\pi$.

The method for sampling points from this distribution was described in sect. (2.4), eq. (2.18). This equation gives the modulus of the displacement. The isotropic direction may be sampled in a very simple way. Draw a three-dimensional vector uniformly distributed in a cube of side 2 and centered at the origin. If this vector is outside the sphere of radius 1, then discard it and draw another vector. When the vector is inside the sphere, normalize it and keep it as the desired direction. This method is not very practical in many dimensions, because most of the vectors tried will be outside the sphere. This is due to the fact that the volume of the hypersphere in much smaller than the volume of the hypercube. Kalos /45/ describes a

```
C       GREEN FUNCTION MONTECARLO: a nucleon in Gaussian potential
        DIMENSION R(3,6000,2),X(3),XI(3)
        COMMON /RAND/ NRAN
        DATA ALAM,NUMBER,EIGM,EIGM2,IN,IO/1., 0, 0., 0., 2, 1/
        TYPE*,' INTEGER FOR RANDOM, INITIAL POP., CRITICAL POP AND NITER'
        ACCEPT*,NRAN,NPOP,ICRIT,NITER
        DO 10 I=1,NPOP            ! ======= STEP 1 =======
        DO 10 J=1,3               ! Starting configuration
10      R(J,I,1)=2*RAN(NRAN)-1.   ! generated at random
        DO 100 ITER=1,NITER       ! ===== LOOP OF ITERATIONS =====
        IN=3-IN                   ! Skip from OLD to
        IO=3-IO                   ! NEW configuration
        JPOP=0
        SW=0
        DO 1000 IPOP=1,NPOP       ! ===== LOOP OVER CONFIG =====
        DO 2000 I=1,3
2000    XI(I)=R(I,IPOP,IN)        ! Old walker
1030    CALL GO (XI,X,POT)        ! Sample from G free, == STEP 2a ==
        IF (ALAM.LT.RAN(NRAN)) GOTO 1020! == Step 2b ==, new point rejected
        JPOP=JPOP+1                         ! Accepted, count walkers of new set
        DO 2010 I=1,3
2010    R(I,JPOP,IO)=X(I)                   ! and save for next iteration
        SW=SW+1.-POT                        ! Sampling of 1-v(z)
1020    IF (POT.LT.RAN(NRAN)) GOTO 1000 ! == Step 2d ==
        DO 2020 I=1,3
2020    XI(I)=X(I)                          ! Rename new walker
        GOTO 1030                           ! Go to step == 2a ==
1000    CONTINUE                            ! Born series ended
        EIG=SW/JPOP                         ! Average of eigenvalue over config.
        PRINT*,ITER,JPOP,EIG                ! Monitor the evolution
        NPOP=JPOP
        ALAM=1.                             ! Keep population increasing
        IF(NPOP.LT.ICRIT) GOTO 100
        EIGM=EIGM+EIG             ! Accumulate statistics
        EIGM2=EIGM2+EIG*EIG
        NUMBER=NUMBER+1
        ALAM=EIG                  ! and stabilize population
100     CONTINUE
        EIGM=EIGM/NUMBER
        EIGM2=EIGM2/NUMBER
        PRINT*,' EIGENVALUE=',EIGM,' ERROR=',SQRT((EIGM2-EIGM*EIGM)/(NUMBER-1))
        END
        SUBROUTINE GO (XO, X1,POT)          !Sample form free GREEN function
        DIMENSION XO(3),X1(3)
        COMMON /RAND/NRAN
        R=-ALOG(RAN(NRAN)*RAN(NRAN))        ! Modulus of displacement
100     DO 200 I=1,3                        ! Direction: sample in a cube
200     X1(I)=2.*RAN(NRAN)-1.               ! centered at (0,0,0), with side = 2
        S=X1(1)**2+X1(2)**2+X1(3)**2        ! Compute norm of vector
        IF (S.GT.1.) GOTO 100               ! Outside sphere of radius one. Repeat
        DO 300 I=1,3                        ! Inside, compute coordinates of
300     X1(I)=XO(I)+X1(I)*R/SQRT(S)         ! generated position
        POT=EXP(-(X1(1)**2+X1(2)**2+X1(3)**2)/9.4973)   ! Compute potential
        RETURN
        END
```

Figure 7.1. GFMC algorithm for a nucleon in a gaussian potential.

method appropriate to many dimensions.

The algorithm previously described has been coded in Fortran, and the program is shown in Figure (7.1). It is a literal translation of the explained method and it does not require further comments.

Figure (7.2) plots the results obtained with this program for the case of a nucleon in the gaussian potential. In general lines the behaviour of the iterations in GFMC is quite similar to the behaviour of the iterations in DFMC. When comparing this figure with Fig. (6.2) note that the scales are different. There is however an important difference, namely the absence of correlations in GFMC between the succesive generations once the system has stabilized. Remember that correlations in DFMC were a direct consequence of the smallness of the time step.

The expectation value of the energy after 200 iterations is $\lambda=0.7594$ and the statistical error is 0.001, which corresponds to $E=-10.83\pm 0.04$. This is a very good result.

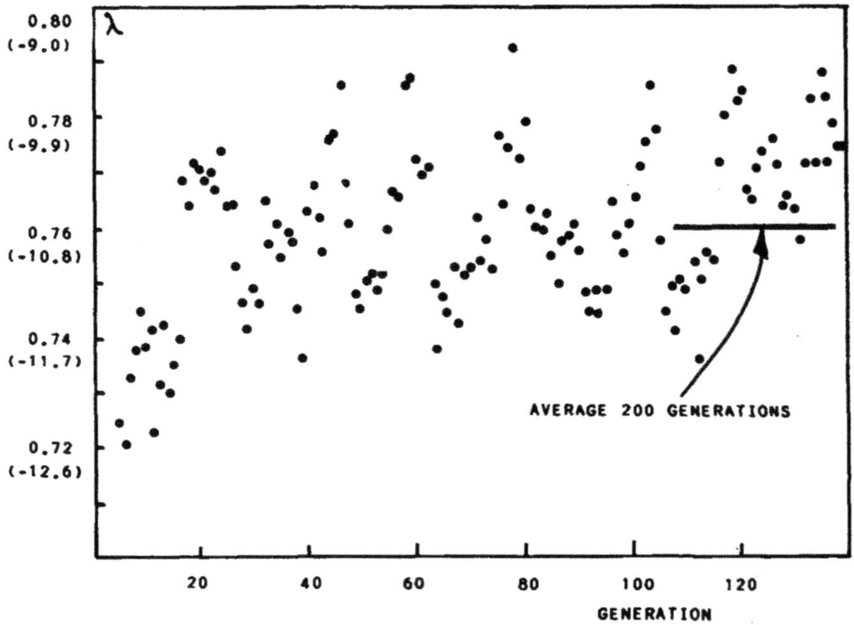

Figure 7.2. The eigenvalue versus the generation number for gaussian potential in GFMC. The values in parenthesis in the vertical scale correspond to the energy in MeV. The horizontal line is the average of 200 generations after the system has stabilized.

7.5 Some complications: unbounded interactions

The previous simple case has shown many of the features of GFMC. However, this method cannot be used with unbounded interactions, so frequent in many body problems.

Kalos, Levesque and Verlet /47/ have developed a method to solve this problem. The basic idea is to establish a Born series expansion for the full Green function in terms of a free Green function defined in a smaller region which does not include the singularities of the potential. Let's describe this method.

Let $G(R,R_0)$ be the full Green function defined in a domain D. Specifically we want to draw random points labelled R from a point labelled R_0. The domain D is the whole physical domain, i.e., all space in finite systems or the fundamental cell with periodic extension in extended systems. G satisfies the equation

$$\begin{cases} \left[-\nabla_1^2 + V(R_1)\right] G(R,R_1) = \delta(R-R_1) \\ \text{in domain D,} \end{cases} \tag{7.23}$$

and is symmetric in R, R_1.

The free or partial Green function is defined in the following form. Let $D(R_0)$ be a subdomain which includes the point R_0, and let U be an upper bound to V(R) in that domain. Note that V(R) is the shifted, always positive potential. The only requirement to define the domain $D(R_0)$ is that it does not include the singularities of the potential. Then define a G_U by the equations

$$\begin{cases} \left[-\nabla_1^2 + U\right] G_U(R_1,R_0) = \delta(R_1-R_0) \\ G_U(R_1,R_0) \text{ is null if } R_1 \text{ is outside } D(R_0) \end{cases} \tag{7.24}$$

Now G_U is no longer symmetric because the dependence of the domain on the point R_0.

One can relate G and G_U by using a familiar technique: multiply on the left eq. (23) times $G_U(R_1,R_0)$, and eq. (24) times $G(R_1,R)$, substract and integrate over R_1 in domain $D(R_0)$. It turns out

$$\begin{aligned} G(R,R_0) = G_U(R,R_0) &+ \int_{\text{Surface}} \left[-\nabla_{n_1} G_U(R_1,R_0)\right] G(R,R_1)\, dR_1 \\ &+ \int_{D(R_0)} \left[U - V(R_1)\right] G_U(R_1,R_0)\, G(R,R_1)\, dR_1 \end{aligned} \tag{7.25}$$

In this equation surface means the hypersurface enclosing domain $D(R_0)$, and the gradient operator is the outwards directed normal derivative on this surface.

Equation (25) is again an integral equation similar to eq. (19), even if quite more complicated. It may be solved by a Born series expansion, as before. This Born series is convergent, because the integral of the Kernel is smaller than one. To ascertain this statement integrate eq. (24) in $D(R_0)$ with respect to R_1 to get

$$\int_{Surf.} \left[- \nabla_{n_1} G_U(R_1,R_0)\right] dR_1 + \int_{D(R_0)} U \; G_U(R_1,R_0) \; dR_1 = 1 \qquad (7.26)$$

This is not the integral of the kernel, because it has the quantity U in the second integral instead of $U-V(R_1)$. However $U-V(R_1)$ is smaller than or equal to U in domain $D(R_0)$, so the integral of the kernel is smaller than 1.

Montecarlo integration of eq. (25) is better described if Born series is explicitly written up. Formally eq. (25) is

$$G(R,R_0)=G_U(R,R_0) + \int G(R,R_1) \; K(R_1,R_0) \; dR_1$$

where K contains both surface and volume terms. Iterating we have

$$\begin{aligned} G(R,R_0) = {} & G_U(R,R_0) \\ & + \int G_U(R,R_1) \; K(R_1,R_0) \; dR_1 \\ & + \int G_U(R,R_1) \; K(R_1,R_2) \; K(R_2,R_0) \; dR_2 dR_1 \\ & + \int G_U(R,R_1) \; K(R_1,R_2) \; K(R_2,R_3) \; K(R_3,R_0) \; dR_3 dR_2 dR_1 \; \ldots \end{aligned}$$

where

$$K(R,R') = - \nabla_n \; G_U(R,R') \; + U \left\{1 - \frac{V(R)}{U}\right\} G_U(R,R')$$

Sampling R given R_0 has two steps. First, sample R from $G_U(R,R_0)$, the first term of Born series. There may result nothing from this sampling, because the volume integral of G_U is smaller than 1, eq. (26). The second step is the sampling from the rest of the series. This may be carried out in a chained way:

1. Put $R_1 = R_0$ (to consider the first term)

2. Put $R_2 = R_1$ (to consider the remaining terms)

3. Define a domain $D(R_2)$, and compute the appropriate constant U for that domain.

4. Sample a new point R_1 form $K(R_1,R_2)$. There are two excluyent cases (see eq. 26):

   a. Sample from the surface part of K, with a probability

   $$\int \left[- \nabla_n \; G_U(R_1,R_2)\right] dR_1 .$$

   Call this point r_1 and go back to step 2, or

   b. Sample a point in the interior of domain $D(R_2)$ and call it R_1. Now consider the part $(1-V(R_1)/U)$ of the kernel, i.e., continue with a probability $(1-V(R_1)/U)$ going back to step 2. Sampling from Born series will only stop when an interior point is selected and afterwards this point is not accepted.

There are still some formidable pending questions to be answered to make this algorithm practical:

1. How to choose the domain $D(R_1)$
2. How to sample from the volume part of the kernel, G_U, and
3. How to sample form the surface part of the kernel.

I will skip these questions. The reader should consult the works of Kalos, Levesque and Verlet /47/ and Zabolitzky /54/ for a deeper discussion.

7.6 A particle in a box

With the purpose of illustrating some of the ideas introduced in section (7.5) it is convenient to consider the simple case of the problem of a particle in a one dimensional box. The particle is confined to move freely between -1 and +1, and the total Green function is given by the differential equation

$$-\frac{d^2}{dx^2}\, G(x,x') = \delta(x-x') \qquad (7.27)$$

with boundary conditions $G(1,x')=G(-1,x')=0$.

This Green function is well known. It has the form of a triangle with a discontinuity at $x=x'$, i.e.

$$G(x,x') = \begin{cases} (1-x')(1+x)/2 & x < x' \\ (1+x')(1-x)/2 & x > x' \end{cases} \qquad (7.28)$$

Define now a G_U Green function and its corresponding domain $D(x')$, see Figure 7.3. There is no potential and the criterium to define $D(x')$ is the size of the box: $D(x')$ extends symmetrically from x' until we arrive to one of the boundaries +1 or -1. G_U will satisfy the same differential equation than G, eq. (27) with the boundary conditions:

$$\left.\begin{aligned} G_U(1,x') &= 0 \\ G_U(2x'-1,x') &= 0 \end{aligned}\right\} \quad \text{if } x' > 0 \qquad (7.29)$$

or

$$\left.\begin{aligned} G_U(-1,x') &= 0 \\ G_U(1+2x',x') &= 0 \end{aligned}\right\} \quad \text{if } x' < 0 \qquad (7.30)$$

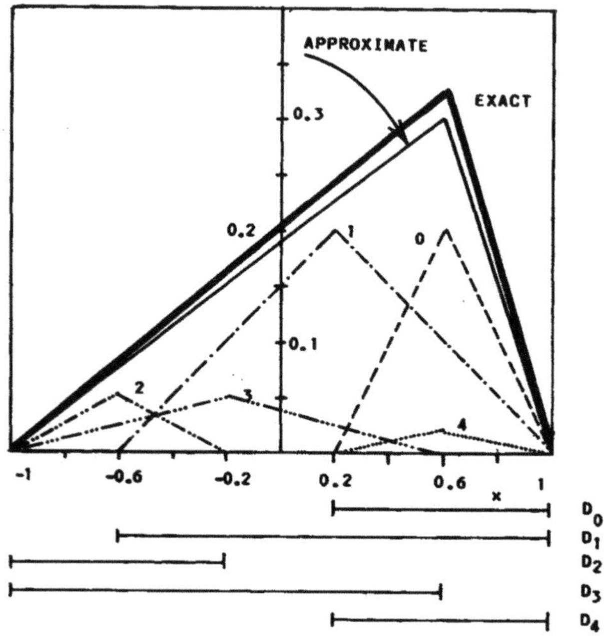

Figure 7.3. Born series for the particle in a box problem. The segments at the bottom of the figure represent the various subdomains. See the text for a full description.

The explicit form of G_U is

$$G_U(x,x') \quad \begin{cases} (1+x-2x')/2 & x < x' \\ (1-x)/2 & x > x' \end{cases} \quad \text{if } x' > 0$$

and (7.31)

$$G_U(x,x') \quad \begin{cases} (1+x)/2 & x < x' \\ (1+2x'-x)/2 & x > x' \end{cases} \quad \text{if } x' < 0$$

Next step is to obtain the integral equation relating G and G_U. Given that there is not a potential, there is not a volume term. Moreover the surface part degenerates to the value at the limits, i.e.,

$$G(x_1,x_0) = G_u(x_1,x_0) + \left\{- \frac{dG_U(x,x_0)}{dx} \quad G(x_1,x)\right\}_{x=x\text{sup}} - \left\{- \frac{dG_U(x,x_0)}{dx} \quad G(x_1,x)\right\}_{x=x\text{inf}} \quad (7.32)$$

```
C       SAMPLING OF GREEN FUNCTION FOR A PARTICLE IN A BOX (-1,1)
        DIMENSION JDIST(20)                ! Histogram
        TYPE*,' NUMBER OF POINTS'          ! Points attempted
        ACCEPT*,NP
        X0=0.6                             ! G(x,0.6)
        DO 1000 ISAMPL=1,NP
        X1=X0
        F=1.                               ! Scale 1/2**n of each iter
100     CALL SAMPLE(X,X1,F,IFLAG)          ! Sample from scaled GU
        IF (IFLAG.EQ.0) GOTO 200           ! No random generated, so skip
        NFP=NFP+1                          ! otherwise, count it and
        I=INT(10.*(1.+X))+1                ! update histogram
        JDIST(I)=JDIST(I)+1                ! size of bin=0.1
200     IF(X1.GT.0.)  X2=2*X1-1            ! Determine next term of
        IF(X1.LE.0.)  X2=1+2*X1            ! Born series
        X1=X2
        F=F/2.                             ! and update coefficient
C       STATEMENT TO END THE ITERATION WHEN PROBABILITY IS SMALL
        IF(NP*F*(1.-ABS(X1))**2/2. .LT. 1) GOTO 1000
        GOTO 100                           ! Sample from next term
1000    CONTINUE
        PRINT*,' FINAL POINTS=', NFP       ! Printout of results
        DO 1010 I=1,20
1010    PRINT*,JDIST(I)
        END
        SUBROUTINE SAMPLE(X,X1,F,IFLAG) ! Sampling from F*GU(x,x1)
        IFLAG=0                            ! When no random number generated
        DATA NRAN/131451/                  ! Start U(0,1) series
        IF (X1 .LE. 0.0) THEN
                XMIN=-1
                XMAX=1+X1+X1
        ELSE                               ! Determine shape of GU
                XMIN=-1+X1+X1
                XMAX=1
        END IF
100     X=2.*RAN(I1)-1.                    ! Von Neumann ACCEPTANCE-REJECTION
        IF(X.LT.XMIN.OR.X.GT.XMAX)RETURN! Outside range: no random
        IF (X1.LE.0) THEN                  ! Negative, at
          IF (X.GT.X1)  THEN
                        G=0.5*(1+2*X1-X)          ! right
                        ELSE                      ! or
                        G=0.5*(1+X)               ! left
          END IF
        ELSE                               ! Positive, at
          IF (X.GT.X1)  THEN
                        G=0.5*(1-X)               ! right
                        ELSE                      ! or
                        G=0.5*(1+X-2*X1)          ! left
          END IF
        END IF
        IF (RAN(I1)/2.GT.G*F) RETURN       ! Outside triangle, reject it.
        IFLAG=1                            ! Inside, accept it.
        RETURN
        END
```

Figure 7.4. Born series for a particle in a box problem.

when x_1 is negative. Moreover, the exact Green function vanishes at +1 and at -1, so that only one of the two integrated terms will survive. Also, the values of the derivatives of G_U may be computed from its explicit form, eq. (31), and the following 'integral' equation is obtained

$$G(x_1,x_0)=G_U(x_1,x_0) + G(x_1,2x_0-1)/2\ \theta(x_0) + G(x_1,1+2x_0)/2\ \theta(-x_0) \quad (7.33)$$

A Born series may be built up by iterating eq. (33). This is described in Figure (7.3) which corresponds to $x_0=0.6$. The first domain extends form 0.2 to 1, and the corresponding contribution is the triangle labelled 0. The next iteration is centered at 0.2 and the new domain extends from -0.6 to 1, the corresponding contribution being the triangle labelled 1 (note the factor 1/2 in eq. (33)). Next domain extends from -1 to -0.2,... and so on. This figure includes up to the fifth contribution (which is labelled 4) and the triangle labelled APPROXIMATE is the sum of all that five triangles. The resulting Green function is very close to the exact Green function and, if the series were continued, the exact result will be obtained.

To sample from this series one has to consider all terms of the expansion. Each term is a triangle like eq. (31) times a factor $1/2^p$ corresponding to the order of the iteration. The area of each triangle is $(1-|x|)^2/2$, where x is the center of the subdomain, and for that reason we will only obtain a random number from each individual term with probability $(1-|x|)^2/2^{p+1}$. The area of the total Green function is $(1-x^2)/2$, so that from a run with N trials only $N(1-x^2)/2$ random numbers will result.

Just to complete the present example I include in Figure (7.4) a Fortran program corresponding to this kind of sampling, and the result for $x_0=0.6$ with 10000 trials is represented in Figure (7.5). A last comment regards to the fact that in this example there is no way to end up, because we always move to a point on the boundary. The Fortran code of figure 4 limits the sampling (statement two lines above statement 1000)

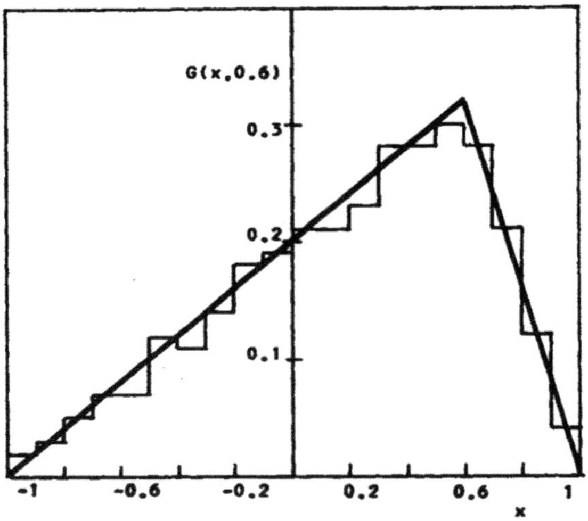

Figure 7.5. Result of Fortran program (7.4) for G(x,0.6)

7.7 Complements: Importance sampling and fermion problems

As in the DFMC case, importance sampling is a very useful tool in GFMC because

i. It is a variance going to zero mechanism, and

ii. It improves the convergence of Born series, by introducing a drift towards the really important regions.

The basic idea is to use the quantity $_T(R)$ (R) as the funtcion to be sampled through the integral equation, like in DFMC case. Actually, importance sampling was firstly developed in GFMC. This new quantity satisfies the integral equation

$$\Psi_T(R)\,\Psi(R) = E \int K(R,R')\ \Psi_T(R')\ \Psi(R')\ dR' \qquad (7.34)$$

where K is a modified kernel

$$K(R,R') = \Psi_T(R)\ G(R,R')\ /\ \Psi_T(R') \qquad (7.35)$$

One can define a free kernel K_U from the free Green function using the same transformation (35), and obtain an integral equation relating the complete and the free kernels,

$$\begin{aligned}\Psi_T(R)\ G(R,R_0)/\Psi_T(R_0) &= \Psi_T(r)\ G_U(R,R_0)/\Psi_T(R_0) \qquad (7.36)\\ &+\int_{Surf}\left[-\Psi_T(R_1)\ \nabla_{n_1} G_U(R_1,R_0)/\Psi_T(R_0)\right]\left[\Psi_T(R)\ G(R,R_1)/\Psi_T(R_1)\right] dR_1\\ &+\int_{Vol}\{U - V(R_1)\}\left[\Psi_T(R_1)\ G_U(R_1,R_0)/\Psi_T(R_0)\right]\left[\Psi_T(R)\ G(R,R_1)/\Psi_T(R_1)\right] dR_1\end{aligned}$$

The stochastic solution of eqs. (35) and (36) follows steps very similar to the steps followed in the non importance sampling method. The more relevant modification is the presence of $\Psi_T(R_1)/\Psi_T(R_0)$ each time a G_U function is involved. The role of this correction is easily understood when R_1 and R_0 are rather close points. A Taylor expansion gives

$$\Psi_T(R_1)/\Psi_T(R_0) = 1 + (\vec{R}_1-\vec{R}_0)\ \vec{\nabla} \log \Psi_T(R_0) \qquad (7.37)$$

which shows the drift effect above mentioned.

With this new function one has the following estimator for the energy

$$E = \frac{\int \Psi(R)\ \Psi_T(R)\ E_L(R)\ dR}{\int \Psi(R)\ \Psi_T(R)\ dR} \qquad (7.38)$$

where E_L is the local energy. As we already know, when the trial function is the exact ground state, then the local energy is a constant and the variance of eq. (38) is zero. In general, we expect to decrease the variance of our estimator by using a reasonable trial function.

As in the case of DFMC, importance sampling helps in dealing with systems of fermions. The troubles related to the antisymmetric nature of the many fermion problem wave function were already briefly described in section (6.5). As in DFMC, the fixed node approximation gives an upper bound to the fermion ground state energy when the mixed estimator eq.(38) is used. The goodness of this upper bound is related to the goodness of the nodal surfaces of the trial function, and not to the function itself.

Fixed node approximation may be followed by nodal relaxation, as in DFMC. This will increase the boson population at each iteration, at the cost of decreasing the fermion population. The consequence is the growing of the variance, so that after some iterations there will remain only fermion noise. Because of that node relaxation can only be carried out a limited number of steps. The energy estimate obtained in this form is called a transient estimate, the reason being obvious.

Two ideas have been put forward to improve the quality of the results for fermions. One of the methods /54,55/ is to keep the total population increasing (using a high value of E in eq. (34)) after relaxing the nodes so as to have a steady fermion population. In practice this means an exponential increase of the population, in other words, an exponential increase of computer time.

Another promising method /53,56,57/ is to include the fermionic character into the Green function itself, instead of being in the wave function. Points are evolved by pairs, one from a positive region and the other from a negative region, and their corresponding Green functions are combined to have a small symmetric component, and an important antisymmetric component. This is a way of compensating the boson background present in the mixed estimator (38). That compensation is exact from the analytic point of view, but only approximate in practice because of the statistical fluctuations. Up to my knowledge, this method has been only used in simple (but not trivial) few body problems /57/.

The description of fermion problems of this section is too short. I suggest the reader to consult Ref. /57/ for a clear and complete analysis of this problem.

7.8 Applications of GFMC: helium systems

Most of GFMC calculations have been devoted to systems of helium atoms. In section (4.4) we have mentioned some results concerning helium-4 with Lennard-Jones potential. The equilibrium energy was found to be a little smaller (absolute energy) than the experimental value, and the discrepancy was associated to the inadequacy of the interaction. Using Aziz /18/ potential, there results a very good value for the binding energy per atom, E(GFMC)=-7.12 0.02 K /17/, to be compared with the experimental value E=-7.14 K.

This result is in a very good agreement with HNC variational calculations /58/ with a wave function including three body correlations, which is -6.96 K. This HNC calculation includes elementary diagrams by means of the so called scaling approximation. A general conclusion of these works is that the description of liquid helium-4 with two- and three- body correlations is appropriate, and that Aziz interaction is a very good interaction to describe liquid helium.

A step forward in the study of helium-4 was the consideration of droplets, i.e., bound clusters of a small number of atoms. A complete an costly GFMC calculation /50/ has been carried out for droplets of varying size, from 3 to 112 atoms per drop. This is an impressive work, both because of the amount of computing time as well as for the beauty of the results. The energy per particle has been fitted to a liquid drop like formula,

$$E(N)/N = E_{vol} + E_{surf}\, N^{-1/3} + E_{curv}\, N^{-2/3}$$

with a volume term (E_{vol}=-7.02 K), a surface term (E_{surf}=18.8 K) and a curvature term (E_{curv}=-11.2 K). The value of the volume term is in a very good agreement with the GFMC calculation mentioned above for the extended system. An important consequence of this work is the confidence it gives to the extrapolation from a finite system to an extended system. Hopefully this extrapolation is also meaningful in nucleon systems.

Another reassuring result is the behaviour of the drop radius with the number of atoms, following the $r=r_0\, N^{1/3}$ low for clusters with 20 or more atoms. Figure (7.6), adapted from Ref. /50/ shows clearly this behaviour.

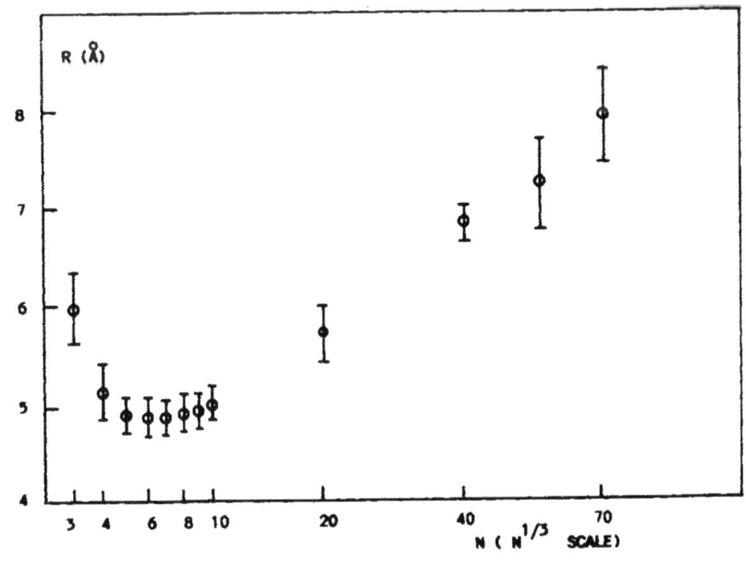

Figure 7.6. Radius of helium-4 droplets versus number of atoms.

Another system which has received very much attention is helium-3, a fermionic system. Variational calculations with Aziz potential have shown the importance of three body correlations, as well as the importance of backflow (formally a momentum dependence correlation). The best variational estimate with VMC method /60/ is $E=-1.91 \pm 0.03$ K, quite low compared with the experimental value of 2.47 ± 0.01 K (I have a hand written number, by Panoff, which is slightly higher, -2.1 K, probably computed with an optimal correlation). This result is also in disagreement with th FHNC+Scaling calculation of Manousakis et al. /61/, which gives -2.36 K. A preliminary, and probably not fully converged, GFMC calculation with fixed node approximation /59/ lowered the VMC result to -2.2 K. This calculation was made with 38 particles only.

The discrepancy between GFMC and FHNC form one side, and with experiment from the other side, has estimulated the improvement of GFMC method. Very recently /51/ some result were presented with a longer GFMC run. The fixed node GFMC obtained was -2.27 ± 0.01 K and -2.32 ± 0.01 K, corresponding to a wave function with the Slater determinant nodes and with the backflow corrected nodes, respectively. A transient estimator lowered the energy to the exciting result of -2.44 ± 0.04 K, but probably this last calculation should be improved.

Attemps have been made to study drops of helium-3 atoms. However, it seems that many atoms are required to form a stable cluster /62/, and there are not still definite results on this system.

8. CLOSING REMARKS

We have arrived to the end of this long trip through Montecarlo forest. Even if the presentation has been somehow superficial in some aspects, I hope that a general view of the power and beauty of Montecarlo method in quantum problems had been reached.

Along the lectures, the emphasis has been mainly directed towards a single quantity, namely the binding energy of the system. Actually with Montecarlo method we obtain much more things than just the energy, for example the wave function. Even if the ground state wave function is obtained in a quite special form, a collection of random points, this information is enough to compute many other properties of the system, particularly the two body distribution function or its Fourier transform, the liquid structure function.

The title of the lectures made a reference to Montecarlo method as a computational technique. Nevertheless, I beleive that one should change this word and write instead theory. In its present form GFMC and DFMC have a solid structure and a form of looking to the problems at hand which convert them in something more than a powerful computational technique.

There are however some disturbing questions behind Montecarlo Many Body theory, and also in other branches of physics. Should one abandone all other theories?. Is Many Body physics reserved to institutions or researchers who have access to really large scale computers?. Even more, some dedicated computers have been constructed (or are being planned) specifically for Montecarlo work, so, are we inventing an Experimental Theoretical Physics?. I beleive I will not be able to answer satisfactorily all these questions. In any case, I feel that feedback among various theories is the most powerful way of progress in physics.

There are still many questions open in Montecarlo study of Many Body systems, and in general in any Many Body theory. Particularly, future work should be focussed in discrete (spin/isospin) degrees of freedom and in the excitation spectrum of extended or finite systems. Probably Montecarlo method has not too much room in these questions, and we have to rely in Correlated Basis Functions or Copuled Cluster theories for the future developments.

ACKNOWLEDGEMENTS

Many people has helped me during the preparation of this work. I would like particularly acknowledge the patience and help of Valdir Aguilera, Ray Bishop, Enrique Buendia, Manuel de Llano, Stefano Fantoni, Bob Panof and Sergio Rosati. The kind hospitality of Leo Violini and the Asociacion pro Centro Internacional de Física made the stay at Bogotá very fruitful and pleasant.

Finantial support from CAICYT (Madrid), under contract 1234/84 is fully acknowledged.

REFERENCES

1. J.M.Hammersley and D.C.Handscomb, "Monte Carlo Methods", Methuen's Monographs (Norwich, U.K., 1967)
2. M.Zelen and N.C.Severo, "Probability Functions" in "Handbook of Mathematical Functions", M.A.Abramowitz and I.A.Stegun Eds. (Dover, New York 1968)
3. G.Marsaglia, "Random Numbers Generation" in "Encyclopaedia of Computer Science", A.Ralston Ed. (Petrocelli/Carter, 1967)

4. K. Binder and D. Stauffer, in "Applications of the Montecarlo Method in Statistical Physics", K.Binder Ed. (Springer, Berlin 1984)
5. M.H.Kalos, Phys. Rev. 128, 1791 (1962)
6. J. von Neumann, Natl. Bur. Stand. Appl. Math. Series 12, 36 (1951)
7. N.Metropolis, A.W.Rosenbluth, M.N.Rosenbluth,A.H.Teller and E.Teller, Jour. Chem. Phys. 21, 1087 (1953)
8. M. Gardner, "Circo Matemático" (Alianza Ed. Madrid 1983)
9. H. Kahn, Rand Corporation Report R-163 (1949)
R.E. Cutkosky, Jour. Res. Natl. Bur. Stand. 47, 113 (1951)
10. R. Jastrow, Phys. Rev. 98, 1479 (1955)
11. W.W.Wood and F.R.Parker, Jour. Chem. Phys. 27, 720 (1957)
12. A. Munster, "Statistical Thermodynamics" (Academic Press, New York 1969)
13. W.L. McMillan, Phys. Rev. A138, 442 (1965)
14. D. Schiff and L. Verlet, Phys. Rev. 160, 208 (1967)
15. D.M. Ceperley and M.H. Kalos, in "Montecarlo Mehtods in Statistical Physics", K.Binder Ed. (Springer, Berlin 1979)
16. K. Schmidt, M.H. Kalos, M.A. Lee and G.V. Chester, Phys. Rev. Lett. 45, 573 (1980)
17. M.H. Kalos, M.A. Lee, P.A. Whitlock and G.V. Chester, Phys. Rev. B24, 115 (1981)
18. R.A. Aziz, V.P.S. Nain, J.S. Carley, W.L. Taylor and G.T. McConville, Jour. Chem. Phys. 70, 4330 (1979)
19. D. Ceperley, G.V. Chester and M.H. Kalos, Phys. Rev. B16, 3081 (1976)
20. J. Lomnitz-Adler, V.R. Pandharipande and R.A. Smith, Nucl. Phys. A361, 399 (1981)
J. Carlson, R.B. Wiringa and V.R. Pandharipande, Nucl. Phys. A371, 301 (1981)
R.B. Wiringa, J.L. Friar, B.F. Gibson, G.L. Payne and C.R. Chen, Phys. Lett. 143B, 273 (1984)
21. J. Carlson ans M.H. Kalos, Phys. Rev. C32, 2105 (1985)
22. N. Metropolis ans S. Ulam, J. A. Stat. Assn. 44, 335 (1949)
23. M.D. Donsker and M. Kac, J. Res. Natl. Bur. Stand. 44, 551 (1950)
24. J.B. Anderson, J. Chem. Phys. 63, 1499 (1975)
25. J.B. Anderson, J. Chem. Phys. 65, 4121 (1976)
26. J.B. Anderson, J. Chem. Phys. 73, 3879 (1980)
27. D.M. Ceperley and B.J. Alder, Phys. Rev. Lett. 45, 566 (1980)
28. D.M. Ceperley, in "Recent Progress in Many Body Theories", J.G. Zabolitzky, M. de Llano, M.Fortes and J.W. Clark Eds. (Springer, Lect.Not.Phys. 142, Berlin 1981)
29. F. Mentch and J.B. Anderson, J. Chem. Phys. 74, 6307 (1981)
30. P.J. Reynolds, D.M. Ceperley, B.J. Alder and W.A. Lester Jr., J. Chem. Phys. 77, 5593 (1982)
31. J.W. Moskowitz, K.E. Schmidt, M.A. Lee and M.H. Kalos, J. Chem. Phys. 77, 349 (1982)
32. J.B. Anderson, J. Chem. Phys. 82, 2662 (1985)
33. J. Vrbik and S.M. Rohtstein, J. Comput. Phys. 63, 130 (1986)
34. F. L. Pilar, "Elementary Quantum Chemistry" (MacGraw Hill, New York 1968)
35. H. Partridge and S.R. Laughoff, J. Chem. Phys. 74, 2361 (1981)
36. D. M. Bishop and L. M. Cheung, J. Chem. Phys. 78, 1369 (1983)
37. B.Jeziorsky, H.J. Monkhorst, K. Szalewicz and J.G. Zabolitzky, J. Chem. Phys. 81, 368 (1984)
38. K. Szalewicz, B. Jeziorsky, H.J. Monkhorst and J.G. Zabolitzky,

J. Chem. Phys. 83, 5543 (1983)
39. E.P. Wigner, Phys. Rev. 46, 1002 (1934)
40. J.G. Zabolitzky, Phys. Rev. B22, 2353 (1980)
41. R.F. Bishop and K.H. Lurmann, Phys. Rev. B17, 3757 (1978)
42. K. Emrich and J.G. Zabolitzky, Phys. Rev. B30, 2040 (1984)
43. L. Lantto and P.J. Siemens, Nucl. Phys. A317, 55 (1979)
44. D.M. Ceperley, Phys. Rev. B18, 3126 (1978)
45. M.H. Kalos, Phys. Rev. 128, 1711 (1962)
46. M.H. Kalos, Ref. /5/. See also
M.H. Kalos, Jour. Comput. Phys. 2, 257 (1967).
47. M.H. Kalos, D. Levesque and L. Verlet, Phys. Rev. A9, 2178 (1974)
48. D. Ceperley, G.V. Chester and M.H. Kalos, Phys. Rev. B16, 3081 (1977)
49. D.M. Arnow, M.H. Kalos, M.A. Lee and K.E. Schmidt, Jour. Chem. Phys. 77, 5562 (1982)
50. V.R. Pandharipande, J.G. Zabolitzky, S.C. Pieper, R.B. Wiringa and U. Helmbrecht, Phys. Rev. Lett. 50, 1676 (1983)
51. R.B. Panoff, work presented at the 10th International Workshop on Condensed Matter Theories (Argonne, 1986), Proc. to be published by Plenum Press.
52. D. Ceperley and M.H. Kalos, in "Montecarlo Methods in Statistical Physics", K. Binder Ed. (Springer, Berlin 1979)
53. K.E. Schmidt and M.H. Kalos, in "Applications of Montecarlo Method in Statistical Physics", K.Binder Ed. (Springer, Berlin 1984)
54. J.G. Zabolitzky, Prog. Part. and Nucl. Phys. 16, 103 (1986)
55. U. Helmbrecht and J.G. Zabolitzky, Nucl. Phys. A442, 109 (1985)
56. M.H. Kalos, Lect. Notes in Phys. 142, 252 (Springer, Berlin 1981)
57. See ref. 49
58. Q.N. Usmani, S. Fantoni and V.R. Pandharipande, Phys. Rev. B26, 6123 (1982)
59. M.A. Lee, K.E. Schmidt, M.H. Kalos and G.V. Chester, Phys. Rev. Lett. 46, 728 (1981)
60. K.E.Schmidt, M.A. Lee, M.H. Kalos and G.V. Chester, Phys. Rev. Lett. 47, 807 (1981)
61. E. Manousakis, S. Fantoni, V.R. Pandharipande and Q.N. Usmani, Phys. Rev. B28, 3770 (1983)
62. See Ref. /55/

MICROSCOPIC VARIATIONAL THEORY FOR BOSE FLUIDS AT ZERO TEMPERATURE

by
S.Rosati[(*)]
Physics Department, University of Pisa
Istituto Nazionale di Fisica Nucleare - Sezione di Pisa (Italy)
M.Viviani[(**)]
Scuola Normale Superiore, Pisa
Istituto Nazionale di Fisica Nucleare - Sezione di Pisa (Italy)

[(*)]Dipartimento di Fisica and INFN, Sezione di Pisa
Piazza Torricelli 2, I-56100 Pisa (Italia)
[(**)]Scuola Normale Superiore and INFN, Sezione di Pisa
Piazza dei Cavalieri, I-56100 Pisa (Italia)

MICROSCOPIC VARIATIONAL THEORY FOR BOSE FLUIDS AT ZERO TEMPERATURE

by S. Rosati and M. Viviani

1. INTRODUCTION.

Only very few systems remain in a gaseous or liquid state as the temperature approaches absolute zero. In correspondence, the particle wave-length associated to the termal motion is of the order of magnitude of the average interparticle distance and the system reveals macroscopic quantum effects. Such "quantum fluids" represent an important test for application and comparison of different many-body theories. Bose and Fermi systems, as well as Bose-Fermi mixtures, are of great experimental and theoretical interest. In the present paper we will focus only on Bose fluids, in particular on liquid ^{4}He. For that system, the interaction energy per particle is in the range $1 \div 10$ K (1K $= 1/11,606$ eV). This energy is extremely smaller than the electronic excitation energy and the atoms of the system can thus be considered as structureless particles. The smallness of the interparticle energy, and the value of the mass which gives a zero point kinetic energy largely dominating the potential energy contribution, are responsible of the peculiar properties of such quantum system, first of all of the superfluidity decovered in 1937 by KAPITZA[(1)]. At present, a wealth of experimental data on the properties of liquid ^{4}He at different values of the temperature T and the external pressure P are available; in most cases extrapolations at $T = 0$ result accurate. In particular at the equilibrium, namely for $P = 0$, one gets[(2)]:

$$\rho_0 = 0.02185\,\text{Å}^{-3} \quad , \qquad E_0/N = -7.14\,\text{K/particle} \quad . \tag{1.1}$$

As a consequence of the Pauli principle on the electron states, two helium atoms strongly repel when come close to each other; for this reason the traditional perturbative approaches cannot be applied to the study of a quantum fluid. Let us consider a large number N of particles in a volume Ω described by the following hamiltonian:

$$H = -\frac{\hbar^2}{2m}\sum_{i=1}^{N}\nabla_i^2 + \sum_{i<j=1}^{N} v_{ij} + \sum_{i<j<k=1}^{N} v_{ijk} + \dots \quad . \tag{1.2}$$

Since we are here interested only in the study of the bulk properties of the system, we will finally take the thermodynamic limit, with both N and Ω going to infinite mantaining a costant value for the density $\rho = N/\Omega$. The knowledge of the eigenvalues and eigenstates of H, allows us to calculate the important macroscopic properties of the system, using the methods of the statistical mechanics. Such a program can be worked out in the low temperature limit when the state of the system is accurately described in terms of the ground-state and elementary weakly interacting excitations.

The variational approach has been widely applied to the study of the ground-state of a quantum fluid. The mean value of the hamiltonian, with respect to a variational wave function, has a minimum around the ground-state wave function (Ritz's principle), then we are faced with the problem of constructing very flexible trial wave function Ψ. With reference to the bulk of a Bose fluid, the function Ψ must be symmetric for the exchange of particles and translationally invariant; moreover it satisfies the condition

$$\Psi \to 0 \ \text{ if } \ r_{ij} \to 0 \quad , \tag{1.3}$$

since the repulsion becomes infinite when the distance between two particles approaches to zero. When the separation between two particle increases, their interaction rapidly decreases. As a consequence, for configuration where the particle $1,\ldots,n$ are sufficiently far from the remaining particles, Ψ must satisfy the cluster property:

$$\Psi(1,\ldots,N) \rightarrow \Psi^{(n)}(1,\ldots,n)\Psi^{(N-n)}(n+1,\ldots,N) \quad . \tag{1.4}$$

It is evident that a symmetrized product of single-particle wave functions violates both conditions (1.3), (1.4); correlated wave functions must therefore be constructed for a succesfull variational theory. This problem will be handled in the chapters 2 – 5.

Useful information on the structure of the system are contained in the radial distribution function $g(r)$, a quantity proportional to the probability of finding two particles at distance r; the relation between $g(r)$ and Ψ is given by:

$$g(r_{12}) = \frac{N(N-1)}{\rho^2} \frac{\int d^3r_3 \ldots d^3r_N \, |\Psi(\vec{r}_1 \ldots \vec{r}_N)|^2}{\int d^3r_1 \ldots d^3r_N \, |\Psi(\vec{r}_1 \ldots \vec{r}_N)|^2} \quad . \tag{1.5}$$

The Fourier transform

$$S(k) = 1 + \rho \int d^3r \, (g(r) - 1) \exp(i \, \mathbf{k} \cdot \mathbf{r}) \quad , \tag{1.6}$$

gives the static structure function as measured from X-ray or neutron diffraction on the fluid; $S(k)$ is defined by the relation[2]

$$S(k) = \frac{\text{rate of diffraction from N atoms in the fluid}}{\text{rate of diffraction from N free atoms}} \quad , \tag{1.7}$$

where $\mathbf{k}$ is the momentum transferred to the system from the incident particle. The static structure function for ^{4}He, at $T = 0$ and at the equilibrium density obtained from experimental data[2], is presented in fig. 1. As indicated by the relation (1.6) it is possible to get information on the inner structure of the fluid from diffraction experiments.

If the measure on the scattered particle is done in function not only of the momentum $\mathbf{k}$ but also of the energy ω trasferred to the system, by the obvious extension of eq. (1.7) one obtains the dynamical structure function $S(k,\omega)$ a quantity that contains[3] all the information on the scattering process.We have thus:

$$S(k) = \int_{-\infty}^{+\infty} S(k,\omega) d\omega \quad . \tag{1.8}$$

In particular, from $S(k,\omega)$ one can derive the excitation spectrum $\epsilon(k)$ and the momentum distribution $n(k)$ of the system. From the knowledge of $n(k)$ it is possible to evaluate the condensate fraction n_0, namely the percentage of particle with zero momentum (see chapter 6). A system of non-interacting particles in the ground state has $n_0 = 1$. Liquid ^{4}He is a strongly interacting system, nevertheless, as it is discussed in sect. 6.2, at low temperature it possesses a rather large condensate fraction.

In the following chapters the variational theory is developped and applied to calculate quantities of interest for the knowledge of the structure of liquid ^{4}He at $T = 0$. The results will be also compared with those obtained by other theoretical approaches and with the experimental data.

2 . JASTROW CORRELATED WAVE FUNCTIONS.

In the first section of this chapter, few examples of accurate interatomic ^{4}He–^{4}He potentials will be presented. The remaining of the chapter is devoted to the discussion of the Jastrow wave function, to the definition and properties of the distribution functions and to the derivation of the expectation value of the hamiltonian.

2.1 INTERACTION ENERGY OF TWO HELIUM ATOMS.

The interaction energy between two helium atoms has received much attention both experimentally and theoretically[4]. The important characteristics of the pair potential are:

i) the strong repulsion for small interparticle distances,

ii) the appreciable attraction at intermediate distances, and

iii) the small attraction for large distances.

The long-range behaviour of the potential is dominated by a tail proportional to r^{-6}, in accordance to a Van der Waals interaction[5]. At very small distances, due to the Pauli principle for the atomic electrons, the interaction between the atoms results strongly repulsive and it is approximated as a hard-core repulsion in some model potentials. A simple and familiar two-body potential is the Lennard-Jones (LJ) 6 − 12 potential

$$v_{LJ}(r) = 4\epsilon\Big[\Big(\frac{\sigma}{r}\Big)^{12} - \Big(\frac{\sigma}{r}\Big)^{6}\Big] \quad , \tag{2.1}$$

$$\epsilon = 10.22 \text{ K}, \qquad \sigma = 2.556 \text{ Å} \quad ,$$

with parameter values determinated by de BOER and MICHELS[6] from second virial coefficient in the temperature range 40 ÷ 200 K. The abundance of experimental data has stimulated the search of ad hoc potentials well suited to reproduce these data. Moreover, the short range part of the potential can be well determined by a self-consistent Hartree-Fock method[7]. Here we limit ourselves to recall the potential proposed by AHLRICHS et al.[8]:

$$v(r) = \epsilon'\Big\{A\exp\Big(-\frac{\alpha r}{r_m}\Big) - [C_6\Big(\frac{r_m}{r}\Big)^6 + C_8\Big(\frac{r_m}{r}\Big)^8 + C_{10}\Big(\frac{r_m}{r}\Big)^{10}]F(r)\Big\} \quad , \tag{2.2}$$

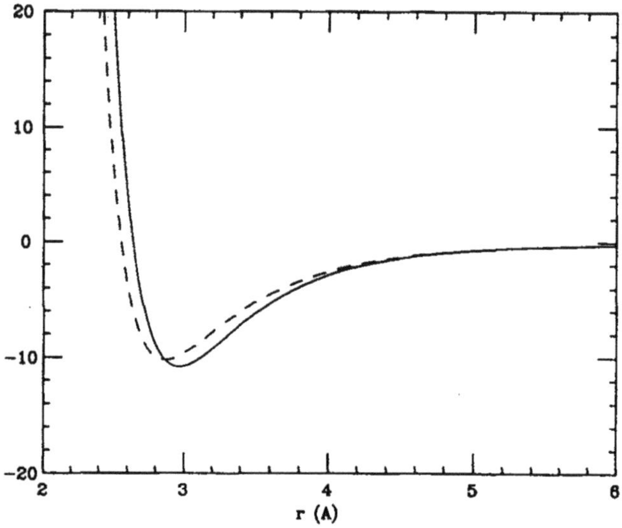

Figure 1. *Comparison of the atomic energy of two helium atoms: solid and dashed lines correspond to HDFHE2 and LJ potentials, respectively.*

where

$$F(r) = \begin{cases} \exp\left[\left(Dr_m/r - 1\right)^2\right], & \frac{r}{r_m} \leq D \ , \\ 1, & \frac{r}{r_m} > D \ . \end{cases}$$

An accurate determination of the parameters of eq. (2.2) has been given by AZIZ et al.[9], and the so-called HFDHE2 resulting potential is specified by the choice:

$$\begin{aligned} A &= 0.5448504\ 10^6, & \epsilon' &= 10.8\ \text{K}, \\ \alpha &= 13.353384, & C_6 &= 1.3732412, \\ D &= 1.241314, & C_8 &= 0.4253785, \\ r_m &= 2.9673\ \text{Å}, & C_{10} &= 0.178100. \end{aligned}$$

AZIZ et al. compared the prediction of several potentials and concluded that the HFDHE2 potential gives the best overall agreement with experiments. Figure 1 shows a comparison between the Lennard-Jones and HFDHE2 potential.

More recently CEPERLEY and PARTRIDGE[10] have suggested that the HFDHE2 potential is too repulsive for separation between the atoms less than 1.8 Å and proposed a modified form of the potential for short distances.

When we consider more than two He atoms also three- and more-body forces should be included. A widely adopted three-particle interaction is the long-range triple dipole interaction as given by AXILROD and TELLER[11]:

$$v_3(r_{12}, r_{13}, r_{23}) = u_0 \frac{1 + 3\cos\theta_1 \cos\theta_2 \cos\theta_3}{r_{12}^3 r_{13}^3 r_{23}^3} \quad . \tag{2.3}$$

The angles θ are the internal angles of the three-particle triangle, and $u_0 = 0.324$ K[12]. The energy due to this term in liquid ^{4}He is a rather small fraction[13] of the total energy per particle. Four-body forces should play a negligible role in liquid ^{4}He.

2.2 CORRELATED WAVE FUNCTIONS.

The simplest trial function for a boson system is a symmetrized product of single-particle wave functions, namely

$$\Psi = S\{\prod_{i=1}^{N} \varphi_{\alpha_i}(i)\} \quad , \tag{2.4}$$

where i and α_i specify the particle and the set of its quantum numbers, respectively; S denotes the symmetrizing operator. Because of the translational invariance, the functions φ_i are plane waves. However, in the case of strong particle-particle interaction, the choice of a variational function as (2.4) is not appropriate. As an example, due to the hard-core repulsion one would obtain infinite expectation value of the hamiltonian; in this case, the wave function must vanish if one or more interparticle distances are smaller than the hard-core radius R_c. To satisfy this condition one can adopt the following form proposed by BIJL[14], DINGLE[15] and JASTROW[16]

$$\Psi = \prod_{i<j=1}^{N} f(r_{ij}) \cdot S\{\prod_{i=1}^{N} \varphi_{\alpha_i}(i)\} \quad , \tag{2.5}$$

where the two-particle correlation factor $f(r)$ is zero for $r \leq R_c$ (or nearly zero in the region of strongly repulsive potential) and goes to unity for large r values:

$$f(r) = 0 \quad \text{if} \quad r \leq R_c \, ; \qquad \lim_{r \to \infty} f(r) = 1 \quad . \tag{2.6}$$

For the ground state of the fluid, the functions $\varphi_{\alpha_i}(i)$ in eq. (2.5) can be taken equal to unity, with the result:

$$\Psi_J \equiv \prod_{i<j=1}^{N} f(r_{ij}) = \exp\Big[-\frac{1}{2} \sum_{i<j=1}^{N} u(r_{ij})\Big] \quad . \tag{2.7}$$

The function $u(r)$ is called pseudopotential and Ψ_J is often cited as Jastrow correlated wave function. For liquid ^{4}He with LJ potential, a simple but reliable form for the trial function $f(r)$ has been suggested by Mc MILLAN[17] and SCHIFF and VERLET[18]

$$f(r) = \exp\Big\{-\frac{1}{2}\Big[\frac{b\sigma}{r}\Big]^m\Big\} \quad , \tag{2.8}$$

where σ is related to the potential, b and m are free parameters (in all cases, the values used for m are in the interval $4 \div 5$). When two particles are close, the pair potential is large with respect to the average potential due to the other ones: this suggests to choose the correlation factor as determined by the condition:

$$-\frac{\hbar^2}{m}\nabla^2 f(r) + (v(r) - \lambda)f(r) = 0, \qquad r < d \ , \tag{2.9}$$

$$f(r) = 1, \qquad f'(r) = 0, \qquad r \geq d \ . \tag{2.10}$$

The "healing" distance d is a trial parameter which can be fixed by a variational calculation on the system. The constant λ takes into account the main effect of the medium on the pair considered and it is determined so as to satisfy the healing conditions (2.10). It can be noticed that in the case of LJ potential, and for the leading order, the analytical form (2.8) of $f(r)$ satisfies eq. (2.9) for small r with $b = 16/25 = 1.134$, independently from the density of the system.

The best choice for $f(r)$ in the variational sense, is the solution of the Euler-Lagrange equation resulting from the stationarity condition:

$$\frac{\delta}{\delta f(r)} \frac{< \Psi_J|H|\Psi_J >}{< \Psi_J|\Psi_J >} = 0 \ . \tag{2.11}$$

This equation will be worked out and its solution discussed in chapter 4.

The behaviour of $f(r)$ at large r values is directly related to measurable quantities: it is known[19] that

$$f(r) = 1 - \frac{mc_s}{4\pi^2 \hbar \rho r^2} \quad , \quad \text{as r is large} \quad , \tag{2.12}$$

where c_s is the sound velocity. It is clear that neither the analytical form (2.8) nor the solution of eq. (2.9) has a tail proportional to r^{-2}. On the other hand (see sect. 4.1), the solution of the Euler-Lagrange equation for the Jastrow ansatz has this behaviour.

2.3 DISTRIBUTION FUNCTIONS.

In the study of a (quantum) system the n-particle distribution functions $\mathrm{P}^{(n)}(1,\ldots,n)$ play an important role. In the case of a boson fluid, the function

$$\mathrm{P}^{(n)}(1,\ldots,n) = \frac{N!}{(N-n)!} \frac{\int d^3r_{n+1}\ldots d^3r_N \, |\Psi(\vec{r}_1\ldots\vec{r}_N)|^2}{\int d^3r_1\ldots d^3r_N \, |\Psi(\vec{r}_1\ldots\vec{r}_N)|^2} \quad , \tag{2.13}$$

depends only on the interparticle distances (in the thermodynamical limit) and gives the probability density of finding n of the N particles with positions $\vec{r}_1\ldots\vec{r}_n$. Some properties of $P^{(n)}(1,\ldots,n)$ which directly follow from general physical considerations are heredown listed:

$$\text{Positiveness} \qquad P^{(n)}(1,\ldots,n) \geq 0 \quad , \tag{2.14}$$

$$\text{Symmetry} \qquad P^{(n)}(\ldots,i,\ldots,j,\ldots) = P^{(n)}(\ldots,j,\ldots,i,\ldots) \quad , \tag{2.15}$$

$$\text{Translational invariance} \qquad P^{(n)}(\vec{r}_1,\ldots,\vec{r}_n) = P^{(n)}(\vec{r}_1+\vec{A},\ldots,\vec{r}_n+\vec{A}) \quad , \tag{2.16}$$

$$\text{Hard-core effect} \qquad P^{(n)}(\ldots,i,\ldots,j,\ldots) = 0, \quad \text{if } r_{ij} \leq R_c \quad , \tag{2.17}$$

$$\text{Sequential relation} \qquad P^{(n-1)}(1,\ldots,n-1) = \frac{1}{N-n+1}\int d^3r_n \, P^{(n)}(1,\ldots,n) \quad , \tag{2.18}$$

$$\text{Cluster property} \qquad P^{(n)}(1,\ldots,i,i+1,\ldots,n) \to P^{(i)}(1,\ldots,i)\, P^{(n-i)}(i+1,\ldots,n) \quad . \tag{2.19}$$

The cluster property refers to configurations where the particles $1,\ldots,i$ are far from the remaining $i+1,\ldots,n$ ones.

From the condition $P^{(1)} = \rho$ and the sequential relation we obtain

$$P^{(1)} = \rho = \frac{1}{N-1}\int d^3r_2 \, P^{(2)}(1,2) \quad . \tag{2.20}$$

The two particle distribution function is the most important one; it is related to the so-called radial distribution function via the relation:

$$g(r_{12}) = \frac{P^{(2)}(r_{12})}{\rho^2} = \frac{N(N-1)}{\rho^2}\frac{\int d^3r_3\ldots d^3r_N \, |\Psi(\vec{r}_1\ldots\vec{r}_N)|^2}{\int d^3r_1\ldots d^3r_N \, |\Psi(\vec{r}_1\ldots\vec{r}_N)|^2} \quad . \tag{2.21}$$

Due to the cluster property, when the particles 1 and 2 are far each from the other,

$$P^{(2)}(1,2) \to P^{(1)}(1)P^{(1)}(2) = \rho^2 \quad , \tag{2.22}$$

and eq. (2.21) implies:

$$\lim_{r\to\infty} g(r) = 1 \quad . \tag{2.23}$$

As stated in chapter 1, the physical observable directly related to $g(r)$ is the static structure function, defined by

$$S(k) = \frac{1}{N}\frac{<\Psi|\rho_k\,\rho_{-k}|\Psi>}{<\Psi|\Psi>} \quad , \tag{2.24}$$

where

$$\rho_k = \sum_{i=1}^{N} \exp(i\,\mathbf{k}\cdot\mathbf{r}_i) = \int d^3r \sum_{i=1}^{N} \delta(\mathbf{r}-\mathbf{r}_i)\exp(i\,\mathbf{k}\cdot\mathbf{r}) \quad , \tag{2.25}$$

is the Fourier transform of the density operator $\sum_{i=1}^{N}\delta(\mathbf{r}-\mathbf{r}_i)$. From the latter two equations it follows

$$S(k) = 1 + \rho \int d^3r\, [g(r)-1]\, \exp(i\,\mathbf{k}\cdot\mathbf{r}) \quad , \tag{2.26}$$

and, in particular:

$$S(k=0) = 0, \qquad \lim_{k\to\infty} S(k) = 1 \quad . \tag{2.27}$$

It can be shown[19] that

$$\lim_{k\to 0} \frac{S(k)}{k} = \frac{\hbar}{2mc_s} \quad , \tag{2.28}$$

c_s being the sound velocity of the system. The behaviour of $S(k)$ for small k values determines the long-range tail of $g(r)$ and $f(r)$ (see sect.4.1):

$$\begin{aligned} \lim_{r\to\infty} r^4\,[g(r)-1] &= -\frac{\hbar}{2\pi^2 m\rho c_s} \quad , \\ \lim_{r\to\infty} r^2\,[f(r)-1] &= -\frac{mc_s}{2\pi^2\hbar\rho} \quad . \end{aligned} \tag{2.29}$$

Many calculations involve $P^{(3)}(1,2,3)$ or, equivalently, the three-particle radial distribution function:

$$g_3(r_{12},r_{13},r_{23}) = \frac{P^{(3)}(1,2,3)}{\rho^3} = \frac{N(N-1)(N-2)}{\rho^3}\frac{\int d^3r_4\ldots d^3r_N|\Psi(\vec{r}_1\ldots\vec{r}_N)|^2}{\int d^3r_1\ldots d^3r_N|\Psi(\vec{r}_1\ldots\vec{r}_N)|^2} \quad . \tag{2.30}$$

In the case of the Jastrow wave function, by applying the operator ∇_1 to both members of eq.(2.21) on gets:

$$\nabla_1 g(r_{12}) = g(r_{12})\nabla_1 u(r_{12}) + \rho\int d^3r_3\, g_3(r_{12},r_{13},r_{23})\nabla_1 u(r_{13}) \quad . \tag{2.31}$$

Analogously, from (2.30) it follows:

$$\nabla_1 g_3 = g_3\big(\nabla_1 u(r_{12}) + \nabla_1 u(r_{13})\big) + \rho\int d^3r_4\, g_4\, \nabla_1 u(r_{14}) \quad . \tag{2.32}$$

The relation (2.31) is known as the BORN–BOGOLIUBOV–GREEN–KIRKWOOD–YVON (BBGKY) equation and it can be used to obtain $g(r)$ when $g_3(r_{12},r_{13},r_{23})$ is known.

In practice, it is necessary to select a suitable approximate expression for $g_3(r_{12}, r_{13}, r_{23})$ as, for instance:

$$\text{KSA} \qquad g_3(r_{12}, r_{13}, r_{23}) = g(r_{12})g(r_{13})g(r_{23}), \tag{2.33}$$

$$\begin{aligned} \text{SA} \qquad g_3(r_{12}, r_{13}, r_{23}) &= 1 + G(r_{12}) + G(r_{13}) + G(r_{23}) \\ &+ G(r_{12})G(r_{13}) + G(r_{12})G(r_{23}) + G(r_{13})G(r_{23}) \\ &+ \rho \int d^3r_4\, G(r_{14})G(r_{24})G(r_{34}) \quad , \end{aligned} \tag{2.34}$$

where $G(r) = g(r) - 1$ and KSA and CA stand for Kirkwood Superposition Approximation and Convolution Approximation, respectively[19].

The expectation value of the hamiltonian can be easily expressed in terms of the two- and three-particle radial distribution functions. The contribution of the potential to the energy per particle is readily shown to be:

$$\left\langle \sum_{i<j=1}^{N} v(r_{ij}) \right\rangle /N = \frac{\rho}{2} \int d^3r\, g(r)\, v(r) \quad . \tag{2.35}$$

The contribution of the kinetic energy $\mathrm{T} = -\hbar^2/(2m)(\nabla_1^2 + \cdots + \nabla_N^2)$ is given by

$$\begin{aligned} <\mathrm{T}> /N =& \tfrac{1}{2}\rho \int d^3r\, g(r) \left(-\frac{\hbar^2}{m} \frac{\nabla^2 f(r)}{f(r)} \right) \\ & - \rho^2 \int d^3r_{12}\, d^3r_{13}\, g_3(r_{12}, r_{13}, r_{23}) \left(-\frac{\hbar^2}{2m} \frac{\nabla_1 f(r_{12})}{f(r_{12})} \frac{\nabla_1 f(r_{13})}{f(r_{13})} \right) \end{aligned} , \tag{2.36}$$

which is known as the PANDHARIPANDE–BETHE (PB) form[20] of the kinetic energy. If we integrate by parts eq. (2.36) it is possible to express <T> in terms of $g(r)$ only: we obtain thus the so-called JACKSON–FEENBERG (JF) form[21] of the kinetic energy:

$$<\mathrm{T}> /N = \frac{\hbar^2}{8m} \rho \int d^3r\, g(r) \nabla^2 \ln(f^2(r)) \quad . \tag{2.37}$$

If triplet correlations are also considered in the trial wave function,

$$\Psi_{JT} = \Psi_J \prod_{i<j<k=1}^{N} f_3(r_{12}, r_{13}, r_{23}) \quad , \tag{2.38}$$

then <T> in the PB form involves also g_4 and g_5, whereas the JF form (2.37) must be implemented by the contribution

$$<\mathrm{T}>_3 /N = -\frac{\hbar^2}{8m} \rho^2 \int d^3r_{12}\, d^3r_{13}\, g_3(r_{12}, r_{13}, r_{23})\, \nabla_1^2 \ln(f_3(r_{12}, r_{13}, r_{23})) \quad . \tag{2.39}$$

3. CLUSTER EXPANSION AND THE METHOD OF INTEGRAL EQUATIONS.

In the first section of this chapter the main elements of the cluster expansion and associated diagramatic analysis are recalled. The method of integral equations to sum up large classes of cluster terms is presented in successive sections where a particular attention is given to techniques useful to approximate the so-called elementary contributions. Few numerical results for liquid helium, in corrispondence to a simple Jastrow ansatz, are also presented.

3.1 CLUSTER EXPANSION.

If the variational wave function adopted is of the Jastrow type, with the position $f(r) = \exp(-\frac{1}{2}u(r))$, the eq. (2.21) for the radial distribution takes the form

$$g(r_{12}) = \frac{N(N-1)}{\rho^2} \frac{\int d^3r_3 \ldots d^3r_N \exp\left(-\sum_{i<j=1}^{N} u(r_{ij})\right)}{\int d^3r_1 \ldots d^3r_N \exp\left(-\sum_{i<j=1}^{N} u(r_{ij})\right)} , \tag{3.1}$$

which formally coincides with the expression for the radial distribution function of a classical fluid if the pseudopotential $u(r_{ij})$ is replaced by $\phi(r_{ij})/k_BT$, $\phi(r_{ij})$ being the interaction potential between the particles i and j. As a consequence, the method devised to calculate $g(r)$ in the classical case can be directly extended to quantum systems.

The correlation factor $f(r)$ is close to unity everywhere, except for small r values: this suggests to introduce a function $h(r)$ as an expansion parameter for the multidimensional integral in eq. (3.1); the function $h(r)$ is defined as follows:

$$f^2(r) = \exp(-u(r)) = 1 + h(r) \quad . \tag{3.2}$$

No theorem exists on the properties of the resulting expansion, but the equations obtained allow significant solutions also when the correlation factor, and consequently $h(r)$, have a long-range tail. Let us write down (3.1) as

$$g(r_{12}) = \frac{N(N-1)}{\rho^2} f^2(r_{12}) \frac{\int d^3r_3 \ldots d^3r_N \left(1 + \sum_{i<j=1}^{N}{}' h(r_{ij}) + \cdots\right)}{\int d^3r_1 \ldots d^3r_N \left(1 + \sum_{i<j=1}^{N} h(r_{ij}) + \cdots\right)} , \tag{3.2}$$

where the prime in the summation indicates that the term $i,j = 1,2$ is excluded. Many terms in the integral at the numerator (denominator) of the latter equation differ only in the particle labels and give the same contribution to the integral. It is useful[22–27] to associate a diagram to a generic term of the sum as follows.

---

1. *points*, corresponding to particles; the points can be of two kinds:
   i) *external points* (1 and 2) depicted by empty circles;
   ii) *internal points*, depicted by full circles; such a circle has associated a factor ρ and an integration over the point coordinates is implied.
2. a *dashed line connecting points i and j*, corresponding to a factor $h(r_{ij})$.

As an example one has:

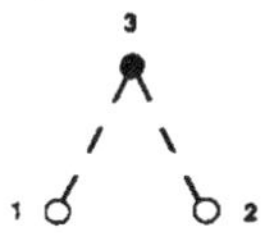

$$\equiv \rho \int d^3r_3\, h(r_{13})\, h(r_{23}) \quad .$$

An $i-j$ *subdiagram* is a part of a diagram which is connected with the rest of the diagram by means of the points i and j, only. Diagrams and subdiagrams are usually classified as follows:

---

1. a diagram can be *linked* or *unlinked*. It is unlinked when it exists at least one group of particles, different from the external particles 1 and 2, which are not connected by any dashed line to the remaining particles;
2. a diagram can be *reducible* or *irreducible*. It is reducible when it possesses at least one point of reducibility, i.e. a point which is the only common point between two parts of the diagram, one of the part containing all the external particles;
3. a subdiagram is called *composite* if it is composed by two or more $i-j$ subdiagrams which form "parallel", indipendent connections between points i and j;
4. any $i-j$ subdiagram is either *nodal* or *non–nodal*; a subdiagram is nodal if it has one or more *nodes*, a node being an internal point through which any path joining the points i and j must pass; by definition a composite diagram is non–nodal;
5. a *chain* is a nodal $i-j$ subdiagram constituted by an unique path of dashed lines;
6. a subdiagram is called *elementary* when it is neither nodal nor composite.

Few examples of diagrams are given in fig.2.

We do not discuss here the details of the cluster expansion which, however, can be found in the literature[25]. We only mention that in the evaluation of the radial distribution function, if the various cluster terms are summed up, one obtains that the contributions from unlinked and reducible diagrams exactly cancel out so that $g(r)$ can be written as a

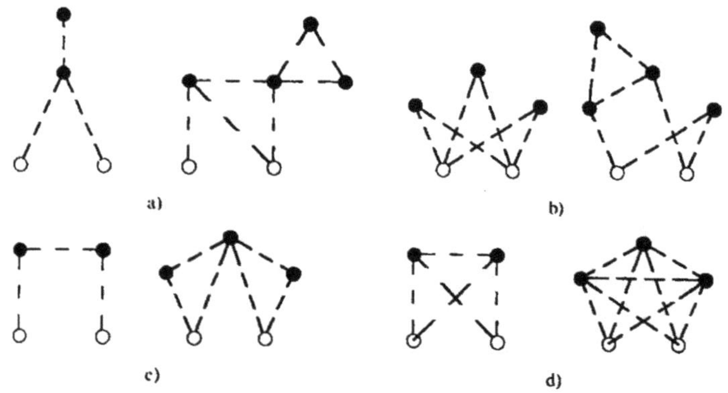

Figure 2. *Some examples of reducible (a), composite (b), nodal (c), and elementary (d) subdiagrams.*

sum of an infinite number of irreducible terms (diagrams) with an increasing number of internal particles (points). The diagrams which contribute to $g(r)$ and involve less than five points are the following

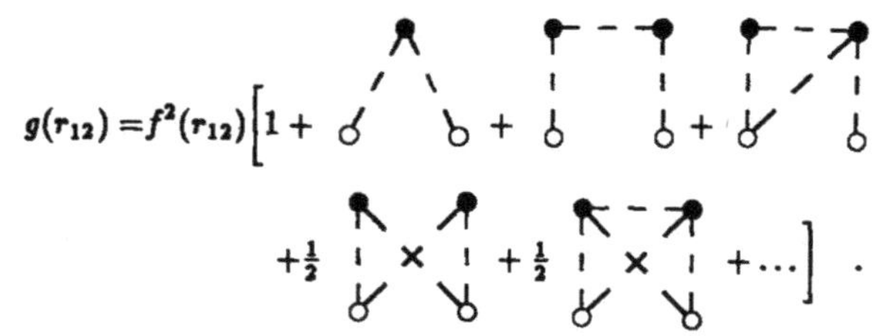

(3.3)

The cluster expansion is an expansion both in $h(r) = f^2(r) - 1$ and in the density ρ. In general, there is not an unique parameter useful to measure the rapidity of the convergence of the expansion, however a significant quantity is

$$\eta = \left| \rho \int d^3r \, (f^2(r) - 1) \right| \quad .$$

For liquid ^{4}He at the equilibrium density $\rho_0 = 0.02185$ Å^{-3}, we obtain

$$\eta = \rho_0 \, \tfrac{4}{3} \pi \, (2\ \text{Å})^3 \approx .8 \quad ;$$

this large value of η is a clear signal that for such a system the convergence of the cluster expansion will result very slow.

3.2 THE METHOD OF INTEGRAL EQUATIONS.

It is possible to derive specific integral equations, with solutions which represent the sum of large classes of terms in eq.(3.3). Let $N(r_{12})$ and $E(r_{12})$ indicate the sum of all nodal and elementary terms (i.e. corresponding to nodal or elementary 1-2 subdiagrams, respectively). It has been shown[26,27] that

$$g(r) = f^2(r)\left[1 + N(r) + E(r) + \frac{(N(r)+E(r))}{2\,!} + \cdots\right]$$
$$= f^2(r)\exp\left[N(r) + E(r)\right] \quad , \tag{3.4}$$

$$N(r_{12}) = \rho \int d^3r_3\,\left[g(r_{13}) - 1\right]\left[g(r_{23}) - N(r_{23}) - 1\right] \quad , \tag{3.5}$$

$E(r_{12}) = \frac{1}{2}$  $+ \cdots$,   (3.6)

where in the expression for $E(r_{12})$ a solid line represents the factor $g(r_{ij}) - 1$. From the preceding equations one obtains:

$$\ln\left\{\frac{g(r_{12})}{f^2(r_{12})\exp(E(r_{12}))}\right\} =$$
$$\rho \int d^3r_3\,\left[g(r_{13}) - 1\right]\left[g(r_{23}) - \ln\left\{\frac{g(r_{23})}{f^2(r_{23})\exp(E(r_{23}))}\right\} - 1\right] , \tag{3.7}$$

which is called the *convolution equation* and gives the relation satisfied by the exact $g(r)$ correspondent to the Jastrow function Ψ_J. In practice, the analytic evaluation of the various diagrams which contribute to $E(r)$ represents a difficult task. In the next three sections we will review some approximate procedures generally used to overcome this difficulty.

3.3 HNC HIERARCHY AND PY EQUATION.

As a first approximation (very crude for dense systems) one can disregard all the elementary contributions and take $E(r) = 0$. This approximation, denoted as HNC/0, leads to the familiar hypernetted chain integral equation, which is readily derived from eq.(3.7) by dropping out the function $E(r)$. To improve the approximation one must include the elementary diagrams involving few particles. The HNC/n approximation is obtained by considering $E(r)$, in eq. (3.7), as the sum of the elementary diagrams with no more than n particles. It must be noticed that the number of elementary diagrams rapidly the three particle distribution function has been calculated with the inclusion of all the elementary diagrams which involve only four particles (see sect. 3.4). The results reported in table I show that the E_{JF} values in the PY and HNC/0 approaches have nearly opposite deviations from the MC value. The PB form together with the HNC treatment produces a good energy estimate due to a sort of cancellation mechanism[31].

	HNC/0	HNC/4	PY
E_{JF}/N	−4.44	−4.76	−7.68
E_{PB}/N	−6.79	−6.27	−6.92
$E(MC)/N$		-5.73	

Table I. *Energy per particle (K) for liquid* ^{4}He *at zero temperature and density* $\rho_0 = 0.365\sigma^{-3}$*, for LJ potential.*

3.4 THE SCALING APPROXIMATION.

By the inspection of fig. 3 it can be verified that the estimates $E_n(r)$, (n=4,5,6) for the elementary function $E(r)$ differ between themselves for (nearly) constant factors. Such circustance is the basis of the suggestion done by USMANI et al.[32] to adopt the expression:

$$E(r) = (1+s)\,E_4(r) \quad . \tag{3.10}$$

This relation is called the scaling approximation for the elementary diagrams, s being the scaling parameter wich can be fixed[32] as follows. The JF and PB expression of the total kinetic energy T, as specified by eqs. (2.36) and (2.37), should furnish coincident results if the exact distribution functions would be used. In general this will not be true when some approximations have been introduced to calculate $g(r)$ and $g_3(r_{12},r_{13},r_{23})$, in particular if the function $E(r)$ is taken as in eq.(3.10): the scaling parameter s can then be fixed so as to have

$$<T>_{JF}=<T>_{PB} \quad . \tag{3.11}$$

However, the evaluation of $<T>_{PB}$ requires the knowledge of the three-body radial distribution function $g_3(r_{12},r_{13},r_{23})$; in correspondence to a Jastrow function Ψ_J, eq. (2.30) takes the form:

$$g_3(r_{12},r_{13},r_{23}) = \frac{N(N-1)(N-2)}{\rho^3} f^2(r_{12})\,f^2(r_{13})\,f^2(r_{23}) \cdot \\ \cdot \frac{\int d^3r_4 \ldots d^3r_N \left(1+\sum_{i<j=1}^{N}{}' h(r_{ij})+\cdots\right)}{\int d^3r_1 \ldots d^3r_N \left(1+\sum_{i<j=1}^{N} h(r_{ij})+\cdots\right)} \quad , \tag{3.12}$$

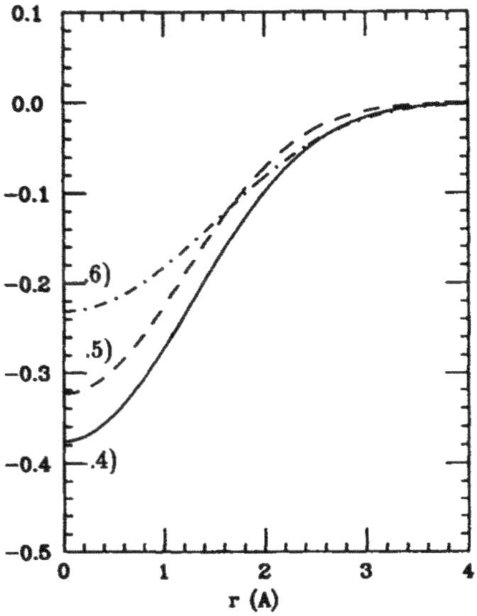

Figure 3 *The elementary function $E_n(r)$ (n=4,5,6) estimated up to inclusion of the n-points elementary diagrams, is shown versus r. The data refer to liquid* ^{4}He *at the equilibrium density and are taken from ref. (28).*

increases with n. From fig. 3 we can realize that for liquid helium the contributions of elementary terms with an increasing number of particles decrease rather slowly[(28)].

The convolution equation can also be written in the form:

$$g(r) = f^2(r)\,[1 + N(r) + E(r) + C(r)] \quad , \tag{3.8}$$

where $C(r)$ represents the contribution from all the composite diagrams. PERCUS and YEVICK[(29)], on the basis of an approximate cancellation mechanism between elementary and composite diagrams, proposed for $g(r)$ the following integral equation

$$g(r_{12})/f^2(r_{12}) = \rho \int d^3r_3\,[g(r_{13}) - 1]\,[g(r_{23}) - g(r_{23})/f^2(r_{23})] \quad , \tag{3.9}$$

which is known as the PY equation; eq. (3.9) is based on the approximate relation $E(r) + C(r) \approx 0$.

To test those different approximations, the energy per particle E/N in liquid ^{4}He has been calculated at $\rho_0 = 0.02185$ Å^{-3} $= 0.365\ \sigma^{-3}$ and compared with the (numerically) exact Monte Carlo (MC) estimate[(30)]. The LJ potential has been used and the correlation factor $f(r)$ has the form (2.8) with $b = 1.17$. The energy has been evaluated both using the JF and PB form for the kinetic energy. In the HNC/4 approximation, with the PB form, where the prime in the summation means that the values $(i=1, j=2)$, $(i=1, j=3)$ and $(i=2, j=3)$ are excluded. It is possible to perform a cluster expansion for $g_3(r_{12}, r_{13}, r_{23})$ and subsequently to sum the cluster terms with the result[33]

$$g_3(r_{12}, r_{13}, r_{23}) = g(r_{12})\, g(r_{13})\, g(r_{23})\, e^{A(r_{12}, r_{13}, r_{23})} \quad . \tag{3.13}$$

The quantity $A(r_{12}, r_{13}, r_{23})$ is expressed as a sum of contributions from the so-called Abe diagrams

$A(r_{12}, r_{13}, r_{23}) =$ 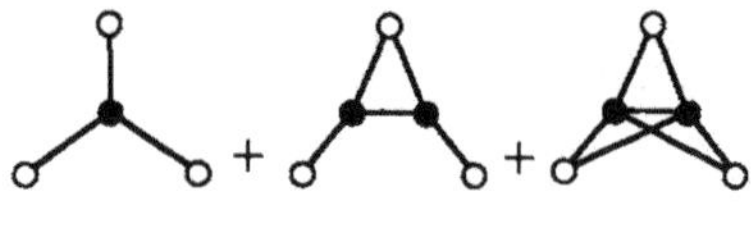 $+ \cdots$, (3.14)

where the solid lines represent the correlation $g(r_{ij})-1$. As for the 1-2 elementary diagrams, also the calculation of the integrals associated to the Abe diagrams has a complexity which strongly increases with the number of points. In ref.(32) it has been suggested to adopt again a scaling procedure via the relation

$$e^{A(r_{12}, r_{13}, r_{23})} = 1 + (1 + s')A_4(r_{12}, r_{13}, r_{23}) \quad . \tag{3.15}$$

The parameter s' is related to s by the approximate relation[32]

$$s \approx 2s' \quad . \tag{3.16}$$

In conclusion, the equations characterizing the scaling approximation are rewritten heredown:

$$\begin{aligned} g_S(r) &= f^2(r)\, \exp\left[N(r) + (1+s)\, E_4(r)\right] \quad , \\ N(r_{12}) &= \rho \int d^3r_3\, \left[g_S(r_{13}) - 1\right] \left[g_S(r_{23}) - N(r_{23}) - 1\right] \quad , \\ g_{3S}(r_{12}, r_{13}, r_{23}) &= g_S(r_{12})\, g_S(r_{13})\, g_S(r_{23})\, \left[1 + (1 + s/2)\, A_4(r_{12}, r_{13}, r_{23})\right] \quad , \end{aligned} \tag{3.17}$$

where

$E_4(r_{12}) =$  , $A_4(r_{12}, r_{13}, r_{23}) =$ 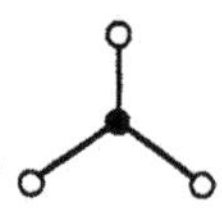 , ( $\equiv g(r_{ij}) - 1$) .

The parameter s should be fixed so as to fulfill the condition $<T>_{JF} = <T>_{PB}$.

3.5 THE METHOD OF INTERPOLATING EQUATION.

If we compare the terms included in the HNC/0 and PY approximations, it is evident that a larger number of diagrams has been disregarded to get the PY equation than in the deduction of the HNC/0 equation. As an example, the neglected diagrams which involve only four particle are:

HNC/0 : (3.18)

PY : (3.19)

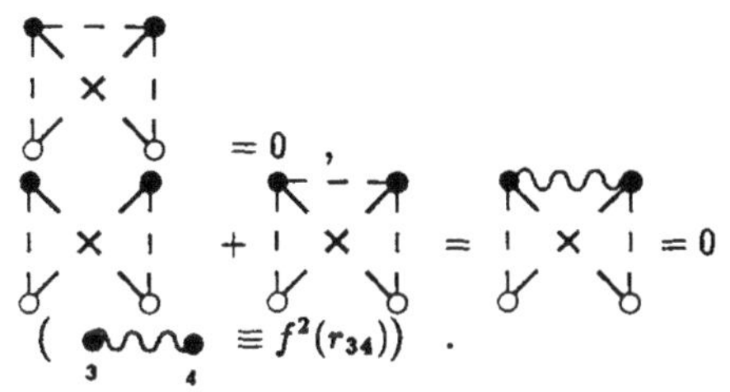

However, in a number of cases the PY approximation appears be superior[34] to the HNC/0 one. This can be easily understood when the interparticle potential contains a strong repulsion and the correlation factor rapidly heals to unity. In this case, $h(r) \neq 0$ only for small interparticle distances what reduces the configurations in diagrams (3.18) and (3.19) to those where the particles 3 and 4 are close each together; for many configurations it happens that $f^2(r_{34}) = 1 + h(r_{34})$ is small (due to the repulsive potential) thus the complessive contribution from the two diagrams in (3.19) is smaller than the one belonging to eq. (3.18). To take into account the fact that in eq. (3.19) the cancellation is only approximate, one can assume[35-38]

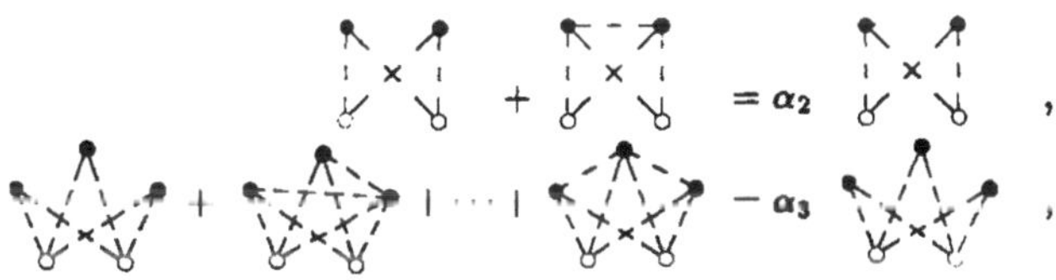

(3.20)

with the analogous relations for elementary and composite diagrams with a larger number of points. The interpolating coefficients α_n are expected to be small numbers ($|\alpha_n| \ll 1$); since the cancellation should be more effective for a large number of correlations, the condition[36,37]

$$\alpha_n = \alpha^{n-1} \quad , \tag{3.21}$$

seems to be appropriate. If all the cluster terms contributing to $g(r)$ are summed up, one gets[36,37]

$$g(r) = f^2(r)\, \frac{e^{\alpha N(r)} - 1 + \alpha}{\alpha} \quad . \tag{3.22}$$

By equating this expression to the exact one (3.4), the following relation for the elementary function is ontained:

$$E_\alpha(r) = \ln\left(\frac{e^{\alpha N(r)} - 1 + \alpha}{\alpha}\right) - N(r) \quad . \tag{3.23}$$

Up to this point α has been taken as a constant parameter, but it is evident that an improved treatment can be realized by introducing an interpolating function $\alpha(r)$, since cancellations as in eq.(3.20) can be larger or smaller depending on r_{12}. FABROCINI and ROSATI[38] have suggested the following form for the interpolating function:

$$\alpha(r) = 1 + \alpha_0 \frac{\ln(1+N_{PY}(r)) - N_{PY}(r)}{N_{PY}(r)} \quad . \tag{3.24}$$

α_0 is a free parameter and $N_{PY}(r)$ is the nodal function calculated in the PY approach to the problem; the expressions (3.22), and (3.23) remain valid if α is substituted by the function $\alpha(r)$.

A possible way for choosing the parameter α_0 (or the constant α in the simpler approach) could be the fulfillement of the condition $<\mathrm{T}>_{JF}=<\mathrm{T}>_{PB}$. This would also require an (approximate) calculation of $g_3(r_{12},r_{13},r_{23})$ and for such a reason in refs. (37,38) a different approach has been adopted. As stated in sect. 3.1, the problem of calculating $g(r)$ for a Bose quantum system at $\mathrm{T}=0$ described by a Jastrow wave function is completely analogous to the treatment of a classical fluid with the correspondence:

$$-\ln(f^2(r)) = u(r) = \phi(r)/K_B\mathrm{T} \quad . \tag{3.25}$$

The exact radial distribution function of a classical fluid satisfies the condition

$$1 - \rho \int d^3r\ [g(r) - 1 - N(r)] = 1 + \rho/3 \int d^3r\ u'(r)\ [g(r) + \rho\frac{\partial g(r)}{\partial \rho}] \quad , \tag{3.26}$$

which is called the Compressibility Consistent Condition (CCC) since it can be obtained by equating two different expressions for the classical compressibility. We can require that $g(r)$ calculated by means of the interpolating procedure satisfies the CCC equation and this can be reached by an appropriate choice of α_0. In conclusion, the equations of the interpolating technique (HNC/α) are:

$$\begin{aligned} g_\alpha(r) &= f^2(r)\,\frac{e^{\alpha(r)N(r)} - 1 + \alpha(r)}{\alpha(r)} \quad , \\ N(r_{12}) &= \rho \int d^3r_3\ [g_\alpha(r_{13}) - 1]\ [g_\alpha(r_{23}) - N(r_{23}) - 1] \quad , \end{aligned} \tag{3.27}$$

with $\alpha(r)$ as in eq.(3.24) and α_0 fixed by the CCC.

3.6 BBGKY HIERARCHY.

The approximation BBGKY/0 to solve eq.(2.31) corresponds to the KSA choice, namely

$$g_3(r_{12},r_{13},r_{23}) = g(r_{12})\,g(r_{13})\,g(r_{23}) \quad . \tag{3.28}$$

As it is apparent from eq.(3.13), successive improvements can be done by including a number of Abe diagrams in the calculation of $g_3(r_{12},r_{13},r_{23})$. As an example, BBGKY/4 is obtained in correspondence to

$$g_3(r_{12},r_{13},r_{23}) = g(r_{12})\,g(r_{13})\,g(r_{23})\,e^{A_4(r_{12},r_{13},r_{23})} \quad . \tag{3.29}$$

$\rho\sigma^3$	$(E/N)_{scal}$	$(E/N)_{int}$	$(E/N)_{MC}$
0.300	−5.77	−5.83	
0.330	−5.79	−5.92	-5.96 ± 0.2
0.350	−5.72	−5.86	-5.90 ± 0.2
0.365	−5.63	−5.76	-5.73 ± 0.1
0.380	−5.43	−5.60	-5.51 ± 0.2
0.401	−5.14	−5.28	-5.25 ± 0.1

Table II. *Variational energy for liquid helium at $T = 0$, with the LJ potential and the correlation factor $f(r) = \exp[-(b\sigma/r)^5]$ ($b = 1.17$). $(E/N)_{scal}$, $(E/N)_{int}$ and $(E/N)_{MC}$ are the energies furnished by the scaling*[32]*, interpolating*[38] *and Monte Carlo technique*[18,30]*, respectively. All the energies are in kelvin per particle.*

$\rho\sigma^3$	b	$(E/N)_{scal}$	$(E/N)_{int}$	$(E/N)_{MC}$
0.328	1.190	−5.91	−6.02	-6.00 ± 0.05
0.347	1.194	−5.83	−5.91	
0.365	1.195	−5.69	−5.71	-5.75 ± 0.05
0.401	1.199	−5.24	−4.96	

Table III. *The same as for table II, but with the the HFDHE2 potential.*

The approximations BBGKY/n with $n > 4$ are defined in a similar way and the corresponding solutions $g(r)$ can be used to calculate the energy per particle . For liquid ^{4}He, the results from BBGKY/5[28] are in good agreement with those furnished by the HNC/s and HNC/α techniques.

3.7 RESULTS FOR SHORT-RANGE CORRELATION FACTOR.

This section presents the results obtained for liquid ^{4}He in the case of a Jastrow variational wave function with a short-range correlation factor $f(r)$ as specified in eq.(2.8) with $m = 5$. The energy per particle calculated using the LJ potential is given in table II for few density values frequently considered in the literature. The value of the parameter b entering in the correlation factor $f(r)$ has been taken equal to 1.17 in the case of Lennard-Jones potential (table II) and in the range 1.19÷1.20 in the case of Aziz potential (table III). When available, also the MC estimates have been reported.

As it can be seen from the presented results, the two techniques for approximating elementary diagrams in the case of a dense system like liquid ^{4}He, give quite similar predictions with a close agreement in correspondence to the equilibrium density.

4. OPTIMUM JASTROW CORRELATED WAVE FUNCTION.

In this chapter the problem of determining the best variational Jastrow correlation factor $f(r)$ is discussed. The equation obeyed by the optimum $f(r)$ is derived in the first section and in the following one a method for the numerical solution in the HNC/0 approximation is presented. In the third section the scaling and the interpolating approximations to that equation are discussed and some numerical results are given in the last section.

4.1 THE EULER–LAGRANGE EQUATION.

If $E[f]$ represents the energy per particle for the ground state of the system corresponding to a Jastrow wave function $\Psi_J = \prod_{i<j=1}^{N} f(r_{ij})$, the variational principle requires

$$\frac{\delta E[f]}{\delta f(r)} = 0 \quad . \tag{4.1}$$

When the JF expression of the kinetic energy given in eq. (2.37) is adopted, $E[f]$ can be written in the form

$$E[f] = \frac{\rho}{2}\int d^3r\; g(r)\left[v(r) - \frac{\hbar^2}{4m}\nabla^2 \ln f^2(r)\right] \quad , \tag{4.2}$$

where the radial distribution function obeys the eq. (3.4),

$$g(r) = f^2(r)\exp\Big(N(r) + E(r)\Big) \quad . \tag{4.3}$$

Therefore, when stationarity of $E[f]$ with respect to $f(r)$ is required, the condition (4.3) must be fulfilled. This can be done as suggested by LANTTO and SIEMENS[39], by minimizing the quantity

$$\frac{\rho}{2}\int d^3r\; g(r)\left[v(r) - \frac{\hbar^2}{4m}\nabla^2 \ln f^2(r)\right] - \int d^3r\; \lambda(r)\left[\ln\Big(g(r)/f^2(r)\Big) - E(r) - N(r)\right] \quad ,$$

where $\lambda(r)$ is a Lagrange multiplier. Alternatively, eq. (4.3) can be used to express $f(r)$ in terms of $g(r)$ and rewrite the stationarity condition as

$$\frac{\delta E[g]}{\delta g(r)} = 0 \quad . \tag{4.4}$$

By Fourier transforming both members of eq. (3.5) for $N(r)$, one gets the equation

$$\widehat{N}(k) = \frac{\Big(S(k) - 1\Big)^2}{S(k)} \quad , \tag{4.5}$$

where $S(k)$ is the static structure function given by eq. (2.26), and

$$\widehat{N}(k) = \rho \int d^3r \, N(r) \exp(i\,\mathbf{k}\cdot\mathbf{r}) \quad .$$

Eq. (4.3) can be put in the form

$$\ln f^2(r) = \ln g(r) - E(r) - \frac{1}{(2\pi)^3\rho} \int d^3k \, \frac{\left(S(k)-1\right)^2}{S(k)} \exp(i\,\mathbf{k}\cdot\mathbf{r}) \quad , \tag{4.6}$$

which, together with eq. (4.2), gives

$$\begin{aligned} E[g] =& \frac{\rho}{2} \int d^3r \; g(r) v(r) + \frac{\hbar^2}{2m}\rho \int d^3r \left[\left(\nabla\sqrt{g(r)}\right)^2 + \tfrac{1}{4} g(r) \nabla^2 E(r) \right] \\ & - \frac{\hbar^2}{8m} \frac{1}{(2\pi^3)\rho} \int d^3k \, k^2 \frac{\left(S(k)-1\right)^3}{S(k)} \quad . \end{aligned} \tag{4.7}$$

The functional derivative of $E[g]$ with respect to $\sqrt{g(r)}$ can be readily obtained and equated to zero with the result:

$$-\frac{\hbar^2}{m}\nabla^2\sqrt{g(r)} + \left[v(r) + w(r)\right]\sqrt{g(r)} = 0 \quad . \tag{4.8}$$

Such an equation, commonly called the Euler-Lagrange (EL) equation, has the form of a Schrödinger equation at zero energy; the potential is a sum of the pair potential $v(r)$ and the "induced" potential $w(r)$ which reflects the effect of the medium on the considered pair and is given by

$$w(r) = w_0(r) + w_B(r) \quad , \tag{4.9}$$

$$\widehat{w}_0(k) = -\frac{\hbar^2}{4m} k^2 \frac{[S(k)-1][2S(k)+1]}{S(k)^2} \quad , \tag{4.10}$$

$$w_B(r) = \frac{\hbar^2}{4m}\left[\nabla^2 E(r) + E^*(r)\right] \quad . \tag{4.11}$$

The function $E^*(r)$ is defined as

$$E^*(r) = \rho \int d^3r' \; \nabla^2 g(r') \frac{\delta E(r)}{\delta g(r')} \quad . \tag{4.12}$$

In the HNC/0 approximation, the term $w_B(r)$ is assumed to be zero, but it is then important to verify that the EL/0 solution corresponds to a minimum of the energy per particle.

To this end, let $g_{EL}(r)$ be the solution of eq. (4.8) and $\delta\sqrt{g(r)}$ an arbitrary variation; from eqs. (4.7),(4.8) one easily gets

$$E[g+\delta g]-E[g]=\frac{\rho}{2}\int d^3r\,\delta\sqrt{g(r)}K\delta\sqrt{g(r)} \quad , \tag{4.13}$$

where the kernel K is defined by

$$Kg(r)=\Big[-\frac{\hbar^2}{m}+v(r)+w_0(r)\Big]g(r)-\sqrt{g(r)}\int d^3r'\,w'(r-r')\sqrt{g(r')}\,g(r') \quad , \tag{4.14}$$

$$\widehat{w'}(k)=k^2\frac{S(k)^3-1}{S(k)} \quad . \tag{4.15}$$

If the right–hand side of eq. (4.13) is a positive quantity for arbitrary $\delta\sqrt{g(r)}$, it can be concluded that $E[g_{EL}]$ is a miminum. This is guaranteed if K has only positive eigenvalues; CASTILLEJO et al.[40] have found that for liquid ^{4}He with LJ potential the minimum eigenvalue is positive for density values $\rho \leq 0.03$ Å^{-3}.

An alternative way to obtain the optimun Jastrow–type trial function for the ground state is to use the paired phonon analysis[41,42,19] (PPA). The method requires the diagonalization of the hamiltonian in the subspace of two–phonons excitation[41], but it can be also stated[42] in a variational form. Let us write the trial wave function as

$$\Psi=\exp\left[-\frac{1}{2}\sum_{i<j=1}^{N}\Big(u_0(r_{ij})+\Delta u(r_{ij})\Big)\right]=\Psi_0\,\exp\left[-\frac{1}{2}\sum_{i<j=1}^{N}\Delta u(r_{ij})\right] \quad . \tag{4.16}$$

The best choice of $\Delta u(r)$ satisfies the condition

$$\frac{\delta}{\delta\Delta u(r)}\left\{\frac{<\Psi|H|\Psi>}{<\Psi|\Psi>}\right\}=0 \quad . \tag{4.17}$$

With reference to a normalization box with periodicity conditions on its walls, the function $\Delta u(r)$ can be expressed in terms of its Fourier transform as

$$\Delta u(r)=\frac{1}{N}\sum_{\mathbf{k}}c_{\mathbf{k}}\exp(i\,\mathbf{k}\cdot\mathbf{r}) \quad . \tag{4.18}$$

Eq. (4.17) can be worked out to get[42]

$$<\Psi|\rho_k\rho_{-k}(H-E)|\Psi>=0 \quad , \tag{4.19}$$

where ρ_k is the operator which induces the excitation of a phonon with momentum $\mathbf{k}$ or, alternatively, the Fourier transform of the density operator (see eq. (2.25)). It can be shown[41,42] that the condition (4.19) is equivalent to the EL equation.

4.2 SOLUTION OF THE EL/0 EQUATION.

Let us now briefly discuss the numerical solution of the Euler–Lagrange equation in the HNC/0 approximation (EL/0). If the pair potential $v(r)$ goes as r^{-6} at large distances, then it can be shown[40] that

$$S(k) \to ak + bk^3 + \ldots \quad , \quad k \to 0 \quad ; \tag{4.20}$$

$$g(r) \to 1 + A/r^4 + \ldots \quad , \quad r \to \infty \quad ; \tag{4.21}$$

$$w_0(r) \to B/r^6 + \ldots \quad , \quad r \to \infty \quad . \tag{4.22}$$

Such behaviours of $S(k)$ and $g(r)$ agree with the ones asserted in sect. (2.3). Since the EL equation is not linear, its numerical solution requires an iterative procedure. In calculating $g(r)$ attention must be paid to both the region of small r values, where the contributions from kinetic and potential energy are important, and the region of large r values where $g(r)$ oscillates around unit. Correspondingly, there can be problems also when one calculates the Fourier transform since i) the behaviour for $k \to 0$ is related to the tail at large r of $g(r)$, and ii) the oscillations at large k values are due to the "hole" produced in $g(r)$ at small r values by the repulsive potential. In ref. (39) it has been suggested to start with some approximate expression $g_0(r)$ and to construct the solution of the EL equation as

$$\sqrt{g(r)} = \sqrt{g_0(r)} + \delta\sqrt{g(r)} \quad . \tag{4.23}$$

If this expression is inserted into eq. (4.8) (when the term $w_E(r)$ is disregarded) and only linear contribution in $\delta\sqrt{g(r)}$ are retained, the result is

$$-\frac{\hbar^2}{m}\nabla^2\delta\sqrt{g(r)} + [v(r) + w_0(r)]\delta\sqrt{g(r)} + \left[\rho\int d^3r'\ \sqrt{g_0(r')}\delta\sqrt{g(r')}w'(r-r')\right]\sqrt{g_0(r)}$$

$$= \left[-\frac{\hbar^2}{m}\nabla^2 + v(r) + w_0(r)\right]\sqrt{g_0(r)} \quad , \tag{4.24}$$

where $w'(r)$ is defined by eq. (4.15). The linear differential equation (4.24) can be solved by means of standard methods and the resulting function $\sqrt{g_0(r)} + \delta\sqrt{g(r)}$ can be used to get a new input function $g_0(r)$ in eq. (4.23). The procedure is then iterated until the required convergence in the solution $g(r)$ and other quantities of interest (energy per particle, behaviour of $S(k)$ at small k values, etc.) has been reached. When the initial choice $g_0(r)$ is not sufficiently close to the solution of the EL/0 equation, the iterative procedure may not converge at all; a possible care to that is to multiply, at each iteration, the obtained $\delta\sqrt{g(r)}$ by a factor (smaller than unit) before summing it to the previously estimated $\sqrt{g_0(r)}$.

4.3 CONTRIBUTION FROM ELEMENTARY DIAGRAMS TO THE EL EQUATION.

The contribution from elementary terms to the Euler–Lagrange equation (4.8) is represented by the "induced" potential $w_E(r)$ given by eq. (4.11). The knowledge of $w_E(r)$ requires the calculation of the functions $E(r)$ and $E^*(r)$ which are related to the elementary diagrams only. The approximations discussed in section 3.4 and 3.5 can be used to evaluate $w_E(r)$ and to improve the HNC/0 solution of the EL equation.

4.3.1 Scaling Approximation.

At each step of the iterative procedure adopted to solve the EL equation one can easily calculate the contribution from the elementary four-particle diagrams. The function $E(r)$ and $E^*(r)$ are approximate as

$E(r_{12}) = (1+s)E_4(r_{12}) = (1+s)$ 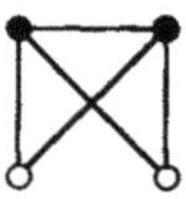 ,

$E^*(r_{12}) = (1+s)\{4$ 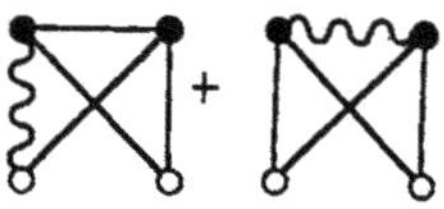 $\}$, (4.25)

where the solid and wiggly lines represent the function $g(r)-1$ and $\nabla^2 g(r)$, respectively.

In ref. (32), where the scaling procedure for evaluating the elementary contributions was firstly introduced, the EL equation was modified as follows. The terms involving $E^*(r)$ was disregarded and the EL equation was solved up to the value $r = d$, with d being $2 \div 3$ times larger than σ. For $r \geq d$, the following conditions were required to be satisfied,

$$f(r) = 1 - \gamma/r^2 \quad , \quad \gamma = \frac{mc_s}{2\pi^2\hbar\rho} \quad , \tag{4.26}$$

where c_s is the sound velocity. Moreover, the potential $w_E(r)$ was approximated with a constant value λ, so that the modified EL equation takes the form

$$-\frac{\hbar^2}{m}\nabla^2\sqrt{g(r)} + [v(r) + w_0(r) + \lambda]\sqrt{g(r)} = 0 \quad , \quad r \leq d \quad . \tag{4.27}$$

For a fixed c_s value, the equation (4.27) is solved to get $g(r)$ and λ is so determined to match the correlation factor

$$f(r) = \sqrt{g(r)}\exp\left[N(r)/2\right] \quad , \tag{4.28}$$

at $r = d$ with the logarithmic derivative of the function given by eq. (4.26). One can then calculate the energy per particle and the sound velocity as

$$c_s = \left(\frac{1}{m}\frac{\partial P[\rho]}{\partial \rho}\right)^{1/2} \quad , \quad P[\rho] = \rho^2\frac{\partial E[\rho]}{\partial \rho} \quad , \tag{4.29}$$

where P is the pressure. The value obtained for c_s can be used to start again iterating from eq. (4.26); the calculations of ref. (32) converge in just two iterations.

Alternatively, the EL equation can be solved as discussed in the preceding section by keeping $w_E(r)$ as

$$w_E(r) = \frac{\hbar^2}{4m}(1+s)\nabla^2 E_4(r) \quad , \tag{4.30}$$

with the scaling parameter s so chosen as to satisfy the condition (3.11) on the kinetic energy. The results obtained[43] are very close to those of ref. (32).

4.3.2 Interpolating Procedure.

In order to evaluate $w_E(r)$, we assume the form (3.23) for $E(r)$, namely

$$E_\alpha(r) = \ln\left(\frac{e^{\alpha N(r)} - 1 + \alpha}{\alpha}\right) - N(r) \quad . \tag{4.31}$$

If we disregard the term $E^*(r)$ in the expression (4.11) for $w_E(r)$, the EL equation can be solved by an iterative procedure in order to obtain $g(r)$ and $N(r)$, which have the proper (long-range) asymptotic behaviour (r^{-4} and r^{-2}, respectively). There, however, remains the question of fixing the parameter α_0. Now the CCC is not useful any longer, since the two integrals in eq. (3.26) are divergent for a long-range function. Anyway, this difficulty can be overcome as follows. First of all, we calculate $g(r)$ and $E_\alpha(r)$ with a short-range correlation factor which gives a reasonable estimate of the energy per particle. The introduction of long-range tails must not change the function $E(r)$ appreciably, in particular for small r values. In solving the Euler equation, the parameter α_0 appearing in the interpolating function $\alpha(r)$ (see eq. (3.24)) is fixed in order to obtain a value of the elementary function $E_\alpha(r)$ at $r = 0$ coincident with the one estimated in correspondence to the short range trial function. Two more remarks are in order: i) changes in α_0 which produce small variation of $E_\alpha(0)$ are nearly irrelevant on the energy estimates; ii) the calculation of the elementary diagram involving four particles shows that the change of $E_4(0)$ with respect to the short-range correlation factor is less than 2 %.

In correspondence with eq. (4.31) for $E_\alpha(r)$, the function $E^*(r)$ takes the form

$$\begin{aligned}
E^*_\alpha(r) &= \rho \int d^3r' \, w_Z(|\mathbf{r} - \mathbf{r}'|) Z(r') \nabla^2 g(r') \quad , \\
Z(r) &= \frac{\alpha(r) \exp\left[\alpha(r) N(r)\right]}{\exp\left[\alpha(r) N(r)\right] - 1 + \alpha(r)} - 1 \quad , \\
\widehat{w_Z}(k) &= \frac{S(k)^2 - 1}{S(k)^2} \quad .
\end{aligned} \tag{4.32}$$

The behaviour at large r values of $E^*_\alpha(r)$ is related to the small k values behaviour of its Fourier transform

$$\widehat{E^*_\alpha} = \left(\widehat{Z\nabla^2 g}\right) \widehat{w_Z} \quad ;$$

for $k \to 0$ it results $\widehat{w_Z}(k) \propto 1/k^2$ and

$$\left(\widehat{Z\nabla^2 g}\right)_{k=0} = \rho \int d^3r\ Z(r)\nabla^2 g(r) \quad .$$

The latter quantity is zero in HNC/0 approximation, when $\alpha = 1$ and $Z(r) = 0$, but it is otherwise different from zero when one has

$$\widehat{E_\alpha^*}(k) \stackrel{k\to 0}{\longrightarrow} 1/k^2 \quad , \quad E_\alpha^*(r) \stackrel{r\to\infty}{\longrightarrow} 1/r \quad .$$

Such a behaviour is not compatible with the one expected for $w(r)$, namely $w(r) \to 1/r^6$ at large r (see sect. 4.2); this in turn implies that no solution is possible for the EL equation; a similar situation has been observed[39] in the case of the PY approximation (namely, when $\alpha = 0$).

It can be noticed that $E^*(r)$ is the sum of the contributions of all the elementary diagrams (those associated with $E(r)$) with one correlation $f(r)^2 - 1$ replaced with by $\nabla^2 g(r)$. Let $g^*(r)$ be the function which corresponds to the diagrams obtained from those which contribute to $g(r)$, after replacing with $\nabla^2 g(r)$ one of the correlation factors connecting two points. The equation obeyed by such a function can be easily derived in the following form

$$\begin{aligned} g^*(r) &= \nabla^2 g(r) + g(r)[N^*(r) + E^*(r)] \quad , \\ \widehat{N^*}(k) &= \widehat{g^*}(k)\frac{S(k)^2 - 1}{S(k)^2} \quad . \end{aligned} \tag{4.33}$$

To evaluate $g^*(r)$, let us introduce an interpolating function $\alpha(r)$:

$$g^*(r) = \nabla^2 g(r) + f(r)^2 N^*(r) \exp\left[\alpha(r)N(r)\right] \quad . \tag{4.34}$$

By comparing eqs. (4.33) and (4.34), it results

$$E^*(r) = \left\{ \frac{\alpha(r)\exp\left[\alpha(r)N(r)\right]}{\exp\left[\alpha(r)N(r)\right] - 1 + \alpha(r)} - 1 \right\} N^*(r) \quad . \tag{4.35}$$

If the interpolating function $\alpha(r)$ is chosen to be the same adopted to calculate $g(r)$ in the HNC/α approximation, then $E^*(r)$ goes rapidly to zero at large r values and consequently, in that region $w_E(r)$ does not reveal anomalous behaviour.

Eq. (4.35) for $E^*(r)$ can be used to calculate $w_E(r)$ and then the EL equation can be solved to obtain an improved approximation to $g(r)$. The inclusion of $E^*(r)$ however, produces a small effect in the energy per particle; for instance, for the ground state of liquid ^{4}He at the experimental equilibrium density, the effect is about 0.02 K.

4.4 RESULTS FOR LONG-RANGE CORRELATION FACTOR.

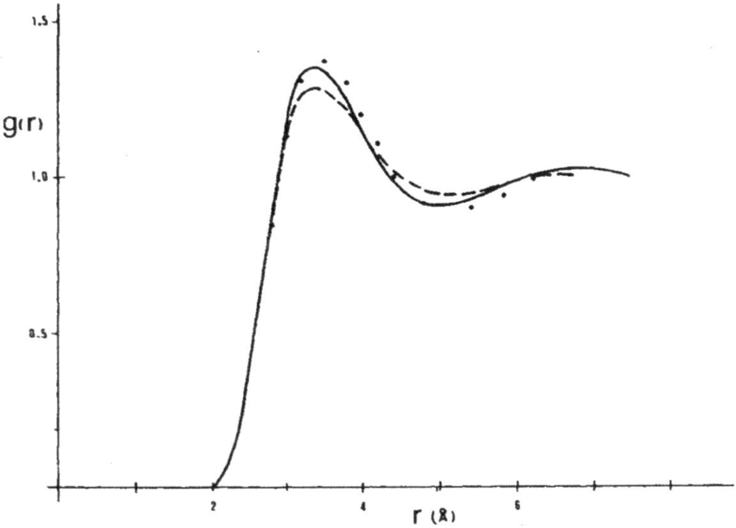

Figure 4. *Radial distribution function for the ground state of the liquid* ^{4}He *calculated at the density* $\rho\sigma^3 = 0.365$ *with the HFDHE2 potential. The solid line reproduces the solution of the EL equation; the dashed line reproduces* $g(r)$ *calculated with the choice (2.8) for the correlation factor. The full circles represent the GFMC results of ref. (13)*

In this section the variational energies obtained for liquid ^{4}He at $T = 0$ by solving the EL equation and applying the HNC/α technique are presented in correspondence to the LJ and HFDHE2 potentials. The short–range correlation factor adopted as reference for the elementary contributions has the form of eq. (2.8); the parameter b of this correlation factor is slowly dependent on the density and the values $b = 1.17$ and $b = 1.19$ are adequate for the LJ anf HFDHE2 potentials, respectively. The results are presented in table IV and V where they are compared with the corresponding estimates obtained in ref. (32) by means of the scaling approximation. The inclusion of medium and long–range tails in the trial wave function produces a lowering of the energy per particle which, although small, is important. For the LJ potential the energy decrease ranges from 0.1 K to 0.14 K depending on to the density value considered. Such variations are in agreement with those predicted by a perturbative calculation, as the one discussed in sect. 6 of ref. (19**). In the case of the HFDHE2 potential the "phonon" corrections are slightly larger than for the LJ potential.

The radial distribution function $g(r)$ obtained at equilibrium density ρ_0 and with the HFDHE2 potential is shown in fig. 4. The dashed line corresponds to the correlation factor given by eq. (2.8) ($b = 1.19$ and $m = 5$); the solid line refers to the solution of the EL equation. For the sake of comparison also the estimates obtained in ref. (13) by means of

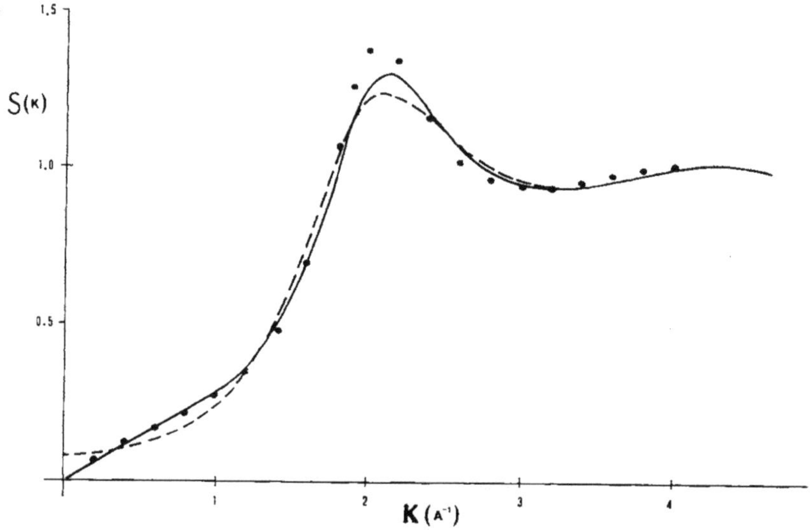

Figure 5. *Static structure function $S(k)$ for liquid ^{4}He at density $\rho\sigma^3 = 0.365$. The solid and dashed lines refer to the radial distribution function reproduced in fig. 5; the full circles represent the experimental data*[2].

$\rho\sigma^3$	b	$E_{sr/\alpha}/N$	$E_{eul/\alpha}/N$	E_{scal}/N
0.300	1.160	−5.84	−5.93	–
0.330	1.160	−5.92	−6.01	−5.93
0.350	1.162	−5.86	−5.95	−5.91
0.365	1.165	−5.78	−5.84	−5.85
0.380	1.168	−5.60	−5.72	−5.75
0.401	1.171	−5.28	−5.42	−5.55

Table IV. *Variational energies per particle for liquid ^{4}He at $T = 0$ with the LJ potential and the "optimum" Jastrow function. Column 1 specifies the density in reduced units. The parameter b of the short-range correlation factor and the corresponding energy $E_{sr/\alpha}$ are given in column 2 and 3 respectively. The energies $E_{eul/\alpha}$ are presented in column 4. The energies obtained with the scaling approximation*[32] *are reported in the last column. All the energies are in kelvin per particle.*

the Green Function Monte Carlo method (GFMC) have been reported (full circles) in the figure.

$\rho\sigma^3$	b	$E_{sr/\alpha}/N$	$E_{eul/\alpha}/N$	E_{scal}/N
0.328	1.191	−6.02	−6.14	−6.12
0.347	1.194	−5.91	−6.06	−6.04
0.365	1.196	−5.71	−5.90	−5.94
0.401	1.199	−4.96	−5.20	−5.46

Table V. *The same as for table IV, but with the the HFDHE2 potential.*

Fig. 5 allows a comparison between the static structure function $S(k)$ calculated with the HFDHE2 potential at the density ρ_0 and the experimental structure function[(2)]. The agreement at small k values is excellent.

On the basis of the results discussed in this section, it can be concluded that the optimum Jastrow ansatz allows a satisfactory description of the properties of the ground state of liquid ^{4}He. In order to get a closer agreement with the experimental data it is necessary to introduce other structural details in the variational wave function, as will be discussed in the next chapter.

5. IMPROVEMENT OF THE JASTROW WAVE FUNCTION.

In the preceding two chapters it has been shown that a variational wave function of the product form

$$\Psi_J = \prod_{i<j=1}^{N} f(r_{ij}) \quad , \tag{5.1}$$

yields an energy per particle of about −6 K for the ground state of liquid ^{4}He at the experimental equilibrium density. The discrepancy between that results and the exact numerical estimate given by the GFMC method[(13)] is about 15 %. Since there is a large cancellation between the potential and kinetic energy, that discrepancy could be judged not very large; on the other side, small errors in the energy correspond to rather large inaccuracies of the trial wave function. For the case we are interested in, the situation is illustrated also in fig. 5 where the experimental structure function $S(k)$ is compared with the one calculated in correspondence with the optimum Ψ_J. The agreement between the experiments and the variational theory is not completely satisfactory and, in conclusion, more flexible trial wave functions are required.

5.1 THREE-BODY CORRELATIONS.

The wave function in the product form given by eq. (5.1) has a number of merits, first of all it can adequately describe the configurations where two particles are very close and the mutual repulsion is large. But it is also evident that many structural details, which can be relevant especially for dense systems, are missing in Ψ_J. As an example, when three particles are close, the trial wave function should include also some dependence on the angular momenta. Moreover, the correlated wave function $\psi(\mathbf{k},\mathbf{r})$ of a pair of particles with relative momentum $\mathbf{k}$, should depend also on $\mathbf{k}$. A complex correlation function $f(\mathbf{k},\mathbf{r})$ can be defined so that

$$\psi(\mathbf{k},\mathbf{r}) = f(\mathbf{k},\mathbf{r}) \exp(i\,\mathbf{k}\cdot\mathbf{r}) \quad . \tag{5.2}$$

For small r values, it results[45,46]

$$\begin{aligned} f(\mathbf{k},\mathbf{r}) &\approx f_0(k=0,r) + i\,\mathbf{k}\cdot\mathbf{r}\Big[f_1(k=0,r) - f_0(k=0,r)\Big] = \\ &\equiv f(r) + \eta(r)\,\mathbf{r}\cdot\nabla \quad . \end{aligned} \tag{5.3}$$

The second term of the correlation describes the Feynman–Cohen[47] back–flow, produced by two particles recoiling from each other in the liquid. A new class of variational wave functions can be defined as

$$\Psi = S\left\{\prod_{i<j=1}^{N}\left[f(r_{ij}) + \frac{1}{2}\eta(r_{ij})\mathbf{r}_{ij}\cdot(\nabla_i - \nabla_j)\right]\right\} \quad . \tag{5.4}$$

The presence of the symmetrizing operator S is due to the fact that correlations do not necessarily commute. Since the function $\eta(r)$ is quite small[45] (approximately 0.2 at the maximum), we may expand (5.4) up to term linear in η, with the result

$$\begin{aligned} \Psi &= \Psi_J \prod_{i<j<k=1}^{N}\Big(1 + X_{ijk} + X_{kij} + X_{jki}\Big) \quad , \\ X_{ijk} &= \frac{1}{4}\left[\eta(r_{ij})\frac{\mathbf{r}_{ij}\cdot\nabla_i f(r_{ik})}{f(r_{ij})f(r_{ik})} + \eta(r_{ik})\frac{\mathbf{r}_{ik}\cdot\nabla_i f(r_{ij})}{f(r_{ik})f(r_{ij})}\right] \quad . \end{aligned} \tag{5.5}$$

It can be noticed[44,45] that $\eta(r)$ and $f'(r)$ have similar shapes; thus, from eq. (5.5) one derives the simpler variational function

$$\Psi = \Psi_J\left[1 + \sum_{\text{cyc}}\xi(r_{ij})\,\xi(r_{ik})\mathbf{r}_{ij}\cdot\mathbf{r}_{ik}\right] \quad , \tag{5.6}$$

where $\xi(r)$ can be considered as a variational function.

The product form (5.1) could be also generalized by including three–body correlation factors depending only on the particles coordinates:

$$\Psi_{JT} = \Psi_J \prod_{i<j<k=1}^{N} f_3(i,j,k) \quad . \tag{5.7}$$

For an infinite and translationally invariant system, f_3 depends only on the interparticle distances so that we can write

$$\Psi_{JT} = \prod_{i<j}^{N} f(r_{ij}) \prod_{i<j<k=1}^{N} f_3(r_{ij}, r_{ik}, r_{jk}) \quad . \tag{5.8}$$

In order to distinguish between different angular momentum states of the three particles, f_3 can be chosen of the form

$$f_3(r_{ij}, r_{ik}, r_{jk}) = \sum_{cyc} \sum_{\ell=0}^{\infty} f_{3\ell}(r_{ij}, R_k)\, P_\ell(\hat{r}_{ij} \cdot \hat{R}_k) \quad , \tag{5.9}$$

where $\mathbf{R}_k = (\mathbf{r}_{ik} + \mathbf{r}_{jk})/2$. Equivalently, one can also consider

$$f_3(r_{ij}, r_{ik}, r_{jk}) = \sum_{cyc} \sum_{\ell=0}^{\infty} F_\ell(r_{ij}, r_{ik})\, P_\ell(\hat{r}_{ij} \cdot \hat{r}_{ik}) \quad . \tag{5.10}$$

The determination of the optimum three–body correlation factor is an interesting problem[42,48], but the necessary numerical procedure is quite difficult. For such reason, up to now the most accurate calculations have been performed in correspondence with simple but flexible analytical choices for f_3.

5.2 CLUSTER EXPANSION FOR JASTROW–TRIPLET CORRELATED WAVE FUNCTIONS.

The cluster expansion of the two– or more–particle distribution function in correspondence with the wave function Ψ_{JT} given by eq. (5.8), can be obtained by a procedure similar to the one adopted for Ψ_J. The expansion involves irreducible diagrams only, but there is now another diagrammatic link, namely

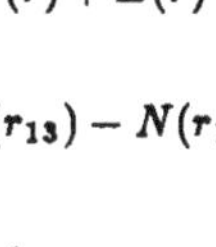

$$= f_3(r_{ij}, r_{ik}, r_{jk})^2 - 1 \quad .$$

The radial two–body distribution function satisfies the following convolution equation

$$g(r) = f(r)^2 \exp\Big[N(r) + E(r) + C(r)\Big] \quad ,$$

$$N(r_{12}) = \rho \int d^3r_3 \Big[g(r_{13}) - N(r_{13}) - 1\Big]\Big[g(r_{23}) - 1\Big] \quad , \tag{5.11}$$

$$C(r_{12}) = \rho \int d^3r_3 \Big[f_3(r_{12}, r_{13}, r_{23})^2 - 1\Big] g(r_{13}) g(r_{23}) \quad ,$$

$E(r_{12}) =$ 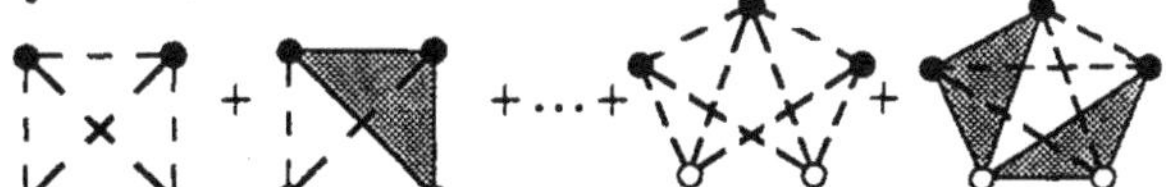 $\dots$.

If the JF expression[21] is adopted for the expectation value of the kinetic energy, the energy per particle can be written in the form.

$$E/N = \frac{\rho}{2} \int d^3r\, g(r) \left[v(r) - \frac{\hbar^2}{4m} \nabla^2 \ln f(r)^2 \right] + \frac{\rho^2}{4} \int d^3r_{12}\, d^3r_{13} g_3(r_{12}, r_{13}, r_{23}) \left[-\frac{\hbar^2}{4m} \nabla_1^2 \ln f_3(r_{12}, r_{13}, r_{23})^2 \right] \quad . \tag{5.12}$$

The difference from eq. (2.37) relative to a Jastrow function, is due to the term containing the radial three–body distribution function. As discussed in sect. 3.4, a possible way to estimate $g_3(r_{12}, r_{13}, r_{23})$ is to use the relation

$$g_3(r_{12}, r_{13}, r_{23}) = f_3(r_{12}, r_{13}, r_{23})^2 g(r_{12}) g(r_{13}) g(r_{23}) \exp\left[A(r_{12}, r_{13}, r_{23}) \right] \quad , \tag{5.13}$$

$A(r_{12}, r_{13}, r_{23}) =$ 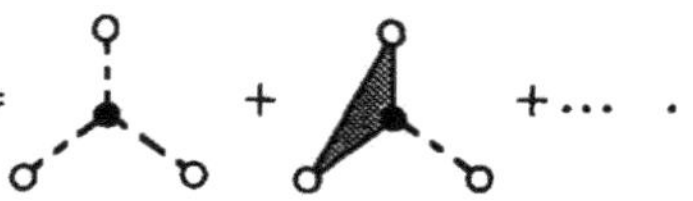 .

where $A(r_{12}, r_{13}, r_{23})$ is the sum of all the Abe diagrams[33]. The numerical calculation of the distribution functions presents difficulties in connection with the elementary and Abe diagrams. Moreover, there are two classes of elementary diagrams containing or not containing triplet correlations, and consequently the number of elementary terms with a given number of particles (or points) increases. As an example, the four-particle elementary diagrams and the four–particle Abe diagrams are the following:

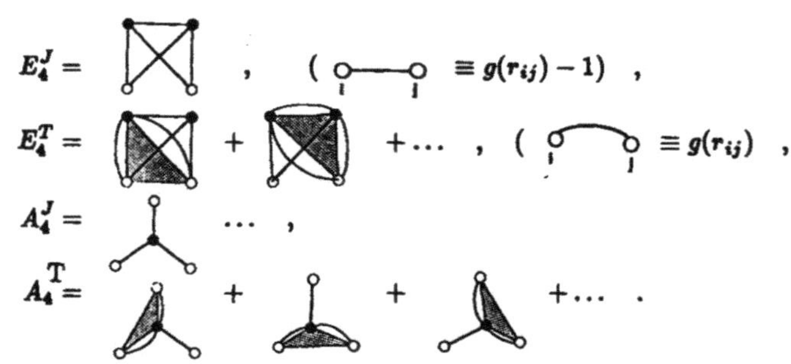

(5.14)

Since $f_3(r_{12}, r_{13}, r_{23})^2 - 1$ is a small quantity, one can adopt the approximation

$$E \approx E^J + E_4^T \quad , \qquad A \approx A^J + A_4^T \quad . \tag{5.15}$$

E^J and A^J should sum up all the terms which do not explicitely contain triplet correlation. In ref. (32), the scaling approximation has been applied to both E^J and A^J,

$$E^J = (1+s)E_4^J \quad ,$$
$$A^J = (1+s/2)A_4^J \quad , \tag{5.16}$$

The values $s/2$ for the scaling parameter relative to A^J seems rather well justified, and the parameter s has been fixed by requiring the condition

$$<T_{JF}>=<T_{PB}> \quad , \tag{5.17}$$

to be satisfied when the triplet correlations are dropped out (Jastrow wave function). If the interpolating equation method is adopted to evaluate E^J as in ref. (44), we write

$$E^J = \ln\left(\frac{e^{\alpha(r)N(r)} - 1 + \alpha(r)}{\alpha(r)}\right) - N(r) \quad ,$$
$$A^J = SA_4^J \quad , \tag{5.18}$$

and S is fixed by means of eq. (5.17). The function $\alpha(r)$ appearing in the preceding expression can be approximated by the one evaluated at the Jastrow level.

5.3 LIQUID ^{4}He *WITH THE "JT" CORRELATED WAVE FUNCTION.*

In this section the results obtained[49] for the ground state of liquid ^{4}He with the variational function Ψ_{JT} given by eq. (5.7) are presented. The elementary and Abe diagrams have been calculated by means of eq. (5.18) and the three-body correlation factor is taken to be

$$f_3(i,j,k) = \exp[-\zeta(i,j,k)] \quad , \quad \zeta(i,j,k) = \sum_{cyc} \xi_1(r_{ij})\xi_1(r_{ik})P_1(\hat{r}_{ij}\cdot\hat{r}_{ik}) \quad . \tag{5.19}$$

Such a form for f_3 has been used in a number of variational Monte Carlo[50] and analytical[51] calculations on quantum fluids. The function ξ_1 is taken as

$$\xi_1(r) = \sqrt{\lambda_1}\exp\left[-\left(\frac{r-r_{t1}}{r_{w1}}\right)^2\right] \quad , \tag{5.20}$$

where λ_1, r_{t1} and r_{w1} are trial parameters. The two-particle distribution function $g(r)$, and the corresponding correlation factor $f(r)$, have been calculated by solving the Euler equation (corresponding to Ψ_{JT}), by a procedure similar to that employed in ref. (44) when only pair correlations were included in the wave function. Table VI presents the energy per particle of liquid ^{4}He described by the HFDHE2 potential. E_J and E_{JT} are the energies estimated with a Jastrow type wave function and with the inclusion of the triplet correlations too, respectively. The parameters of the three-body correlation (see

$\rho\sigma^3$	E_J/N	E_{JT}/N	GFMC
0.328	−6.14(−6.12)	−6.75(−6.75)	-7.03 ± 0.04
0.347	−6.06(−6.04)	−6.78(−6.77)	
0.365	−5.90(−5.94)	−6.78(−6.81)	-7.12 ± 0.02
0.401	−5.20(−5.46)	−6.50(−6.65)	-6.89 ± 0.05

Table VI. *Energy per particle (kelvin) for the liquid phase of ^{4}He at $T = 0$ with the HFDHE2 potential in the J and $J+T$ approximation. The numbers quoted in parenthesis are taken from refs.(32,50). In the last column the GFMC results[(13)] are reported.*

eq. (5.20)) have been varied to minimize the energy with the resulting values $\lambda_1 = -1.08$, $r_{w1} = 0.41$ and $r_{t1} = 0.80$, for all the densities considered; these values coincide with those obtained in ref.(49). The numbers quoted in parentheses in table VI have been taken from refs. (32,51), where a different (approximate) Euler equation and the scaling approximation were adopted. The agreement between the results yielded by the two different approaches (the interpolating and the scaling approximation) is satisfactory, in particular at the equilibrium density. For the sake of comparison, in the last column of the table, the GFMC[(13)] estimates have been reported. An overall satisfactory agreement is evident, but small differences remain between the (supposed exact) GFMC results and the variational ones.

As regards the variational approach, one could criticize on the following points:

(i) *the approximate treatment of the elementary and Abe diagram contributions.* As previously stated, two different approximation procedures gives close estimates. In particular, at the equilibrium density the results nearly coincide and we can therefore be rather confident with them;

(ii) *the approximate form adopted for $f_3(r_{12}, r_{13}, r_{23})$.* In ref. (51), a more flexible three-body correlation factor has been used, namely

$$f_3(i,j,k) = \sum_{\ell=0}^{2} \sum_{cyc} \xi_\ell(r_{ij})\xi_\ell(r_{ik})P_\ell(\hat{r}_{ij} \cdot \hat{r}_{ik}) \quad , \tag{5.21}$$

$\xi_0(r)$, $\xi_1(r)$ and $\xi_2(r)$ being trial functions. The energy gain obtained is small but of interest; as an example, at the experimental equilibrium density the energy per particle is decreased by 0.15 K. In conclusion, small discrepancies between the variational and GFMC results survive and, perhaps, a more flexible f_3 correlation factor is required;

(iii) *a four-body correlation factor has been completely ignored.* It is believable that including f_4 would produce small effects, however the system under consideration is very dense and a sure conclusion cannot be drawn.

As final remark, we observe that, with the exception of the precedent points which require some better understanding, the variational wave functions discussed in this section allow a satisfactory description of the properties of liquid ^{4}He and, in particular, of the effects related to long-range tails of the wave function.

6. CALCULATION OF THE MOMENTUM DISTRIBUTION OF LIQUID ^{4}He AT ZERO TEMPERATURE.

The occupation probability $n(k)$ of a state with momentum k is an important property of a quantum fluid. In the case of a non-interacting Bose system at temperature T, for $k > 0$ one has the well known Bose-Einstein distribution

$$n(k) = \frac{1}{\exp\left[\epsilon(k)/k_B T\right] - 1} \quad , \quad \epsilon(k) = \frac{\hbar^2 k^2}{2m} \quad . \tag{6.1}$$

At $T = 0$ the condensate fraction, namely the probability of having particles with $k = 0$, is one and $n(k) = 0$ for $k > 0$; when the temperature increases the condensate fraction rapidly decreases. In a real system, the interparticle interactions are responsable for correlations among the particles. and at $T = 0$ the condensate fraction n_0 is no longer equal to unity. Strong interactions cause large reductions of n_0 which turns out to be an important parameter characterizing the structure of the system.

The momentum distribution can be derived in the limit of high momentum transfer (inpulse approximation) from X-rays or neutron scattering experiments[52]. For an isotropic liquid, $n(k)$ depends only on $k = |\mathbf{k}|$ and it can be calculated as the Fourier transform of the one-body density matrix $\rho(\mathbf{r}_1, \mathbf{r}_{1'})$. Moreover, for a homogeneous liquid $\rho(\mathbf{r}_1, \mathbf{r}_{1'})$ depends only on $r_{11'} = |\mathbf{r}_1 - \mathbf{r}_{1'}|$ and one obtains the equations

$$\rho(r_{11'}) = N \frac{\int d^3r_2 \dots d^3r_N \; \Psi^*(1,2,\dots,N)\Psi(1',2,\dots,N)}{\int d^3r_1 \dots d^3r_N |\Psi(1,\dots,N)|^2} \quad , \tag{6.2}$$

$$n(k) = n_0 \delta(k) + \int d^3r \left(\rho(r) - n_0\rho\right) \exp(i\,\mathbf{k}\cdot\mathbf{r}) \quad , \tag{6.3}$$

$$\lim_{r\to\infty} \frac{\rho(r)}{\rho} = n_0 \quad , \quad \rho(r=0) = \rho \quad . \tag{6.4}$$

The average kinetic energy is expressed in terms of $n(k)$, by the following equation:

$$<T> = \int \frac{d^3k}{(2\pi)^3\rho} \frac{\hbar^2 k^2}{2m} n(k) \quad . \tag{6.5}$$

The momentum distribution for a quantum fluid as liquid ^{4}He at $T = 0$ can be calculated from variational wave function by means of hypernetted chain methods or numerical Monte

Carlo method[17,18,30]. The basis of the analytical variational approach is discussed in the next section, whilst the remaining of the chapter is devoted to numerical results for the ground state of liquid ^{4}He.

6.1 HYPERNETTED CHAIN TECHNIQUE FOR THE MOMENTUM DISTRIBUTION.

The equations needed to calculate $n(k)$ in correspondence with trial functions containing only two–body correlations have been derived by FANTONI[53] in the frame of a HNC technique. The starting point is the cluster expansion for the one–body density matrix which is worked out by means of a procedure similar to the one for $g(r)$. It is convenient, in order to expand the numerator in eq. (6.2), to introduce the following functions:

 $\equiv f(r_{ij})^2 - 1 = h(r_{ij})$,  $\equiv f(r_{ij}) - 1 = \xi(r_{ij})$. (6.6)

It must be noticed that one of the indices in $\xi(r_{ij})$ must be 1 or 1′ and moreover no link involving $r_{11'}$ exists. In terms of the expansion parameters $h(r)$ and $\xi(r)$ one gets:

$\rho(r_{11'}) = N$ 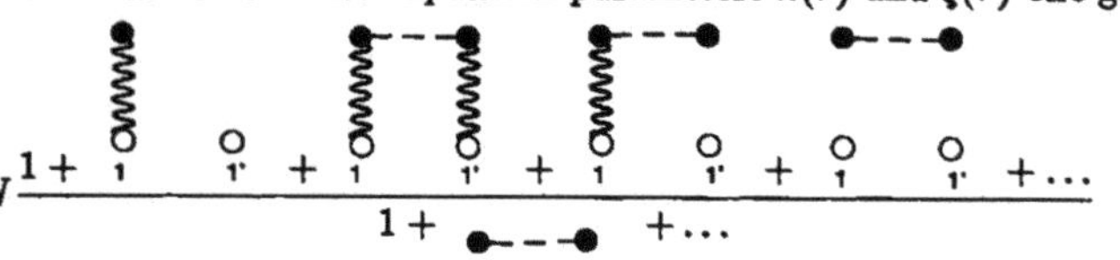 . (6.7)

The denominator of the latter equation cancel out with the unlinked diagrams and a large part of the reducible diagrams, but not all, which appear at the numerator. The terms associated with the remaining diagrams can be summed up by means of integral equations to get the following equations:

$$\rho(r) = \rho\, n_0 \exp\Big[N_{ww}(r) + E_{ww}(r)\Big] \quad , \tag{6.8}$$

$$N_{ww}(r_{12}) = \rho \int d^3r_3 \, \Big[g_{wd}(r_{23}) - 1\Big]\Big[g_{wd}(r_{13}) - N_{wd}(r_{13}) - 1\Big] \quad , \tag{6.9}$$

$$g_{wd}(r) = f(r) \exp\Big[N_{wd}(r) + E_{wd}(r)\Big] \quad , \tag{6.10}$$

$$N_{wd}(r_{12}) = \rho \int d^3r_3 \, \Big[g_{wd}(r_{23}) - 1\Big]\Big[g_{dd}(r_{13}) - N_{dd}(r_{13}) - 1\Big] \quad . \tag{6.11}$$

The function $g_{dd}(r)$ and $N_{dd}(r)$ coincide with the radial two–body distribution function and the associated nodal function already discussed in sects. (3.2) and (5.2). The quantity $E_{ww}(r_{12})$ ($E_{wd}(r_{12})$) results from the sum of all elementary diagrams where both the external points 1 and 2 (only one) are extremities of wiggly lines. The condensate fraction is given by

$$n_0 = \exp\Big[2R_w - R_d\Big] \quad , \tag{6.12}$$

with

$$R_w = \rho \int d^3r \left[g_{wd}(r) - N_{wd}(r) - E_{wd}(r) - 1\right] - \frac{\rho}{2}\int d^3r \left[g_{wd}(r) - 1\right]\left[N_{wd}(r) + 2E_{wd}(r)\right] + E_w \quad , \tag{6.13}$$

and R_d is given by an analogous equation after changing the subscript w with d. The quantity E_w and E_d correspond to particular elementary diagrams[53].

The above equations can be rather easily modified[54] for the case of trial wave functions including three-body correlations, too. The function $E_{wd}(r)$ must be replaced by $E_{wd}(r) + C_{wd}(r)$, where

$$C_{wd}(r) = \rho \int d^3r_3 \, \left(f_3(r_{12}, r_{13}, r_{23}) - 1\right) g_{wd}(r_{13}) g_{dd}(r_{23}) \quad , \tag{6.14}$$

and, of course, the radial distribution function is the one pertinent to Ψ_{JT}. Moreover, the functions $E_{wd}(r)$ and $E_{ww}(r)$ sum up also all the elementary diagrams containing explicit factors like $f_3(r_{ij}, r_{ik}, r_{jk})^2 - 1$; if one of the indices i, j, k is 1 or 1′, the factor is $f_3(r_{ij}, r_{ik}, r_{jk}) - 1$ instead of $f_3(r_{ij}, r_{ik}, r_{jk})^2 - 1$. The expression for R_w is now as follows[54]

$$R_w = \rho \int d^3r \left[g_{wd}(r) - N_{wd}(r) - E_{wd}(r) - 1\right] - \frac{\rho}{2}\int d^3r \left[g_{wd}(r) - 1\right]\left[N_{wd}(r) + 2E_{wd}(r)\right] - \frac{\rho}{2}\int d^3r \left[g_{wd}(r) C_{wd}(r)\right] + E_w \quad . \tag{6.15}$$

R_d is again obtained by replacing the subscript w with d in the last equation. In ref. (54) it was observed that the function $g_{wd}(r)$ does not largely deviate from $g(r)$ and therefore the associated elementary functions could be approximated with a scaling procedure, namely:

$$\begin{aligned} E_{ww}(r) &= (1 + s_{ww}) E^J_{4ww}(r) + E^T_{4ww}(r) \quad , \\ E_{wd}(r) &= (1 + s_{wd}) E^J_{4wd}(r) + E^T_{4wd}(r) \quad , \end{aligned} \tag{6.16}$$

where $E^g_{4xy}(r)$ correspond to the four-particle elementary diagrams with only $(g(r_{ij}) - 1)$ links, whereas the four-particle elementary terms $E^t_{4xy}(r)$ contain at least one explicit three-body correlation. In ref. (54) the quantities E_w and E_d have been scaled as

$$\begin{aligned} E_w &= (1 + \frac{2}{3} s_{ww}) E^J_{4w} + E^T_{4w} \quad , \\ E_d &= (1 + \frac{2}{3} s_{wd}) E^J_{4d} + E^T_{4d} \quad . \end{aligned} \tag{6.17}$$

The afore–mentioned elementary contributions can be approximated also by using the interpolating technique, as done in refs. (44) and (49),

$$E_{ww}(r) = \ln\Big[\frac{\exp\big[\gamma(r)N_{ww}(r)\big] - 1 + \gamma(r)}{\gamma(r)}\Big] - N_{ww}(r) + E^{T}_{4ww}(r) \quad ,$$
$$E_{wd}(r) = \ln\Big[\frac{\exp\big[\beta(r) + N_{wd}(r)\big] - 1 + \beta(r)}{\beta(r)}\Big] - N_{wd}(r) + E^{T}_{4wd}(r) \quad , \tag{6.18}$$

where the interpolating functions $\beta(r)$ and $\gamma(r)$ contain both one free parameter, namely β_0 and γ_0 respectively.

In conclusion, both the scaling and the interpolating approximation contain two parameters which can be fixed in order to satisfy the normalization condition given by eq. (6.4), namely

$$\rho(0)/\rho = n_0 + \int \frac{d^3k}{(2\pi)^3\rho} n(k) = 1 \quad , \tag{6.19}$$

and eq. (6.5) where $<T>$ is preliminarily evaluated by means of the JF form of the kinetic energy given by eq. (2.37).

6.2 *MOMENTUM DISTRIBUTION FOR LIQUID* ^{4}He.

The calculation of the momentum distribution for a strongly interacting quantum fluid is a rather complicate task which causes some troubles to both the variational and GFMC approach. For such a reason, the comparison between the results yielded by different techniques can be very enlightening. First of all, let us consider Jastrow wave functions with the "optimum" correlation factor as discussed in chapter 4. The results obtained in various papers for the condensed fraction n_0 of the ground state of liquid ^{4}He are presented in table VII both for the HFDHE2 and LJ potential. The techniques adopted are based on the scaling approximation[54], the mixture formalism of ref. (55), and the interpolating procedure[44], respectively. In the latter case, the elementary contribution E_w and E_d of eq. (6.13) have been approximated as

$$E_w = \frac{\rho}{3}\int d^3r \,\Big[g_{wd}(r) - 1\Big]E_{wd}(r) \quad , \quad E_d = \frac{\rho}{3}\int d^3r \,\Big[g_{dd}(r) - 1\Big]E_{dd}(r) \quad . \tag{6.20}$$

In such a way, wrong symmetry numbers are attributed to diagrams involving more than four particles. This is not a severe restriction since the quantity $2E_w - E_d$ is expected to be small. In ref. (55) this quantity was disregarded in calculating n_0, which produces the small differences between the values given there and those obtained by the interpolating procedure.

A more accurate description of the system requires the inclusion of explicit three–body correlation factors, as it has been discussed in chapter 5. The corresponding wave function Ψ_{JT} can be used for calculating the momentum distribution. The condensate

$\rho\sigma^3$	HFDHE2 potential Scal.appr.(54)	Inter.appr.(44)	LJ potential Mixture form.(55)	Inter.appr.(44)
0.365	0.098	0.115	0.139	0.134
0.380	0.087*	0.101	0.124	0.120
0.401	0.071	0.086	0.103*	0.105

Table VII. *Condensate fraction n_0 for the ground state of* ^{4}He *at $T = 0$ described by a "optimum" Jastrow wave function. The values marked by an asterisk have been obtained by linear interpolation of the data available in the original paper.*

$\rho\sigma^3$	JT wave function Scal.appr.(54)	Inter.appr.(49)	GFMC
0.365	0.092	0.108	0.092 ± 0.001
0.380	0.082*	0.093	0.075*
0.401	0.065	0.075	0.052 ± 0.001

Table VIII. *Condensate fraction n_0 for the ground state of* ^{4}He *at $T = 0$ described by a wave function Ψ_{JT}. The values marked by an asterisk have been obtained by linear interpolation from the data available in the original papers.*

fraction obtained for the HFDHE2 potential is given in table VIII. The estimates based on the interpolating or scaling approximation have been taken from refs. (49) and (54). In the last column of table VIII, the Green Function Monte Carlo results[56] are reported.

The quantity $kn(k)$, as evaluated for ^{4}He, by using the interpolating[49] or the scaling[54] approximation for Ψ_{JT}, at the experimental equilibrium density is represented in fig. 6, together with the GFMC estimate[56] and the experimental data quoted in ref. (54).

In conclusion, it must be noticed that the momentum distribution and, in particular, the value of n_0 are strongly dependent on the elementary contributions and the improvements on the variational wave function. The GFMC method should give the exact values (provided the effects due to the finite box size in the GFMC simulation are negligible). The variational and GFMC predictions for n_0 and $n(k)$ differ to some extent and at present the reasons of such discrepancies are not completely clear.

7. CONCLUSIONS.

The study of strongly interacting quantum fluids has received large attention in the

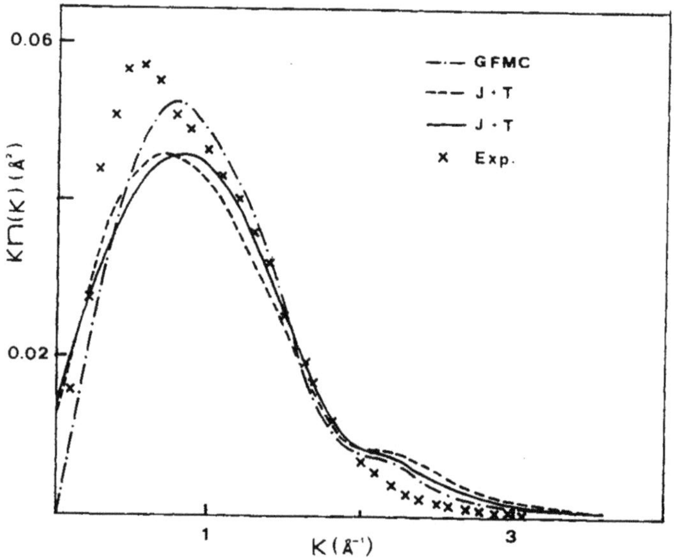

Figure 6. *The function kn(k), as calculated for the ground state of liquid ^{4}He at the experimental equilibrium density. The solid and dashed lines correspond to the results obtained with the interpolating and the scaling approximation, respectively. The dot–dashed line represents the GFMC results. The crosses are the experimental data.*

last few years. A number of techniques have been widely investigated and applied to that problem with the aim of obtaining, with different procedures, results in close agreement. At present, the most popular approaches are the variational ones, in both the analytical and Monte Carlo version, the perturbative theory in conjunction with the variational method, and the exact numerical solution of the problem in the frame of the Green Function Monte Carlo method. As a product of many converging efforts, it is now possible to perform accurate calculations on the quantum fluids of interest, as liquid ^{4}He and $\uparrow$H, liquid ^{3}He, nuclear and neutron matter and electron gas.

This paper has been devoted to a discussion of the analytical variational theory for Bose fluids on the basis of the hypernetted chain summation method. The variational and perturbative theory for Fermi systems is accurately discussed in the paper of Fabrocini and Fantoni[57]. There are few points of the variational approach which deserve some final comments. First of all, for dense systems, like liquid ^{4}He and ^{3}He, the elementary contributions are quite substantial, and crude approximations as the HNC/0 and PY approximations, are not reliable. Moreover, the convergence of the elementary contributions in terms of the number of particles involved is slow and therefore methods as accurate as possible for evaluating the elementary contributions have been devised and carefully tested. The scaling and interpolating procedures discussed in chapter 3 are reliable, in particular at the equilibrium density.

The long–range behaviour of the system can be carefully described at the level of pair correlations, which allows reliable calculations of the properties of the system at small k values.

The pair correlations have a predominant role but as the density of the system increases, also effective three–body correlations must be included in the trial wave function. The inclusion of four–body correlations has not received attention up to now, but it is presumable that, for the physical systems of interest, their contribution, if not at all negligible, should be small.

In conclusion, the variational theory allows a detailed description of the structure of a strongly interacting quantum fluid and gives accurate estimates of the important properties of the system.

8. REFERENCES.

1 – P.KAPITZA, Nature **141**, 74 (1937);

2 – E.K.ACHTER and L.MEYER, Phys.Rev. **188**, 291 (1969); R.B.HALLOCK Phys.Rev. **A5**, 320 (1972); E.C.SVENSSON, Phys.Rev. **B21**, 3638 (1980);

3 – L.VAN HOVE, Phys.Rev. **95**, 249, 1374 (1954);

4 – R.FELTGEN, H.KIRST, K.A.KÖHLER, H.PAULY and F.TORELLO; J.Chem.Phys. **76**, 2360 (1985);

5 – LONDON; Z.Phys.Chem **B11**, 222 (1930);

6 – J.de BOER and A.MICHELS; Physica (Utrecht) **5**, 945 (1938);

7 – D.R.McLAUGHLIN and H.F.SHAEFER III, Chem.Phys.Lett. **12**, 244 (1971);

8 – R.AHLRICHS, P.PENCO and G.SCOLES, Chem.Phys. **19**, 119 (1976);

9 – R.A.AZIZ, V.P.S.NAIN, J.S.CARLEY, W.L.TAYLOR and G.T.McCONVILLE, J.Chem.Phys. **70**, 4330 (1979);

10 – D.M.CEPERLEY and H.PARTRIDGE, J.Chem.Phys. **84**, 820 (1986);

11 – M.AXILROD and E.TELLER, J.Chem.Phys. **11**, 293 (1943);

12 – R.D.MURPHY and J.A.BAKER, Phys.Rev. **A3**, 1037 (1971);

13 – M.H.KALOS, M.A.LEE, P.A.WHITLOCK and G.V.CHESTER. Phys.Rev. **B24**, 115 (1981);

14 – A.BIJL, Physica **7**, 869 (1940);

15 – R.B.DINGLE, Phil.Mag. **40**, 573 (1949);

16 – R.JASTROW, Phys.Rev. **98** 1479 (1955);

17 – W.L.McMILLAN, Phys.Rev. **138**, 442 (1965);

18 – D.SCHIFF and L.VERLET, Phys.Rev. **160**, 208 (1967);

19 – E.FEENBERG, **THEORY OF QUANTUM FLUIDS**, (Academic Press, New York, 1969);

20 – V.R.PANDHARIPANDE and H.A.BETHE, Phys.Rev. **C7**, 1312 (1973);

21 – H.W.JACKSON and E.FEENBERG, Ann. of Phys. (N.Y.) **15**, 266 (1961);

22 – J.E.MAYER and M.G.MAYER, **STATISTICAL MECHANICS**, (New York, 1940);

23 – G.E.UNLHEBECK and G.W.FORD, **STUDIES IN STATISTICAL MECHANICS**, vol. 1 part B (Amsterdam, 1962);

24 – M.GAUDIN, J.GILLESPIE and G.RIPKA, Nucl.Phys. **A176**, 237 (1971);

25 – see, for instance, T.H.HILL, **AN INTRODUCTION TO STATISTICAL MECHANICS**, (Reading, Mass., 1960);

26 – J.M.J.VAN LEEUWEN, J.GROENEWELD and J.DE BOER, Physica **25**, 792 (1959);

27 – T.MORITA, Progr.Theor.Phys. **20**, 920 (1958);

28 – R.A.SMITH, AS.KALLIO, M.PUOSKARI and P.TOROPAINEN, Nucl.Phys. A328, 186 (1979);

29 – J.K.PERCUS and G.J.YEVIK, Phys.Rev. **110**, 1 (1958);

30 – M.H.KALOS, D.LEVESQUE and L.VERLET, Phys.Rev. **A9** 2178 (1974);

31 – V.R.PANDHARIPANDE and K.E.SCHIMIDT, Phys.Rev. **B19**, 1189 (1978);

32 – Q.N.USMANI, B.FRIEDMAN and V.R.PANDHARIPANDE, Phys.Rev. **B25**, 4502 (1982);

33 – R.ABE, Progr.Theor.Phys. **21**, 421 (1959);

34 – See, for instance, A.MÜNSTER, **STATISTICAL THERMODYNAMICS**, (Academic Press, New York, 1969);

35 – J.S.ROWLINSON, Mol.Phys. **9**, 217 (1965); D.D.CARLEY and F.LADO, Phys.Rev. **137**, 42 (1965);

36 – T.MORITA, Progr.Theor.Phys., **41**, 339 (1969);

37 – A.FABROCINI and S.ROSATI, Nuovo Cimento **D1**, 567 (1982);

38 – A.FABROCINI and S.ROSATI, Nuovo Cimento **D1**, 615 (1982);

39 – L.J.LANTTO and P.J.SIEMENS, Phys.Lett. **B68**, 308 (1977), L.J.LANTTO, A.D.JACKSON and P.J.SIEMENS, Phys.Lett. **B68**, 311 (1977); A.D.JACKSON, A.LANDE and L.J.LANTTO, Nucl.Phys. **A317**, 70 (1979);

40 – L.CASTILLEJO, A.D.JACKSON, B.K.JENNINGS and R.A.SMITH, Phys.Rev. **B20**, 3631 (1979);

41 – C.E.CAMPBELL and E.FEENBERG, Phys.Rev. **188**, 396 (1969);

42 – C.C.CHANG and C.E.CAMPBELL, Phys.Rev. **B15**, 4238 (1977);

43 – A.FABROCINI and A.POLLS, Phys.Rev. **B30**, 1200 (1984);

44 – M.VIVIANI, E.BUENDIA, A.FABROCINI and S.ROSATI, Nuovo Cimento **D8**, 561 (1986);

45 – V.R.PANDHARIPANDE and N.ITOH, Phys.Rev. **A8**, 2564 (1979);

46 – V.R.PANDHARIPANDE, Phys.Rev. **B18**, 218 (1978);

47 – R.P.FEYNMAN and M.COHEN, Phys.Rev. **102**, 1189 (1956);

48 – E.KROTSCHECK, Phys.Rev. **B33**, 3158 (1986);

49 – S.ROSATI, E.BUENDIA and M.VIVIANI, contribution to the 4th *International Conference on Recent Progress in Many–Body Theories*, San Francisco, 1985, unpublished;

50 – K.SCHMIDT, M.H.KALOS, M.A.LEE and G.V.CHESTER, Phys.Rev.Lett. **45**, 5731 (1980);

51 – Q.N.USMANI, S.FANTONI and V.R.PANDHARIPANDE, Phys.Rev. **B26**, 6123 (1982);

52 – V.F.SEARS, Phys.Rev. **B28**, 5109 (1983);

53 – S.FANTONI, Nuovo Cimento **A44**, 191 (1978);

54 – E.MANOUSAKIS, V.R.PANDHARIPANDE and Q.N.USMANI, Phys.Rev. **B31**, 7022 (1985);

55 – M.PUOSKARI and A.KALLIO, Phys.Rev. **B30**, 152 (1984);

56 – P.A.WHITLOCK and R.M.PANOFF (unpublished);

57 – A.FABROCINI and S.FANTONI, see contribution to this volume.

HARTREE-FOCK METHODS AND SOLUBLE MODELS IN NUCLEAR PHYSICS

by
M.C.Cambiaggio(*)
Member of CONICET, Argentina

(*)Departamento de Física, CNEA,
Av. Libertador 8250 - 1429 Buenos Aires (Argentina)

HARTREE-FOCK METHODS AND SOLUBLE MODELS IN NUCLEAR PHYSICS

by M. C. Cambiaggio

1.- Introduction

In these lectures we shall talk about two well-known methods in nuclear physics: the Hartree-Fock (HF) and the Time Dependent Hartree-Fock (TDHF) methods. Both are approximations to the exact many body problem obtained considering an independent particle model. The basic idea is that the interaction among the nucleons creates an average potential which is felt by each nucleon. Consequently, a good starting point for the approximate description of the system takes into account this average potential and ignores all the other effects. Of course, there are methods for including these effects, at least partially, but we will not talk about them here. The idea of independent particles is justified by the success of the phenomenologically introduced shell model. To implement it one looks for the best function within a set of trial functions formed with independent particle wave functions in an appropriate average potential. As one studies a fermion system the trial functions are antisymmetric Slater determinants.

Moreover, both methods are variational approaches, i.e., the best wave function is determined variationally. In the time independent case the variational principle states that the exact Schrödinger equation is equivalent to the variational equation $\delta E(\psi) = 0$

$$\delta E = \delta \frac{\langle\psi|\hat{H}|\psi\rangle}{\langle\psi|\psi\rangle} = 0 . \tag{1.1}$$

Calculating the variation one obtains

$$\langle\delta\psi|\hat{H} - E|\psi\rangle + \langle\psi|\hat{H} - E|\delta\psi\rangle = 0 . \tag{1.2}$$

$|\psi\rangle$ is in general a complex wave function and as the equation must be valid for any infinitesimal $|\delta\psi\rangle$, we replace $|\delta\psi\rangle$ by $i|\delta\psi\rangle$ and get

$$-i\langle\delta\psi|\hat{H} - E|\psi\rangle + i\langle\psi|\hat{H} - E|\delta\psi\rangle = 0 . \tag{1.3}$$

Eqs. (1.2) and (1.3) imply

$$\langle\delta\psi|\hat{H} - E|\psi\rangle = 0 \tag{1.4}$$

and the complex conjugate equation. This means that the variation may be performed with respect to $|\delta\psi\rangle$ and $\langle\delta\psi|$ independently, instead of with respect to the real and imaginary parts. As $|\delta\psi\rangle$ is arbitrary, the Schrödinger equation

$$\hat{H}|\psi\rangle = E|\psi\rangle \tag{1.5}$$

follows immediately.

The demonstration is based on the usage of the most general variation but usually $|\psi\rangle$ is restricted to a set of mathematically simple trial wave functions. If this set does not contain the exact function the obtained solution is an approximate one. An important characteristic of the variational method is that the approximate ground state energy turns out to be an upper limit to the exact energy. To prove it we expand the trial function in terms of the exact eigenfunctions

$$|\psi\rangle = \sum_n a_n \; |\psi_n\rangle \tag{1.6}$$

with

$$\hat{H}|\psi_n\rangle = E_n|\psi_n\rangle . \tag{1.7}$$

Then,

$$E(\psi) = \frac{\sum_n |a_n|^2 \; E_n}{\sum_n |a_n|^2} \geq \frac{\sum_n |a_n|^2 \; E_o}{\sum_n |a_n|^2} = E_o \tag{1.8}$$

where Eo is the exact ground state energy.

Once an approximate ground state $|\phi_o\rangle$ has been obtained it is possible to calculate an approximate function $|\phi_1\rangle$, for the first excited state solving the variational equation with the additional condition of $|\phi_1\rangle$ being orthogonal to $|\phi_o\rangle$, $\langle\phi_1|\phi_o\rangle = 0$. For the second excited state one has two additional conditions: $\langle\phi_2|\phi_1\rangle = 0$ and $\langle\phi_2|\phi_o\rangle = 0$. These constraints are coupled to the problem through Lagrange multipliers. It is clearly seen that the method complicates itself very rapidly as the excitation energy increases. It has been thus applied mainly to the calculation of the ground state. However, in some cases these conditions are easily satisfied due to symmetry properties and then a whole band may be calculated. Another point to be mentioned is that when doing realistic calculations one has to be sure about reaching a minimum and not only an extremum through, for example, the calculation of the second derivative of the energy functional.

2.- Hartree-Fock equations

To derive explicitly the HF equations we shall use the second quantization formalism. We have a system of N particles (fermions) represented by the hamiltonian

$$\hat{H}=\sum_{\nu\nu'}\varepsilon_{\nu\nu'}a_{\nu}^{+}a_{\nu'}+\frac{1}{4}\sum_{\nu_1\nu_2\nu_1'\nu_2'}\bar{v}_{\nu_1\nu_2\nu_1'\nu_2'}\,a_{\nu_1}^{+}a_{\nu_2}^{+}a_{\nu_2'}a_{\nu_1'} \qquad (2.1)$$

where $\bar{v}_{\nu_1\nu_2\nu_1'\nu_2'}$ is the antisymmetrized matrix element of the interaction. The HF approximation consists in finding that N particle wave function

$$|\phi\rangle=a_{\nu_1}^{+}\;a_{\nu_2}^{+}\;\ldots\;a_{\nu_N}^{+}|0\rangle \qquad (2.2)$$

which minimizes the expectation value of the hamiltonian:

$$\delta\langle\phi|\hat{H}|\phi\rangle=0\,. \qquad (2.3)$$

The trial functions are Slater determinants and their variation is performed varying the single particle states. Consequently the new single particle states correspond to new creation and annihilation operators b^{+}_{ν} and b_{ν} which may be written in terms of the previous ones

$$b_{\nu}^{+}=\sum_{\mu}C_{\mu\nu}\,a_{\mu}^{+}\,. \qquad (2.4)$$

Both sets of states are complete and orthogonal and as a consequence the transformation C must be unitary

$$C^{+}C=CC^{+}=1\,. \qquad (2.5)$$

This condition also guarantees that the new operators satisfy Fermi commutation relations. With the b_{ν}^{+} operators a new Slater determinant is formed and the energy is minimized with respect to the variational parameters $C_{\mu\nu}$.

Now we see how to obtain an expression for the most general non orthogonal variation of $|\phi'\rangle$ derived by Thouless

$$|\phi'\rangle=b_{\nu_1}^{+}\;b_{\nu_2}^{+}\;\ldots\;b_{\nu_N}^{+}\;|0\rangle\;. \qquad (2.6)$$

The Slater determinant is constructed ordering the single particle states $\nu_1<\nu_2<\ldots<\nu_N<\ldots$ and occupying the N lowest ones. ν_N = is called Fermi momentum even if one does not work with momentum eigenstates. We represent by ν_i, ν_j the states below the Fermi level, $1\le(i,j)\le N$, and by ν_m , ν_n the ones above, $(n, m)>N$. Then,

$$b_{\nu_i}^{+}=\sum_{j}C_{ji}\;a_{\nu_j}^{+}+\sum_{m}C_{mi}\;a_{\nu_m}^{+}\,. \qquad (2.7)$$

But the first part of the transformation that mixes occupied states, also called hole states, does not contribute to the variation of the Slater determinant. This may be shown through the well-known theorem that states that the sum of any two rows of a determinant does not change the value of the determinant. Consequently, we may write

$$b^{+}_{v_i} = a^{+}_{v_i} + \sum_{m} C_{mi}\, a^{+}_{v_m} \tag{2.8}$$

and calculate $|\phi'\rangle$

$$\begin{aligned} |\phi'\rangle &= \prod_i b^{+}_{v_i} |0\rangle = \prod_i (a^{+}_{v_i} + \sum_m C_{mi} a^{+}_{v_m}) |0\rangle = \\ &= \prod_i (a^{+}_{v_i} + \sum_m C_{mi}\, a^{+}_{v_m}\, a_{v_i}\, a^{+}_{v_i}) |0\rangle = \\ &= \prod_i (1 + \sum_m C_{mi}\, a^{+}_{v_m}\, a_{v_i}) \; |\phi\rangle . \end{aligned} \tag{2.9}$$

Using that $a^2_{vi} = 0$ and that the sums commute, i.e.,

$$\left[\sum_m C_{mi}\, a^{+}_{v_m}\, a_{v_i} , \sum_{m'} C_{m'j}\, a^{+}_{v_{m'}}\, a_{v_j} \right] = 0 \tag{2.10}$$

one gets

$$\begin{aligned} |\phi'\rangle &= \prod_i \exp\left(\sum_m C_{mi}\, a^{+}_{v_m}\, a_{v_i} \right) |\phi\rangle = \\ &= \exp\left(\sum_{i=1}^{N} \sum_{m=N+1}^{\infty} C_{mi}\, a^{+}_{v_m}\, a_{v_i} \right) |\phi\rangle . \end{aligned} \tag{2.11}$$

This last expression is known as the Thouless variational wave function. It is to be noted that $|\phi'\rangle$ is not normalized. Instead one has $\langle\phi|\phi'\rangle = 1$.

To perform the variation of the expectation value of the hamiltonian one must take the C_{mi} infinitesimal. Then,

$$|\delta\phi\rangle = \left[\exp\left(\sum_{im} C_{m_i} a^+_{v_m} a_{v_i}\right) - 1\right]|\phi\rangle$$

$$= \sum_{im} C_{mi}\, a^+_{v_m} a_{v_i}\, |\phi\rangle \tag{2.12}$$

and

$$\langle\delta\phi| = \langle\phi|\sum_{im} C^*_{mi}\, a^+_{v_i}\, a_{v_m}\ . \tag{2.13}$$

As all the C_{mi} and C_{mi} are independent the variation $\delta\langle\phi|\hat{H}|\phi\rangle = 0$ gives

$$\langle\phi|\, a^+_{v_i}\, a_{v_m}\, \hat{H}\, |\phi\rangle = 0\,. \tag{2.14}$$

The operator $a^+_{vm}\, a_{vi}$ is called a particle-hole operator because it destroys a particle in the occupied state v_i (it creates a hole) and creates a particle in the unoccupied state v_m. The particle-hole excitations constitute an important and usual class of nuclear excitations. According to the result obtained in (2.14) the HF wave function is such that the hamiltonian does not connect it with one particle- one hole states.

Taking into account that $a_{vm}|\phi\rangle = 0$ one gets

$$\langle\phi|a^+_{v_i} a_{v_m} \hat{H}\,|\phi\rangle = \langle\phi|\left[a^+_{v_i} a_{v_m}\ ,\ \hat{H}\right]\phi\rangle$$

$$= \langle\phi|a^+_{v_i}\left[a_{v_m}\ ,\ \hat{H}\right]|\phi\rangle\ . \tag{2.15}$$

The commutator may be easily calculated with (2.1) and the Fermi commutation relations among the operators

$$\left[\hat{H}, a_v\right] = -\sum_{v'} \varepsilon_{vv'}\, a_{v'} - \frac{1}{2}\sum_{v_2 v_1' v_2'} \bar{v}_{v v_2 v_1' v_2'}\, a^+_{v_2}\, a_{v_2'}\, a_{v_1'} \tag{2.16}$$

where we have used the antisymmetry of the interaction matrix element. Then, (2.15) gives

$$\langle\phi|a^+_{v_i} a_{v_m}\, \hat{H}\,|\phi\rangle = \varepsilon_{v_m v_i} + \sum_j \bar{v}_{v_m v_j v_i v_j} = h_{mi} = 0\ . \tag{2.17}$$

The matrix elements h_{mi} have two terms: the first one, ε_{mi} is the matrix element of the kinetic energy; the second term,

$$U_{mi} = \sum_j \bar{v}_{v_m v_j v_i v_j} ,$$

may be considered as the matrix element of a one body operator U which is called the HF potential. Then, one has a HF hamiltonian h such that the best single particle states from the variational point of view are those that define an h which does not connect occupied states with unoccupied ones. There is no condition with respect to the matrix elements which connect two occupied (or unoccupied) states because any linear combination of occupied states does not change the Slater determinant Therefore one may choose the single particle states in such a way as to diagonalize separately the submatrices $h_{ii'}$ and $h_{mm'}$

$$h_{kk'} = \varepsilon_{v_k v_{k'}} + \sum_{j=1}^{N} \bar{v}_{v_k v_j v_{k'} v_j} = \eta_k \delta_{kk'} . \tag{2.18}$$

The HF hamiltonian depends on the states taken as occupied because the sum is restricted to them. We may thus find a self-consistent solution. We begin with a certain set of single particle states, with the N lowest of them we construct the HF potential U, diagonalize h and obtain a new set of single particle states. We repeat the procedure until getting consistency. The self-consistent solution will have physical meaning if the corresponding energy is a minimum and not only an extremum. I will not derive the condition to be satisfied, I will only say that it may be derived calculating $<\phi'|H|\phi'>$ up to second order in the coefficients C_{mi}.

Another interesting point is the following. Calculating the HF energy one obtains

$$E_{HF} = <\phi|\hat{H}|\phi> = \sum_i \varepsilon_{v_i v_i} + \frac{1}{2}\sum_{ij} \bar{v}_{v_i v_j v_i v_j}$$

$$= \sum_i \eta_i - \frac{1}{2}\sum_{ij} \bar{v}_{v_i v_j v_i v_j} . \tag{2.19}$$

This means that in spite of being $|\phi>$ an eigenfunction of the self-consistent potential its eigenvalue, which is the sum of the single particle energies, is not the nuclear energy. The difference may be understood thinking that the best potential U is the average field felt by a determined particle i caused by the interaction with all the others. Then η_i also reflects this interaction and on summing all the η_i one takes into account each interaction twice and so overestimates the nuclear energy.

The HF equations are non linear due to the fact that the self-consistent potential U depends not only on the original hamiltonian but also on the solution. Therefore, this potential does not necessarily have the same symmetries than the hamiltonian. Usually the solution has a broken symmetry that may be for example the translational or the rotational invariance or the conservation of the number of particles. The transition from a solution which conserves the symmetry to another one with the symmetry broken is called a phase transition. The great advantage of this symmetry breaking is that allows us to take into account, in an approximate way, many body correlations without losing the simple model of independent particles. Obviously, the exact wave function must have the symmetries of the hamiltonian and their violation is a matter of convenience. However we should try to restore these violated symmetries. We shall see something about this problem in the examples. On the other hand, if the initial solution has a certain symmetry of the hamiltonian it will be conserved by the HF theory. These are called self-consistent symmetries. Consequently if we begin with a certain symmetry the energy minimum will necessarily be found within the wave functions with that symmetry.

3. - The Lipkin model

The HF method has been thoroughly treated in several books (see for example refs. 1 and 2). There also exist numerous applications whose enumeration would take too long. A few of them are given in refs.3 6.

On the other hand, it is illuminating to apply it to a soluble model. In general exactly soluble models have proved to be extremely useful in theoretical studies concerning the validity and/or usefulness of diverse theoretical approximations to the quantum many-body problem. Not only one has exact solutions for comparing with the approximate ones but also the models are simple enough to make transparent the application of the approximations. It is clear that for being meaningful a model must include the essential features of the problem to be studied.

Of these models, the most widely known is perhaps the so-called Lipkin Model[7]. It deals with N particles distributed in two N-fold degenerate single particle (sp) levels, separated by the sp energy ε. Two quantum numbers characterize a given sp state. One of them adopts the value $\sigma = -1$ (lower level) and +1 (upper level). The other, which may be called the p-spin, singles out a state within the N-fold degeneracy (p = 1, 2, ..., N).

Lipkin et al introduced the operators

$$\hat{J}_+ = \hat{J}_-^+ = \sum_p a_{p+}^+ \, a_{p-}$$

$$\hat{J}_z = \frac{1}{2} \sum_{p\sigma} \sigma \, a_{p\sigma}^+ \, a_{p\sigma} \qquad (3.1)$$

which satisfy commutation relations identical to those for angular momentum operators

$$\left[\hat{J}_+ , \hat{J}_-\right] = 2\hat{J}_z \; ; \quad \left[\hat{J}_z , \hat{J}_\pm\right] = \pm \hat{J}_\pm \; .$$

The operators (3.1) are then called quasi-spin operators, we have to deal with an SU(2) algebra and there exists a Casimir operator

$$\hat{J}^2 = \hat{J}_z^2 + \frac{1}{2}\left(\hat{J}_+ \hat{J}_- + \hat{J}_- \hat{J}_+\right) \tag{3.3}$$

which commutes with the three operators J_+, J_-, J_z. Some of the 2^N many body states of the system may be classified as eigenstates of J^2 and J_z labelled by quantum numbers J, M such that

$$\begin{aligned} \hat{J}^2 \, |J, M\rangle &= J(J+1) \, |J, M\rangle \\ \hat{J}_z \, |J, M\rangle &= M |J, M\rangle \; . \end{aligned} \tag{3.4}$$

Lipkin's hamiltonian reads

$$\hat{H} = \varepsilon \hat{J}_z + \frac{V}{2}\left(\hat{J}_+^2 + \hat{J}_-^2\right) . \tag{3.5}$$

When the interaction, that does not affect the p-quantum number and is accordingly referred to as a monopole one, is switched off, the system will be found in its unperturbed ground state (ugs)

$$|u\,g\,s\rangle = |J = \frac{N}{2} \, , \; M = -\frac{N}{2}\rangle \tag{3.6}$$

as it is easily seen from the definition (3.1), i.e., the lowest possible value of J_z is obtained when all particles accomodate themselves in the lower level. Obviously, H commutes with J^2 which means that J is a good quantum number and that the exact solution is obtained by diagonalizing H within a given J-multiplet. Since the maximum value of J is N/2, the size of the matrix to be diagonalized is at most (N+1)x(N+1), a simple task if N is not too large.

Physically one may think on the two levels as representing subshells in the jj-coupling nuclear shell model having the same value of j and opposite parity. Whenever states have the same value of the quantum number p and opposite values of the shell quantum number σ, they must have opposite parity. In the non-interacting system the lower level is filled with 2j+1 nucleons, i.e., it is below the Fermi level whereas the upper level is above the Fermi level, it is empty. This fixed sp potential may be considered as generated by a very stable core formed by the rest of the nucleons. On applying HF we shall calculate, in a self-consistent way, only that part of the average potential that comes from the nucleons in the last shell j, neglecting the influence between the core and the last shell.

The hamiltonian (3.5) commutes not only with $\hat{J}^2$ but also with the parity operator because the interaction always involves a simultaneous change in parity of two particles, and with the number of particles in each p value

$$\hat{n}_p = \sum_{\sigma} a^{+}_{p\sigma}\, a_{p\sigma} \; . \tag{3.7}$$

In searching for the HF solution one tries to retain in it as many symmetries as possible. Agassi et al.[8] used a transformation that violates parity conservation. The requirement that the trial function remains an eigenfunction of $\hat{n}_p$ restricts it to a very simple form. One needs only to consider sp states obtained by mixing a particular state having a given value of the quantum numbers p and σ with its partner in the other shell having the same p-value and opposite σ. A mixing of these two states corresponds simply to a rotation in quasi-spin space. The HF approach consists then in looking for the zero particle-zero hole wave function in the rotated frame. The corresponding angle is that which minimizes the expectation value of the hamiltonian.

The HF transformation is

$$\begin{pmatrix} b_{p-} \\ b_{p+} \end{pmatrix} = \begin{pmatrix} \cos\beta/2 & -i\sin\beta/2 \\ -i\sin v/2 & \cos\beta/2 \end{pmatrix} \begin{pmatrix} a_{p-} \\ a_{p+} \end{pmatrix} \tag{3.8}$$

and inverting it we easily relate the J-operators in the old basis to those in the new one

$$\begin{aligned} \hat{J}_x(a) &= \hat{J}_x(b) \\ \hat{J}_y(a) &= \cos\beta\,\hat{J}_y(b) - \sin\beta\,\hat{J}_z(b) \\ \hat{J}_z(a) &= \cos\beta\,\hat{J}_z(b) + \sin\beta\,\hat{J}_y(b) \end{aligned} \tag{3.9}$$

Then, we calculate the expectation value of the hamiltonian working in the new basis and remembering that in it the HF wave function is

$$|J = \frac{N}{2}\,,\; M = -\frac{N}{2}>$$

$$<\phi|\hat{H}|\phi> = -\,\varepsilon\frac{N}{2}\left[\cos\beta + \frac{V}{2\varepsilon}\;(N-1)\;\;(1-\cos^2\beta)\right]. \tag{3.10}$$

Minimizing (3.10) with respect to β and using the usual parameters

$\chi = \frac{v(N-1)}{\varepsilon}$ and $\varepsilon = 1$, we get

$$\beta = 0; \qquad E_{HF} = -\frac{N}{2} \qquad \text{if } \chi \leq 1,$$
$$\cos\beta = \frac{1}{\chi}; \quad E_{HF} = -\frac{N}{4}\left(\chi + \frac{1}{\chi}\right) \qquad \text{if } \chi > 1 \,. \tag{3.11}$$

This means that there is a phase transition at $\chi = 1$. For coupling constants smaller than the critical value the solution is the trivial one and the HF wave function coincides with the unperturbed ground state which has good parity. For $\chi > 1$ a deformed solution exists and the parity symmetry is broken. Studying the exact solution in the thermodynamical limit ($N \to \infty$, $\chi =$ constant) one actuallly sees a second-order phase transition in the ground state energy.

Comparison of the HF ground state energy with the exact result shows a reasonable agreement which improves with increasing number of particles and coupling strength. For $\chi \simeq 2$ the difference is of the order of 20% and for $\chi \simeq 5$ it is only a few percent[8].

4. - An SU(2)*SU(2) model

As already mentioned in sect. 2 the HF method violates some of the symmetries of the hamiltonian. This has been explicitly shown in the application to the Lipkin model of the preceding section. One thinks then in restoring these symmetries by means of projection techniques.

Now we shall see how this works in a model specially suited to the study of projection techniques. This is also a two level model with identical sp states as before. For convenience, however, we allow the p-spin to adopt the values ±1, ±2,..., ±N/2 and define further, the quasi-spin ooerators[9]

$$\hat{S}_+ = \hat{S}_-^+ = \sum_p a_{p,+}\, a_{-p,-}$$
$$\hat{S}_z = \hat{J}_z = \frac{1}{2}\sum_{p\sigma} a^+_{p\sigma}\, a_{p\sigma} \,. \tag{4.1}$$

The S-operators commute among themselves like angular momentum ones. If one wants to work simultaneously with these S-operators and the J-ones given in (3.1), it is necessary to introduce

$$\hat{U} = \frac{1}{2} \sum_{\sigma} \sigma a^{+}_{p,\sigma} a_{-p,\sigma} \tag{4.2}$$

in order to close the algebra. Moreover, we define

$$\begin{aligned}
\hat{V}_{+} &= \hat{V}_{-}^{+} = \frac{1}{2}(\hat{J}_{+} + \hat{S}_{+}) \\
\hat{W}_{+} &= \hat{W}_{-}^{+} = \frac{1}{2}(\hat{J}_{+} - \hat{S}_{+}) \\
\hat{V}_{z} &= \frac{1}{2}(\hat{J}_{z} + \hat{U}) \\
\hat{W}_{z} &= \frac{1}{2}(\hat{J}_{z} - \hat{U})
\end{aligned} \tag{4.3}$$

and obtain two quasi-spin sets: $\hat{V}_{+}$, $\hat{V}_{-}$, $\hat{V}_{z}$ and $\hat{W}_{+}$, $\hat{W}_{-}$ and $\hat{W}_{z}$. Any given V-operator commutes with all W-ones and therefore the corresponding algebra structure is SU(2)*SU(2). $\hat{V}^2$ and $\hat{W}^2$ are the relevant Casimir operators and the hamiltonian reads (with $\varepsilon = 1$ as before)[9]

$$\begin{aligned}
\hat{H} &= \hat{J}_{z} + \frac{v}{2} (2\hat{V}_{z}\hat{W}_{z} + \hat{W}_{+}\hat{V}_{-} + \hat{W}_{-}\hat{V}_{+}) \\
&= \hat{J}_{z} + \frac{v}{2} (\hat{J}^{2} - \hat{V}^{2} - \hat{W}^{2})
\end{aligned} \tag{4.4}$$

which commutes not only with $\hat{V}^2$ and $\hat{W}^2$ but also with $\hat{J}^2$, so that its exact eigenstates may also be labelled by J. As the unperturbed ground state is characterized by $V = W = N/4$ and $V_Z = W_Z = -N/4$, the energy of the lowest state in each J- multiplet is

$$E = -J + \frac{v}{2} \left[J(J+1) - \frac{N}{2} (\frac{N}{4} + 1) \right] \tag{4.5}$$

with $J \leq \frac{N}{2}$.

Studying the energy difference between two states of different J one finds a set of critical coupling constants $v_J = 1/J$. At $v = v_J$, the Lipkin quasi-spin of the ground state changes from J to J-1. So, the ground state of the system which for $v = 0$ possesses the maximum possible J compatible with N(J=N/2), will run downwards in J as v grows, passing through all possible values until it reaches J=0 at v=1. This is illustrated in fig.1 for N=6.

In order to obtain the HF transformation for this model one must mix the sp orbital $|p,\sigma\rangle$ hot only with $|-p,\sigma\rangle$ as in sect.3 but also with $|-p,\sigma\rangle$. This is conveniently done in two steps, i.e., one performs the two successive rotations

$$\begin{pmatrix} a_{-p,-} \\ a_{p,+} \end{pmatrix} = \begin{pmatrix} \cos\beta/2 & -i\sin\beta/2 \\ -i\sin\beta/2 & \cos\beta/2 \end{pmatrix} \begin{pmatrix} c_{-p,-} \\ c_{p,+} \end{pmatrix}$$

and (4.6)

$$\begin{pmatrix} c_{p,-} \\ c_{p,+} \end{pmatrix} = \begin{pmatrix} \cos\alpha/2 & -i\sin\alpha/2 \\ -i\sin\alpha/2 & \cos\alpha/2 \end{pmatrix} \begin{pmatrix} b_{p,-} \\ b_{p,+} \end{pmatrix} .$$

The former is a rotation in quasi-spin S-space which commutes with $\hat{S}^2$ but not with $\hat{J}^2$, while the latter is a rotation in quasi-spin J-space, with the opposite behaviour. As a result both the V-quasi-spin and the W-one rotate around the x-axis, the former in an angle $(\alpha+\beta)$, the latter, according to $(\alpha-\beta)$. The eigenvalues of both $\hat{V}^2$ and $\hat{W}^2$ are preserved in this transformation whereas the conservation of J^2 is violated.

Calculating the expectation value of the hamiltonian (4.4) with the same procedure as in sect.3 and minimizing it with respect to α and β one obtains the HF solution

$$\alpha=\beta=0; \qquad E_{HF} = -\frac{N}{2}\left(1-\frac{vN}{8}\right) \qquad \text{if } v \le 2/N ,$$

$$\alpha=0, \cos\beta=\frac{2}{vN}; \quad E_{HF} = -\frac{1}{2N}-\frac{vN^2}{16} \qquad \text{if } v > 2/N . \tag{4.7}$$

As in the application to the Lipkin model a phase transition is found at a critical coupling constant $v_c = 2/N$. For small enough values of the interaction strength the HF solution is the trivial one and matches the exact result. Then, a phase transition is reached which is connected with the fact that, at that critical value, the attractive terms of the interaction overcome the effect of the repulsive ones. Comparison between the HF and the exact ground state energy is given in fig.1 for N = 6.

The non-trivial HF wave function for $v > 2/N$ given in (4.7) does not possess good quasi-spin J although this is a symmetry of the hamiltonian (4.4). In order to project out this quantum number we diagonalize J^2 within our relevant multiplet, this procedure yielding the set of wave functions

$$|V, W, J, n\rangle = \sum_{V_z W_z} C_{J,n}(V_z, W_z)\, |V, W, V_z, W_z\rangle . \tag{4.8}$$

Here n distinguishes among states of the same J. The set (4.8) allows us to give the projection operator $\hat{P}_J$ an explicit expression

$$\hat{P}_J = \sum_n |V, W, J, n\rangle \langle V, W, J, n| . \qquad (4.9)$$

The projected energy E_J is

$$E_J = \frac{\langle\phi|\hat{P}_J^+ \hat{H} \hat{P}_J|\phi\rangle}{\langle\phi|\hat{P}_J^+ \hat{P}_J|\phi\rangle} = \frac{\langle\phi|\hat{H}\hat{P}_J|\phi\rangle}{\langle\phi|\hat{P}_J|\phi\rangle} \qquad (4.10)$$

where we have used the properties of the projector $P^+_J = P_J$, $P^2_J = P_J$ and $[\hat{H}, \hat{P}_J] = 0$. To obtain an exact expression for E_J in terms of the wave functions of the multiplet $|V,W,V_z,W_z\rangle$, i.e. in terms of the matrix elements $\langle V,W,V'_z|\hat{H}|V,W,V_z,W_z\rangle$ which have a very simple form, we only need to express the HF wave function $|\phi\rangle$ in terms of these states. This may be done taking into account that the HF transformation (4.6) is the product of two independent rotations and using the rotation matrices

$$|\phi\rangle = \sum_{V_z W_z} D(V_z, W_z)\, |V, W, V_z, W_z\rangle \qquad (4.11)$$

with

$$D(V_z, W_z) = d^V_{-V,V_z}(\beta)\, d^W_{-W,W_z}(-\beta) . \qquad (4.12)$$

The d functions are defined by Edmonds[10] in eq. (4.1.15). It is to be noted that our phase convention requires the replacement of $\sin\beta/2$ by $i\sin\beta/2$. Moreover, we have fixed $\alpha = 0$.

Projection after variation is performed by using in (4.1.2), for β, the corresponding HF value given at (4.7). Projection before variation is achieved selecting β in (4.12) so as to minimize (4.10).

Fig.1 illustrates the results for the case N = 6. For J = 3 the exact solution is matched by the HF one. For J = 2,1 one can appreciate the fact that projecting out J considerably improves the HF solution. When J = 0 the projected HF solution coincides with the exact one while the HF energy lies considerably higher. Projection before variation yields the same results as the exact treatment and is thus seen to be a much more powerful method than projection after variation.

A numerical estimation of the correlations introduced by the projection procedure is obtained by evaluation of the overlap between either the exact of the projected HF wave function and the state built upon the HF one by creating n particles and n holes (np-nh) with n = 0,1,2...N. This can be accomplished by expansion of all relevant wave functions in terms of our basic set $|V,W,V_z,W_z\rangle$ as in (4.11). We present in Table 1 an example for the case N = 4, J = 1.

An important conclusion that can be drawn from this study is that, qualitatively, projecting either before or after variation introduces the same kind of correlations. Only the relative weights of the different np-nh correlations are different (remember that in this model projection before variation is equivalent to the exact solution).

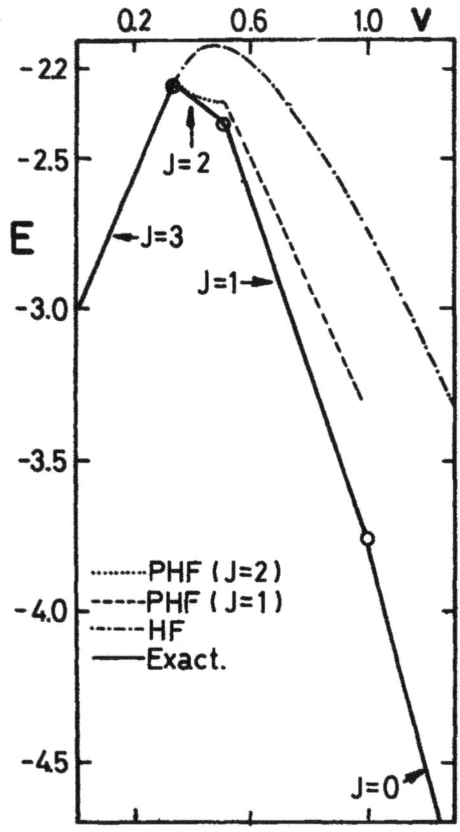

- Fig.1 -

Exact, J.projected and HF energies for N=6 in the SU(2) * SU(2) model. For J=3 the exact and HF energies coincide. For J=0 the exact and the PHF energies are equal. For J=1,2 projection before variation yields the same result as the exact treatment.

Table 1

Overlaps between either the exact or the projected HF wave function and|np-nh>, for N=4, J=1.

v	$\vert\langle HF\vert Exact\rangle\vert^2$	$\vert\langle 1p1h\vert Exact\rangle\vert^2$	$\vert\langle 2p2h\vert Exact\rangle\vert^2$	$\vert\langle 3p3h\vert Exact\rangle\vert^2$	$\vert\langle 4p4h\vert Exact\rangle\vert^2$
0.55	0.158	0.609	0.215	0.017	0.000
0.60	0.257	0.373	0.318	0.049	0.002
0.65	0.319	0.227	0.363	0.086	0.005
0.70	0.360	0.135	0.375	0.120	0.010
0.75	0.386	0.077	0.370	0.151	0.016
0.80	0.402	0.041	0.358	0.178	0.021
0.85	0.412	0.020	0.340	0.201	0.028
0.90	0.419	0.007	0.320	0.220	0.034
0.95	0.421	0.002	0.301	0.237	0.040
1.00	0.422	0.000	0.282	0.250	0.047

v	$\vert\langle HF\vert PHF\rangle\vert^2$	$\vert\langle 1p1h\vert PHF\rangle\vert^2$	$\vert\langle 2p2h\vert PHF\rangle\vert^2$	$\vert\langle 3p3h\vert PHF\rangle\vert^2$	$\vert\langle 4p4h\vert PHF\rangle\vert^2$
0.55	0.158	0.618	0.194	0.027	0.001
0.60	0.259	0.396	0.261	0.077	0.008
0.65	0.325	0.261	0.270	0.124	0.021
0.70	0.370	0.176	0.253	0.162	0.039
0.75	0.401	0.122	0.228	0.190	0.059
0.80	0.424	0.086	0.201	0.209	0.081
0.85	0.440	0.062	0.175	0.220	0.104
0.90	0.453	0.045	0.151	0.226	0.126
0.95	0.461	0.033	0.130	0.227	0.148
1.00	0.469	0.025	0.113	0.224	0.169

5.- The self-consistent cranking method

Exact projection calculations as the one performed in the previous section are easy only in soluble models. In most practical calculations approximations are made in the application of projection techniques. A widely applied approach is the self-consistent cranking method (SCC) which may be derived as a first approximation to an angular momentum projection before variation.

The general expression for the angular momentum projector is[2)]

$$\hat{P}^M_J = \sum_K \int D^{*J}_{MK} \, (\Omega) \hat{R} \, (\Omega) \, d\Omega \tag{5.1}$$

where the $D^J{}_{KK'}$ (Ω) are Wigner functions and $\hat{R}(\Omega)$ is the rotational operator

$$\hat{R}(\Omega) = \exp(i\alpha\hat{J}_z) \; \exp(i\beta\hat{J}_y) \; \exp(i\gamma\hat{J}_z) \tag{5.2}$$

in the three Euler angles α, β, γ.

Inserting (5.1) in the expression (4.10) for the projected energy E_J and using some properties of Wigner functions, one obtains

$$E_J = \frac{\sum_{KK'} \int D^{*J}_{KK'} \, \langle\phi|\hat{H}\hat{R}(\Omega)|\phi\rangle \, d\Omega}{\sum_{KK'} \int D^{*J}_{KK'} \, \langle\phi|\hat{R}(\Omega)|\phi\rangle \, d\Omega} \, . \tag{5.3}$$

Kamlah[11)] has shown that the energy overlap function $\langle\phi|\hat{H}\hat{R}|\phi\rangle$ may be expanded in derivatives of the norm overlap function $\langle\phi|\hat{R}|\phi\rangle$. These derivatives are produced by the operators[10)]

$$L_x \pm iL_y = i \, \exp(\mp i\alpha)\left[-\text{ctg}\,\beta\frac{\partial}{\partial\alpha} \mp \frac{\partial}{\partial\beta} + \frac{1}{\sin\beta}\frac{\partial}{\partial\gamma}\right] \tag{5.4}$$

$$L_z = -i\frac{\partial}{\partial\alpha}.$$

Thus we write

$$\langle\phi|\hat{H}\hat{R}|\phi\rangle = \left[a + bL_x + CL^2 + \ldots\right]\langle f|\hat{R}|\phi\rangle. \tag{5.5}$$

This expansion converges rapidly if the following conditions are satisfied[2,11,12]: i) strong deformation, i.e., large angular momentum fluctuations $\langle\phi|\Delta\hat{J}^2{}_z|\phi\rangle \gg 1$; ii) good signature, $\exp(i\pi J_x|\phi\rangle \alpha|\phi\rangle$; iii) nearly axial symmetry, $\langle\phi|\Delta J^2{}_z|\phi\rangle \ll J(J+1)$. In this case we truncate the expansion after the second term, i.e., we retain only the coefficients a and b. These are determined by requiring that the approximate expressions for the energy overlap and for its first derivative (obtained applying L_x to (5.5)) equal the exact expressions for no rotation, i.e. for $\Omega=(0,0,0)$. Using the relation

$$L_i\langle\phi|\hat{R}|\phi\rangle = \langle\phi|\hat{J}_i\hat{R}|\phi\rangle; \quad i = x,\ y,\ z \tag{5.6}$$

the two conditions result

$$\begin{aligned} &\langle\phi|\hat{H}|\phi\rangle = a + b\ \langle\phi|\hat{J}_x|\phi\rangle \\ &\langle\phi|\hat{H}\hat{J}_x|\phi\rangle = a\langle\phi|\hat{J}_x|\phi\rangle + b\langle\phi|\hat{J}_x^2|\phi\rangle\,. \end{aligned} \tag{5.7}$$

In addition, replacing (5.5) in (5.3) we have

$$E_J = a+b\,\frac{\sum_{KK'}\int D^{*J}_{KK'}L_x\langle\phi|\hat{R}|\phi\rangle d\Omega}{\sum_{KK'}\int D^{*J}_{KK'}\langle\phi|\hat{R}|\phi\rangle d\Omega} \tag{5.8}$$

and considering approximate axial symmetry[13]

$$E_J \approx a + b\sqrt{J(J+1)}\,. \tag{5.9}$$

Determining the coefficients a and b from (5.7) and replacing them in (5.9) we get

$$E_J = \langle\phi|\hat{H}|\phi\rangle + b\left(\sqrt{J(J+1)} - \langle\phi|\hat{J}_x|\phi\rangle\right) \tag{5.10}$$

with

$$b = \frac{\langle\phi|\hat{H}\,\Delta\hat{J}_x|\phi\rangle}{\langle\phi|\Delta\hat{J}_x^2|\phi\rangle}\,;\quad \Delta\hat{J}_x = \hat{J}_x - \langle\phi|\hat{J}_x|\phi\rangle\,. \tag{5.11}$$

Varying (5.10) with respect to ϕ and performing a special variation[13]

$$|\delta\phi\rangle = \eta\,\Delta\,\hat{J}_x|\phi\rangle\,; \qquad \eta \ll 1 \tag{5.12}$$

we obtain

$$\sqrt{J(J+1)} - \langle\phi|\hat{J}_x|\phi\rangle = 0\ . \tag{5.13}$$

Therefore, this first order approximation of the projection before variation method is equivalent to the cranking model of Inglis[14] in which one minimizes the expectation value of the hamiltonian with the constraint (5.13). This is solved introducing a Lagrange multiplier W, considering the modified hamiltonian $\hat{H}' = \hat{H} - W\hat{J}_x$ and minimizing the expectation value of $\hat{H}'$, with the constraint being used to eliminate the Lagrange multiplier W in the usual way. The set of equarions obtained

$$\begin{aligned} &\delta\frac{\langle\phi|\hat{H}'|\phi\rangle}{\langle\phi|\phi\rangle} = \delta\frac{\langle\phi|\hat{H} - W\hat{J}_x|\phi\rangle}{\langle\phi|\phi\rangle} = 0 \\ &\langle\phi|\hat{J}_x|\phi\rangle = \sqrt{J(J+1)} \end{aligned} \tag{5.14}$$

may be solved self-consistently as in the HF case. However, problems may arise in this constrained case (see sect.7.6 of Ref.2). Finally, it is to be noted that no assumptions concerning the set of variational wave functions have been made. Consequently, one may use any set that satisfies the conditions established after eq.(5.5).

6. - An application of SCC

The SSC has been extensively used to describe the rotationa spectra of well deformed nuclei. Only a few of these applications are given in refs. 15-17. However the physical meaning of the condition referred to the strong deformation needed for the validity of the approximation (see previous section) is not clear because large angular momentum fluctuations do not necessarily imply large deformation parameters of the corresponding mean field. Consequently, we shall study through the application to a soluble model, how much may the deformation be decreased before the cranking approximation turns meaningless.

With this objective in mind we consider the interacting boson model (IBA)[18] which may be used as a soluble model for testing many-body approximations due to the fact that is is simple enough to be solved exactly and yet it includes the essential features of nuclear collective phenomena. In particular, it is simple within this model to write a one-parameter hamiltonian[19] to describe the transition from an axial symmetric rotor (SU(3) limit) to a spherical vibrator (U(5) limit).

The many boson hamiltonian considered is [19,20]

$$\hat{H} = \mathcal{E}(1-x)\ \hat{n}_d - K\ x\ (\hat{Q}_2 \cdot \hat{Q}_2) \tag{6.1}$$

where

$$\hat{Q}_2 \cdot \hat{Q}_2 = \sum_{\mu} (-)^{\mu}\ \hat{Q}_{2\mu}\ \hat{Q}_{2-\mu} \tag{6.2}$$

and

$$\hat{Q}_{2\mu} = \left(\gamma_0^+\gamma_2 + \gamma_2^+\gamma_0\right)^2_{\mu} - \frac{\sqrt{7}}{2}\left(\gamma_2^+\gamma_2\right)^2_{\mu}\ . \tag{6.3}$$

The parameter X varies between X = 0 (U(5) limit) and X = 1 (SU(3) limit). $\gamma_\ell^+ (\gamma_\ell)$ creates (destroys) one boson with angular momentum ℓ. Only s(l=0) and d(l=2) bosons are included. $\hat{n}_d$ is the d-boson number operator. The interaction strengths $\mathcal{E}$ and K are so chosen as to locate the phase transition from the vibrational phase to the rotational one at X $\simeq$ 0.5. The corresponding values are $\mathcal{E}$ = 1 MeV, K = 0.030 MeV. The exact results for different values of the parameter X are obtained through the code PHINT[21].

For applying the SCC we use as the trial wave function $|\phi\rangle$ the one provided by the Hartree Bose approximation, i.e., a condensate of only one kind of bosons

$$|\phi\rangle = \frac{1}{\sqrt{N!}} (\Gamma_0^+)^N\ |0\rangle \tag{6.4}$$

with

$$\Gamma_0^+ = \frac{1}{\mathcal{N}} \sum_i \eta_{io}\ \gamma_i^+\ . \tag{6.5}$$

The coefficients η_{io} are the variational parameters, $\mathcal{N}$ is the norm, $|0\rangle$ is the boson vacuum and N the number of bosons. If we minimize the expectation value of the hamiltonian we obtain the Hartree ground state. A phase transition is observed, as in the exact solution, at X_c = 0.46. For X > X_c the Hartree ground state is a condensate of s bosons which means that it is in the spherical phase. For X > X_c the system is in the deformed phase and the deformation increases with X.

In the SCC approach we must vary the modified hamiltonian $\hat{H}'$ as indicated in (5.14). The angular momentum operators for the particular case considered are

$$\hat{J}_\mu = \sqrt{10}\ (\gamma_2^+\gamma_2)^1_\mu . \qquad (6.6)$$

On the other hand, if we do not neglect the mean value of $J^2{}_z$ in the derivation of (5.13) because we are not sure if this is right near the spherical phase, we get a more complete constraint

$$<\phi|\hat{J}_x|\phi> = \sqrt{J(J+1) - <\phi|\hat{J}_z^2|\phi>} . \qquad (6.7)$$

For explicitly calculating the mean values of $\hat{H}$, $\hat{J}_x$ and $\hat{J}^2{}_z$ one must decouple the boson operators in (6.3) and (6.6) by means of Clebsch-Gordan coefficients and use the boson commutation relations instead of the Fermi anticommutation ones employed in the previous examples. Moreover, one must remember on performing the variation that the norm $<\phi|\phi>$ depends on the variational parameters η_{io} and that only five of the six η_{io} are independent.

The SCC results for the excitation energies in the ground state band are shown in fig.2 for N = 10 bosons, as a function of the parameter X. As it was to be expected the SCC results are very good in, or near to, the rotational limit, i.e., for X > 0.5. But what is unexpected is that they are also good through the phase transition and in the spherical phase, even just in the U(5) limit (X=0). The only difference is that in the deformed phase the agreement in the excitation energies gets worse when the angular momentum J increases, whereas in the spherical phase the difference between the approximate and the exact results remains almost constant throughout the band. The J=20 level is a special case due to a truncation effect as 20 is the maximum angular momentum possible with 10 bosons. Moreover the SCC results are below the exact ones in the deformed phase because we represent excitation energies instead of absolute ones.

To explain the validity of the SCC in the spherical phase we recall that, as a matter of fact, a strong deformation is not required in the quantum mechanical derivation of the method. The condition is, instead, to have large angular momentum fluctuations in the intrinsic wave function $|\phi>$. Certainly, this is not the case with the Hartree ground state for $X<X_c$. This is a condensate of s bosons and has zero fluctuations. But it must be remembered that in the SCC, as in any projection before variation approach, the intrinsic wave function is different for each J value and, consequently, the fluctuations also change. Taking this into account we calculate the fluctuations throughout the band for different X values. Some of these results are shown in table 2. It is seen that for X<Xc the fluctuations increase very rapidly with J and from J=8 on they stabilize in values of an order of magnitude comparable with that obtained near the rotational limit. These results explain the goodness of the agreement between approximate and exact energies. Moreover, it is interesting to notice that whereas in the deformed phase the fluctuations are effectively due to the deformation, in the spherical case they are generated by the SCC method

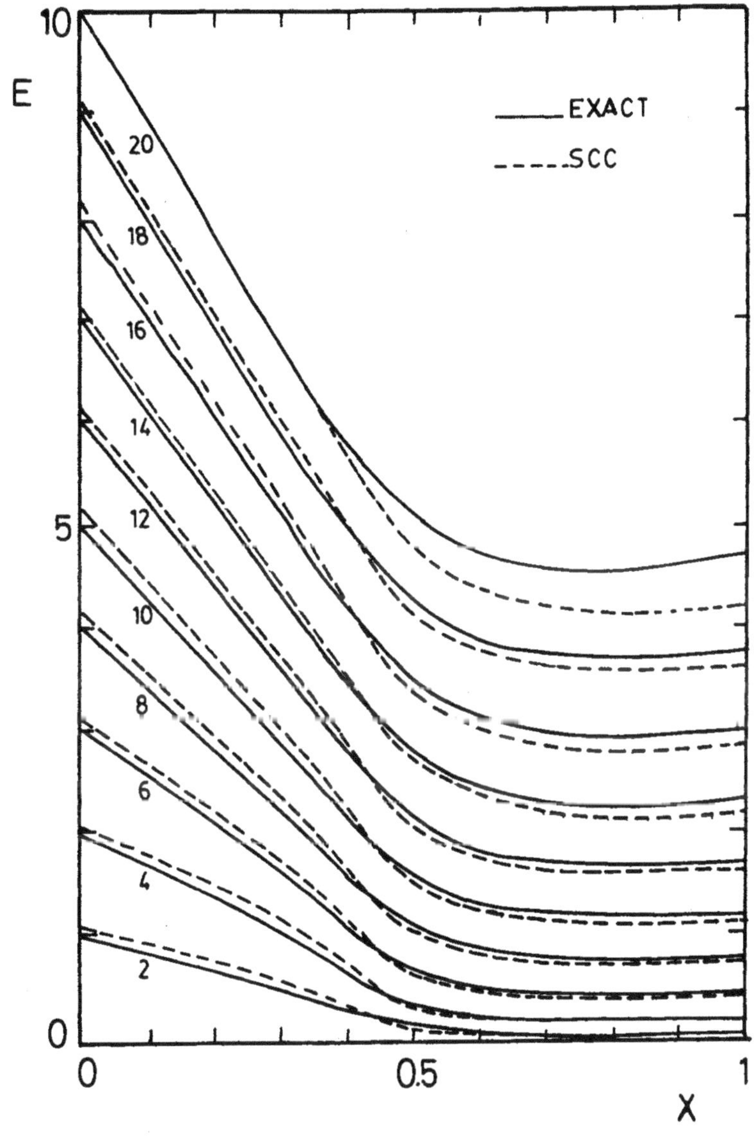

- Fig.2 -

Excitation energy E versus the parameter X, for N=10. The full lines correspond to the exact results in the ground state band and the dashed lines to the corresponding SCC calculations. The number in each curve gives the value of the angular momentum J.

Table 2

Angular momentum fluctuation $\langle\phi|\Delta\hat{J}^2|\phi\rangle$ for various values of the angular momentum J as a function of the parameter X, for N = 10.

	X						
J	0	0.2	0.4	0.5	0.6	0.8	1
0	0	0	0	14.3	24.7	34.2	38.9
2	6.1	6.4	8.9	16.8	25.1	34.1	38.6
4	10.9	11.2	14.3	19.8	25.9	33.8	38.0
6	14.8	15.2	18.3	22.2	26.7	33.2	37.0
8	17.9	18.4	21.1	24.0	27.2	32.4	35.6
10	20.2	20.7	23.0	25.1	27.4	31.3	33.9
18	21.4	21.5	21.9	22.1	22.4	22.8	23.1

itself through the addition of the cranking term $\omega\hat{J}_x$ to the hamiltonian.

The behaviour of the Lagrange multiplier ω as a function of J and X is shown in fig.3. In the rotational case ω changes with J and one may obtain a self-consistent moment of inertia through

$$\mathcal{J}_{SC} = \frac{<f|\hat{J}_x|\phi>}{\omega} \tag{6.8}$$

which is almost constant throughout the band. This is usually interpreted as having a nucleus with a fixed deformation that rotates more and more rapidly, i.e. with increasing angular velocity ω. In the vibrational case one has the opposite situation. The value of w is the same for every level and if one insists in calculating $\mathcal{J}_{SC}$ by means of (6.8). this turns out to be strongly variable. The corresponding interpretation would be to have a nucleus rotating with a fixed angular velocity w but which has an increasing deformation.

As the quantum mechanical derivation of the SCC does not depend on the bosonic or fermionic nature of the system considered the above conclusions are also valid for fermionic systems. One may then state that the SCC could be used for studying high spin phenomena in vibrational nuclei. This conclusion agrees with the suggestions of ref. 22.

7. - The Time Dependent Hartree-Fock method

For describing some nuclear processes that involve large amplitudes such as anharmonic vibrations or fusion and fission processes, it seems convenient to formulate the problem as time dependent. We begin at time t=0 with a wave function $|\psi(0)>$ that is not an exact solution of the stationary Schrödinger equation. Such state represents a wave packet and its behaviour with increasing time is given by the time evolution operator

$$|\psi(t)> = e^{-i\hat{H}t}|\psi(0)> \tag{7.1}$$

where $\hat{H}$ is the complete many-body hamiltonian. The information contained in the time dependent wave function $|\psi(t)>$ is, in principle, equivalent to the solution of the stationary Schrödinger equation. This is due to the fact that a Fourier analysis of $|\psi(t)>$ provides the spectrum of $\hat{H}$ and the corresponding eigenstates. To see this we take a solution of the time dependent Schrödinger equation

$$\hat{H}\psi(t) = i\,\frac{\partial\psi(t)}{\partial t} \tag{7.2}$$

and perform a Fourier analysis

$$\psi(t) = \int e^{-i\omega t}\psi(\omega)\,d\omega\,. \tag{7.3}$$

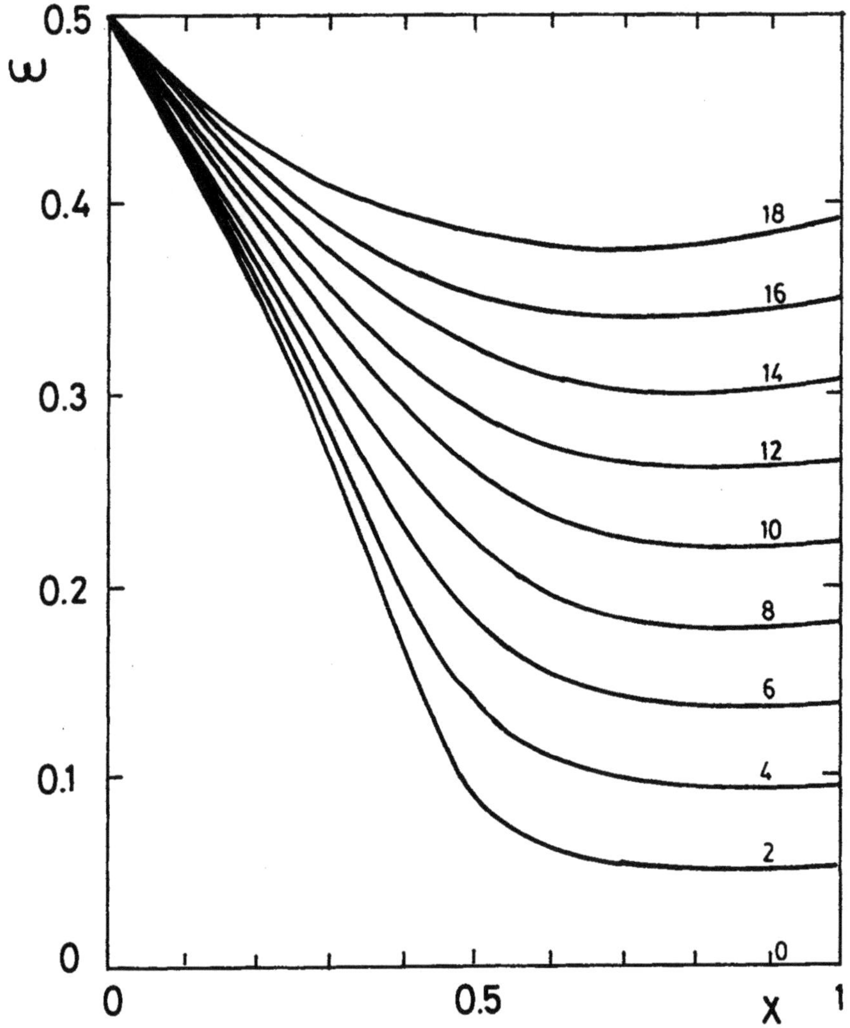

- Fig.3 -

Lagrange multiplier w versus the parameter X, for N=10. The number in each curve gives the value of the angular momentum J.

It is clear that any ω such that $\psi(\omega) \neq 0$ satisfies

$$\hat{H}\psi(\omega) = \omega\psi(\omega) \tag{7.4}$$

and consequently is an eigenvalue of $\hat{H}$.

The general expression of $\psi(t)$ is too complicated for practical calculations. One possible approximation is the one utilized in TDHF in which one begins with a Slater determinant $|\psi(0)\rangle$ and assumes that the time dependent wave function is a Slater determinant for all times. This means that one assumes, as in HF, that each particle does not interact with the others directly but only with the wall of a self-consistent field. This approach seems reasonable for several phenomena such as low energy collective excitations, nuclear fission and heavy ion collisions at moderate energies.

It may be shown, similarly to the HF case, that the time dependent Schrödinger equation is derived from the time dependent variational principle. One defines a Lagrangian as

$$\mathcal{L} = \langle\psi(t)|\, i\frac{\partial}{\partial t} - \hat{H}\,|\psi(t)\rangle \tag{7.5}$$

and consequently the action is

$$S = \int \langle\psi(t)|\, i\frac{\partial}{\partial t} - \hat{H}\,|\psi(t)\rangle\, dt \;. \tag{7.6}$$

Varying the action with respect to ψ and ψ^* independently and equating the variation to zero one easily obtains the time dependent Schrödinger equation.

In the TDHF approximation the many-body wave function is parametrized as a time dependent Slater determinant

$$\Phi(1,\ldots,N,t) = \frac{1}{\sqrt{N!}} \begin{vmatrix} \phi_1(1,t) & \phi_1(2,t) & \ldots & \phi_1(N,t) \\ \phi_2(1,t) & \phi_2(2,t) & \ldots & \phi_2(N,t) \\ \ldots & \ldots & \ldots & \ldots \\ \phi_N(1,t) & \phi_N(2,t) & \ldots & \phi_N(N,t) \end{vmatrix} \tag{7.7}$$

where the single particle (sp) wave functions $\phi_\nu(r)$ are orthonormal.

Then the action is

$$S = \int dt \left[\sum_n \int dr\, \phi_\nu^*(r)\; i\frac{\partial}{\partial t}\phi_\nu(r) \;- \mathcal{H}(\phi,\phi^*) \right] \tag{7.8}$$

where

$$\mathcal{H}(\phi,\phi^*) = \langle\phi|\hat{H}|\phi\rangle . \tag{7.9}$$

Varying with respect to ϕ_ν^* and ϕ_ν one obtains the TDHF equations of motion

$$i\frac{\partial}{\partial t}\phi_\nu = \frac{\partial\mathcal{H}}{\partial\phi_\nu^*} ; \quad -i\frac{\partial}{\partial t}\phi_\nu^* = \frac{\partial\mathcal{H}}{\partial\phi_\nu} . \tag{7.10}$$

Therefore, TDHF looks like a classical field theory with the hamiltonian function $\mathcal{H}$ obeying Hamilton's equations. It is to be noted that the degrees of freedom are actually less as any unitary transformation among the ϕ_ν gives the same determinant.

The variational trial function is the same as in HF but time dependent. The new Slater determinant is formed with new sp states given by the transformation

$$b_\nu^+ = \sum_\mu C_{\mu\nu}(t)\ a_\mu^+ \quad . \tag{7.11}$$

Now we shall see an alternative representation of TDHF equations that is very useful because the equations are expressed in a representation independent way by means of the one-body density matrix. We work with the general wave function $|\phi(t)\rangle$ and calculate its one-body density at time t

$$\begin{aligned} \rho_{\mu\nu}(t) &= \langle\psi(t)|a_\nu^+ a_\mu|\psi(t)\rangle \\ &= \langle\psi(0)|e^{i\hat{H}t}|a_\nu^+ a_\mu\ e^{-i\hat{H}t}|\psi(0)\rangle . \end{aligned} \tag{7.12}$$

For getting an equation of motion we calculate the derivative

$$\begin{aligned} i\dot{\rho}_{\mu\nu} &= \langle\psi(t)|\left[a_\nu^+ a_\mu, \hat{H}\right]|\psi(t)\rangle = \sum_{\nu'}\left(\varepsilon_{\mu\nu'}\ \rho_{\nu'\nu} - \varepsilon_{\nu'\nu}\,\rho_{\mu\nu'}\right) + \\ &+\frac{1}{2}\sum_{\nu'\nu_1\nu_2}\left(\bar{v}_{\mu\nu'\nu_1\nu_2}\,\rho^{(2)}_{\nu_1\nu_2\nu\nu'} - \bar{v}_{\nu_1\nu_2\nu\nu'}\,\rho^{(2)}_{\mu\nu'\nu_1\nu_2}\right) \end{aligned} \tag{7.13}$$

using the general hamiltonian given by (2.1), the commutator (2.16) and the corresponding one with a_ν^+, and the two-body density

$$\rho^{(2)}_{klpq}(t) = \langle\psi(t)|a^+_p a^+_q a_l a_k|\psi(t)\rangle \quad . \qquad (7.14)$$

Equation (7.13) is exact. For approximating it we introduce the correlation function $g^{(2)}$ that is what remains of $\rho^{(2)}$ after extracting the non-correlated pairs

$$\rho^{(2)}_{klpq}(t) = \rho_{lq}\rho_{kp} - \rho_{lp}\rho_{kq} + g^{(2)}_{klpq} \quad . \qquad (7.15)$$

Then we have

$$i\dot{\rho}_{\mu\nu} = [h,\rho]_{\mu\nu} + \frac{1}{2}\sum_{\nu'\nu_1\nu_2}(\bar{v}_{\mu\nu'\nu_1\nu_2}\, g^{(2)}_{\nu_1\nu_2\nu\nu'} - \bar{v}_{\nu_1\nu_2\nu\nu'}\, g^{(2)}_{\mu\nu'\nu_1\nu_2}) \qquad (7.16)$$

where $h_{\mu\nu} = \varepsilon_{\mu\nu} + U_{\mu\nu}$ is the HF hamiltonian defined in (2.17). The only difference is that now the HF potential U is time dependent. This means that the particles may have two types of interactions: collisions with the moving walls of the self-consistent field or collisions with the other particles which cause the two-body correlations. In TDHF the wave function is approximated by a Slater determinant $|\phi(t)\rangle$ for all times. Therefore $g^{(2)} = 0$ as in the static case and TDHF equation results

$$i\dot{\rho} = [h,\rho]. \qquad (7.17)$$

This is a non linear matrix equation for the one-body density matrix ρ. Physically the nuclear dynamics is described in this approximation through the simultaneous evolution of N sp wave functions in the non local mean field U. Moreover, one has an initial value problem because the first order differential equation requires the information about the initial wave function.

Now we see several properties of TDHF equation.

i) First we note that if we have a Slater determinant

$$|\phi\rangle = b^+_1 b^+_2 \ \ldots\ b^+_N|0\rangle \qquad (7.18)$$

the one-body density matrix turns out to be diagonal in this basis

$$\rho_{kl} = \langle\phi|b^+_l b_k|\phi\rangle = 0 \qquad \text{if } l \neq k$$

$$(7.19)$$

$$\rho_{11} = \langle\phi|b_1^+ b_1|\phi\rangle = \begin{cases} 0 & \text{if } 1 > N \\ 1 & \text{if } 1 \leq N \end{cases}$$

and consequently

$$\rho^2 = \rho; \quad \text{Tr } \rho = \sum_1 \rho_{11} = N . \tag{7.20}$$

We study the variation of $\rho^2 - \rho$ with the time

$$i\frac{\partial}{\partial t}(\rho^2 - \rho) = i(\dot{\rho}\rho + \rho\dot{\rho} - \dot{\rho}) = [h, \rho^2 - \rho] . \tag{7.21}$$

Therefore, if we begin with a Slater determinant, $\rho^2(0) = \rho(0)$, this property is conserved and the theory is consistent.
ii) Another property that is immediately derived is the conservation of the overlap matrix of the sp wave functions

$$i\frac{\partial}{\partial t}\langle\phi_i|\phi_k\rangle = i\langle\dot{\phi}_i|\phi_k\rangle + i\langle\phi_i|\dot{\phi}_k\rangle = \langle\phi_i|h - h^+|\phi_k\rangle = 0 \tag{7.22}$$

iii) Now we study the conservation of the expectation values of one-body operators. If $\hat{O}$ is any one-body time-independent operator

$$\hat{O} = \sum_{lm} O_{lm}\, a_l^+ a_m , \tag{7.23}$$

the TDHF expectation value is

$$\langle\phi|\hat{O}|\phi\rangle = \sum_{lm} O_{lm}\, \rho_{ml} = \text{Tr}(O\rho) . \tag{7.24}$$

Its time derivative is

$$i\frac{d}{dt}\langle\phi|\hat{O}|\phi\rangle = \text{T}\rho(O.[h,\rho]) . \tag{7.25}$$

On the other hand, it is straightforward to show that

$$\langle\phi|[\hat{O},\hat{H}]|\phi\rangle = \text{Tr}(O.[h,\rho]). \tag{7.26}$$

Consequently one gets

$$<\phi|[\hat{O},\hat{H}]|G\phi> = \mathrm{Tr}(O\cdot[h,\rho]. \qquad (7.27)$$

and concludes that the expectation values of those one-body operators that commute with the hamiltonian (i.e., that represent symmetries) are conserved. Examples of them are the number of particles, the total linear momentum, the total angular momentum. This does not mean that these symmetries are not violated as only the mean values are conserved. Apart from this property there also are self-consistent symmetries as in the static case.

iv) With respect to the energy we have

$$\frac{d}{dt}<\phi|\hat{H}|\phi> = \sum_{v}\left(\frac{\partial \mathcal{H}}{\partial \phi_v}\frac{\partial \phi_v}{\partial t}+\frac{\partial \mathcal{H}}{\partial \phi_v^*}\frac{\partial \phi_v^*}{\partial t}\right)=0 \qquad (7.28)$$

using (7.9) and (7.10). Consequently the energy is conserved. This means that the movement is along constant energy lines on the energy surface of the Slater determinants.

In summary, the TDHF method seems to have great advantages. It provides an unified description of every sort of nuclear collective movements, beginning with the static solution for the ground state, going on with vibrational and rotational excitations and up to the large amplitude collective movements such as fission, fusion, compound nucleus formation and deep inelastic reactions. It is not necessary to previously choose some collective coordinates. On the other hand, being the wave function a Slater determinant it takes into account aspects of the sp structure. Moreover, the TDHF solutions include damping produced by the interaction between the particles and the wall of a time dependent mean field, called one-body friction. But it has to be remembered that neglecting $g^{(2)}$ is meaningless when the energy per particle is of the order of the Fermi energy.

However there also are important limitations. One is the enormous numerical effort that solving the equations implies in realistic cases. Apart from the fact that the physical interpretation of the results is not always clear. Additionally, the theory is in a sense classical, for example there is no tunnel effect through a potential barrier. Therefore one needs a requantization method for calculating any observable. On the other hand, the two-body correlations are not well treated. Finally, there is no obvious way of systematically improving the variational wave function.

8.- Calculations in TDHF

The TDHF method has been extensively studied, see for example refs. 2, 23 and 24. There also exist numerous applications, a few of which are given in refs. 25-29.

As in the HF case, it is illuminating to apply this method to a soluble model and we choose the Lipkin model[7] which was reviewed in sect.3. Instead of (3.8) we use a more general transformation[30]

$$\begin{pmatrix} b^+_{p+} \\ b^+_{p-} \end{pmatrix} = \begin{pmatrix} \cos\alpha & e^{i\psi}\sin\alpha \\ -\varepsilon^{-i\psi}\sin\alpha & \cos\alpha \end{pmatrix} \begin{pmatrix} a^+_{p-} \\ a^+_{p+} \end{pmatrix} \tag{8.1}$$

in which the parameters α and ψ depend on time. This transformation provides a trial wave function which remains an eigenfunction of n_p, the number of particles with each p-value defined in (3.7). Consequently, the one-body density matrix ρ divides itself in blocks of 2 x 2, one for each p value. Any of these blocks is

$$\rho = \begin{pmatrix} \sin^2\alpha & e^{i\psi}\sin\alpha\cos\alpha \\ e^{-i\psi}\sin\alpha\cos\alpha & \cos^2\alpha \end{pmatrix} \tag{8.2}$$

with Tr ρ =1 instead of N because we work within the p-block.

In addition, the HF hamiltonian h defined in (2.18) is also broken in blocks and one gets

$$h = \begin{pmatrix} \frac{1}{2} & \chi\sin\alpha\cos\alpha\, e^{-i\psi} \\ \chi\sin\alpha\cos\alpha\, e^{i\psi} & -\frac{1}{2} \end{pmatrix} \tag{8.3}$$

with $\chi = \frac{v(N-1)}{\varepsilon}$ and ε = 1 as before.

With (8.2) and (8.3) and using new variables

$$\beta = \cos 2\alpha \quad ; \quad \gamma = \sin 2\psi \tag{8.4}$$

the TDHF equation (7.17) yields[30]

$$\begin{aligned} \dot\beta &= \chi(1-\beta^2)\gamma \\ \dot\gamma &= -2\left[\sqrt{1-\gamma^2} + \chi\beta(1-\gamma^2)\right]. \end{aligned} \tag{8.5}$$

These coupled equations may be solved numerically providing the TDHF solution β(t) and γ(t). One obtains trajectories on the β–γ plane along which the energy per particle

$$\frac{\mathcal{H}}{N} = \frac{E}{N} = -\frac{1}{2}\left[\beta - \frac{1}{2}\chi(1-\beta^2)\sqrt{1-\gamma^2}\right] \tag{8.6}$$

remains constant. These trajectories turn out to be periodic with a period T which depends on the energy E.

The problem of extracting quantum information from a time dependent variational solution has attracted much attention. In particular is has been shown[31,32] that, from a family of periodic solutions of the time dependent variational principle, quantized energy levels E_n can be extracted by the Bohr-Sommerfeld like condition on the variational action

$$S_o(E) = \frac{i}{2}\int_0^T \frac{<\phi|\dot{\phi}> - <\dot{\phi}|\phi>}{<\phi|\phi>}\, dt = 2\pi n \tag{8.7}$$

where n is integer. Notice that as $\mathcal{H}$ is constant in TDHF one uses the reduced action, i.e., the first term in (7.8).

For calculating (8.7) in the Lipkin model one uses the general expression for the trial function

$$|\phi> = \prod_{p=1}^{N} b^+_{p-}\ |0> \quad , \tag{8.8}$$

derivates it and obtains the overlap $<\phi|\dot{\phi}>$ through the anticommutator

$$\left\{b_i,\ \frac{d}{dt}\, b^+_j\right\} = b_i\dot{b}^+_j + \dot{b}^i_j b_i = -i\dot{\psi}\sin^2\alpha \tag{8.9}$$

in which (8.1) has been used. Consequently the action is

$$S_o(E) = N\int_0^T \dot{\psi}\sin^2\alpha\, dt \quad . \tag{8.10}$$

Then one may utilize (8.4) and (8.5) for replacing $\dot{\psi}$ and (8.6) for establishing a relation between ψ and α and writing explicitly the dependence on E. The results for $\chi = 0.5$ are shown in fig.4A. The requantized TDHF levels for a certain N are obtained dividing the interval $-1 \leq S_o/\pi N \leq 1$ into N equal parts. Notice that in the figure only the negative energies are represented as the positive ones are symmetric.

For comparison the exact energy levels are also included in fig.4A, for different N values. The ordinates are obtained by assigning an integer ordering number m ranging from -N/2 for the lowest energy eigenstate to +N/2 for the highest one. It is clear that in the limit N $\rightarrow \infty$ the exact spectrum crowds along the universal curve provided by TDHF. The scaling adopted is indispensable to bring out this feature. Similar results are obtained after the phase transition[33], i.e., for χ >1.

Taking this limit into account we may assume an expansion of the exact action[34]

$$S = S_o(E,\chi) + \frac{1}{N}\, S_1(E,\chi) + \ldots \tag{8.11}$$

$S_0(E,\chi)$ is the TDHF action divided by πN, $S_1(E,\chi)$ is again an universal curve and it can be obtained from the exact results by considering the limit

$$S_1 = N(S - S_0) \ ; \quad N \rightarrow \infty \ . \tag{8.12}$$

Fig.4.B shows the resulting curve for $\chi = 0.5$. The existence of this new universal curve justifies the expansion (8.11). Moreover, this proves that N is, in this model, the relevant expansion parameter on which to base a systematic analysis of higher order corrections to TDHF taken as a zero order approximation.

Although (8.7) takes care of one aspect of requantization, namely the selection of quantized energies, it does not give any information on the quality of the wave function or on matrix elements of relevant operators. We may extend this requantization formula to the calculation of matrix elements by extending a simple argument used in the semiclassical limit of quantum mechanics in one dimension.

It is well known from elementary quantum mechanics that the semiclassical matrix elements of any operator are given by the Fourier components of the classical periodic time dependent trajectory. Based on this fact, we propose to compute matrix elements of a given operator $\hat{\theta}$ as[33]

$$\tilde{\theta}_k = \frac{1}{T}\int_0^T e^{ik2\pi t/T}\,\theta(t)\,dt \tag{8.13}$$

where

$$\theta(t) = \frac{\langle\phi|\hat{\theta}|\phi\rangle}{\langle\phi|\phi\rangle} \ . \tag{8.14}$$

The quantity θ_k is to be identified with the matrix elements of the operator θ according to

$$\tilde{\theta}_k(E) = \langle n + k|\hat{\theta}|n\rangle \tag{8.15}$$

and the relationship between n and E is given by (8.7). Of course this definite prescription is based on the assumption that a family of periodic solutions to the variational equations is known.

Applying this prescription to the Lipkin model we compute the matrix elements of $\hat{J}_z$ (defined in (3.1.)) using the mean value

$$J_z(t) = \frac{\langle\phi|\hat{J}_z|\phi\rangle}{\langle\phi|\phi\rangle} = -\frac{N}{2}\,\beta(t) \tag{8.16}$$

calculated with (8.8), (8.1) and (8.4). The first few Fourier components

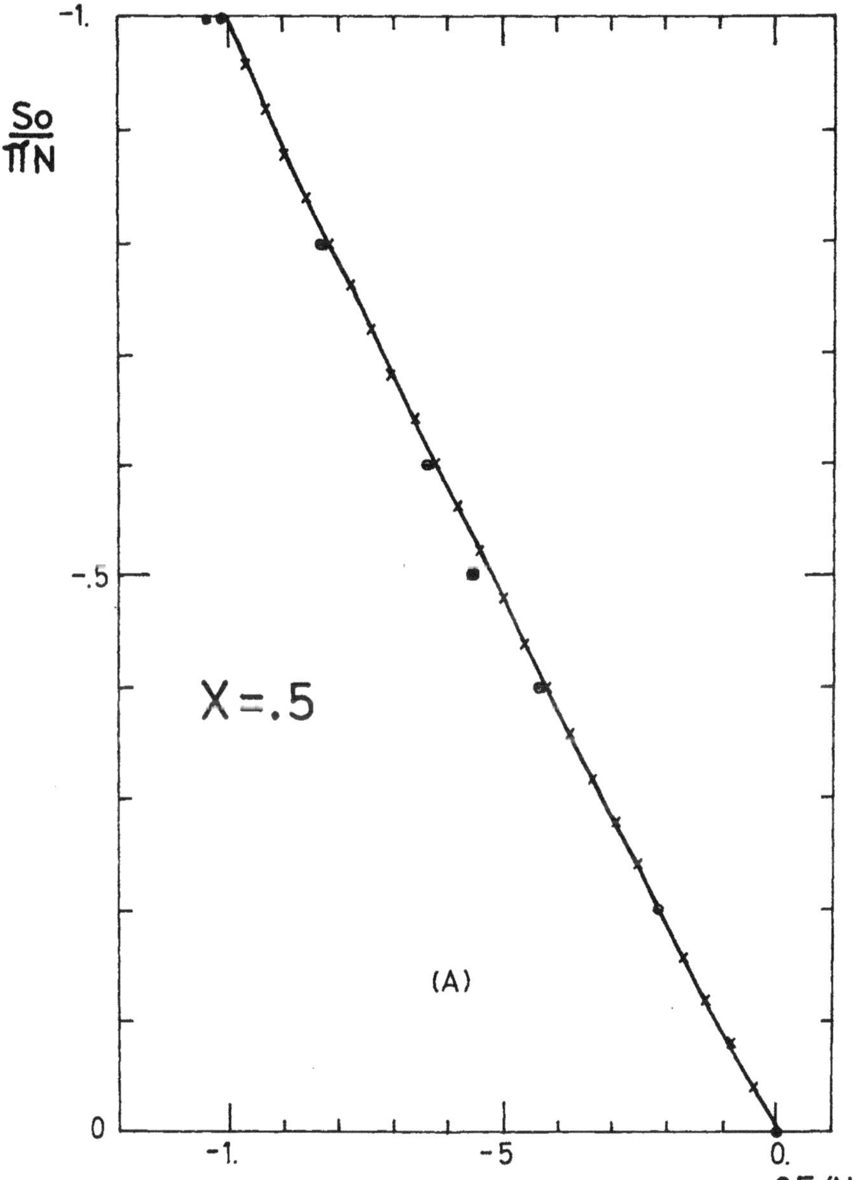

- Fig.4A -

Action as a function of energy for $\chi=0.5$. The full line represents the TDHF result. The exact results correspond to different values of N.

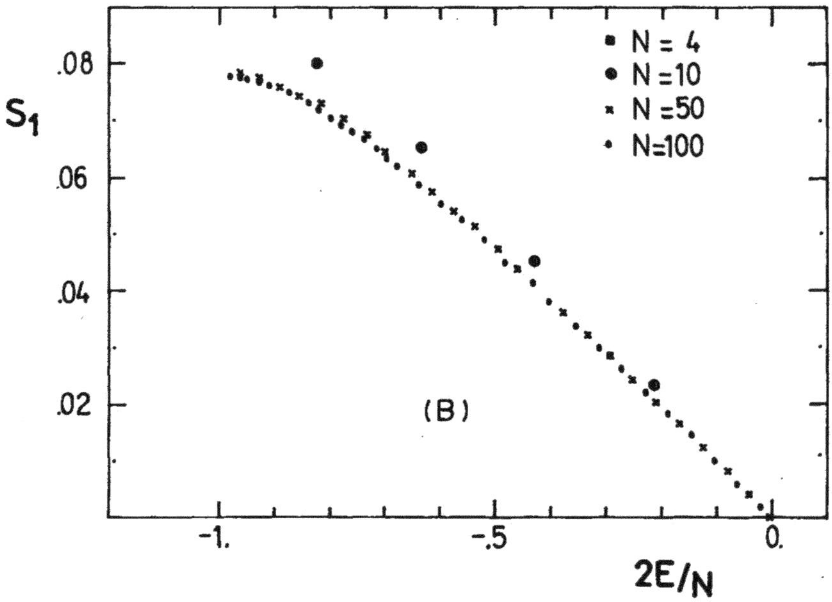

- Fig.4B -

1/N correction to the TDHF action as a function of energy for different values of N.

found numerically, are shown in fig.5 for $\chi = 0.5$ and $N = 30$. We have scaled the exact results, both the energy and the matrix elements, by $N/2$.

The first striking characteristic is the absence of the $k = 1$ component. This is a very welcome reflection of the well known parity alternation between eigenstates in this model, i.e., the mixing of only even, or odd, values of M by the interaction. The agreement with the exact results for the $k = 0$ and $k = 2$ components is very good and improves with increasing N. Similar results are got for $\chi > 0.5$ (see ref.33). Therefore the prescription (8.15) provides for this model the $N \to \infty$ limit for matrix elements.

Finally we investigate the time evolution of single particle properties (i.e. the one body density matrix ρ) in the TDHF approximation and compare it with the exact evolution[30,34]. We choose as a representative quantity to study the following

$$k = \sqrt{2\frac{\mathrm{tr}\,(\rho^2 - \rho)}{\mathrm{tr}\ \rho} + 1} \quad . \tag{8.17}$$

It is obvious that for TDHF, $\rho^2 = \mathrm{tr}\rho$ and therefore $k = 1$ for all times. This can be restated saying that in some single particle basis the occupation of the levels is such that all particles are in one state and the diagonal form of ρ is $\begin{pmatrix}1&0\\0&0\end{pmatrix}$. At the other extreme is the totally "thermalized" situation in which the occupation of the levels is equal with a diagonal form $\begin{pmatrix}1/2&0\\0&1/2\end{pmatrix}$. For this case $\mathrm{tr}\ \rho^2 = 1/2$, $\mathrm{tr}\rho = 1$ and therefore $k = 0$. Thus the way in which k deviates form unity signals the growth of correlations in the evolution and one can check whether and how the evolution of an initial determinant state yields thermalization.
In order to obtain the exact $\rho(t)$ for replacing it in (8.17) we expand the exact time dependent wave function in the basis $|J,M\rangle$ given in (3.4),

$$|\psi(t)\rangle = e^{-i\hat{H}t}\,|\phi(0)\rangle = \sum_M A_M |J,M\rangle \tag{8.18}$$

and use the following matrix elements

$$\begin{aligned}\langle J,M|a^+_{p+}a_{p+}|J,M'\rangle &= \frac{1}{N}\langle J,M|\sum_p a^+_{p+}\ a_{p+}|J,M'\rangle \\ &= \frac{1}{N}\langle J,M|\frac{\hat{N}}{2} + \hat{J}_z|J,M'\rangle = \left(\frac{1}{2} + \frac{M}{N}\right) S_{MM'}\end{aligned} \tag{8.19}$$

and

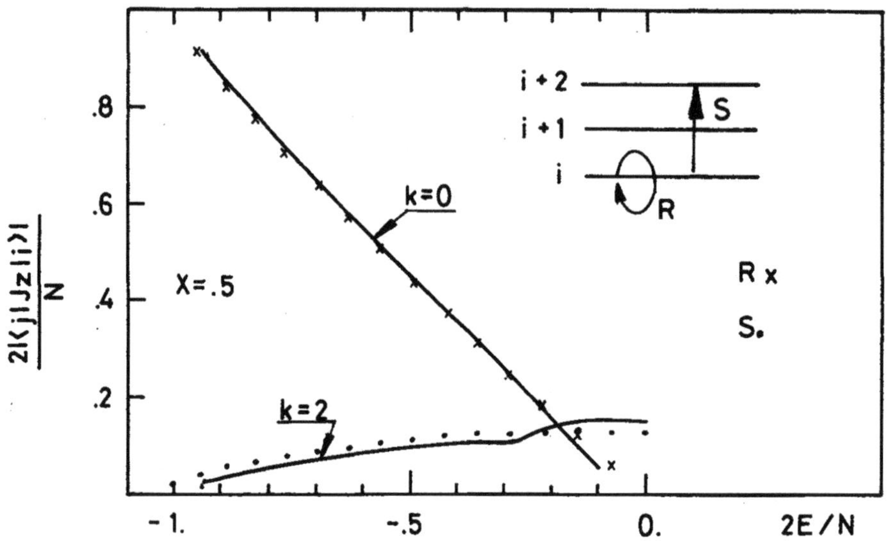

- Fig.5 -

Matrix elements of the operator $\hat{J}_z$ for χ=.5 and N=30. The transitions shown in the inset are compared with Fourier components (k=0 and 2).

$$\langle J,M|a^{+}_{p+}a_{p-}|J,M\rangle = \frac{1}{N}\langle J,M|\sum_{p} a^{+}_{p+}a_{p-}|J,M'\rangle$$

$$= \frac{1}{N}\langle J,M|\hat{J}_{+}|J,M'\rangle = \frac{1}{N}\sqrt{\frac{N}{2}(\frac{N}{2}+1) - M'(M'+1)}\,\delta_{M,M'+1} \tag{8.20}$$

where (3.1) has been used. (See also (7.12) for the general expression of $\rho(t)$).

For getting the expansion coefficients A_M we first expand the initial Slater determinant $|\phi(0)\rangle$ in the basis of the eigenstates $|\psi_i\rangle$

$$|\phi(0)\rangle = \sum_{i} c_i |\phi_i\rangle \tag{8.21}$$

with

$$\hat{H}|\psi_i\rangle = E_i|\psi_i\rangle \tag{8.22}$$

obtaining

$$|\psi(t)\rangle = \sum_{i} c_i e^{-iE_i t}|\psi_i\rangle . \tag{8.23}$$

The diagonalization of the hamiltonian provides

$$|\psi_i\rangle = \sum_{M} B^{i}_{M}|J,M\rangle . \tag{8.24}$$

We may also expand $|\phi(0)\rangle$ in the basis $|J,M\rangle$

$$|\phi(0)\rangle = \sum_{M} D_M|J,M\rangle = \sum_{M} d^{N/2}_{M,-\frac{N}{2}}(2\alpha)\,e^{i(\frac{N}{2}+M)\psi}|J,M\rangle \tag{8.25}$$

similarly to (4.12). Comparing (8.21). (8.24) and (8.25) one gets

$$c_j = \sum_{M} D_M B^{j*}_{M} \tag{8.26}$$

and finally

$$|\psi(t)\rangle = \sum_{iMM'} D_M B^{i*}_{M} B^{i}_{M'} e^{-iE_i t}|M'\rangle . \tag{8.27}$$

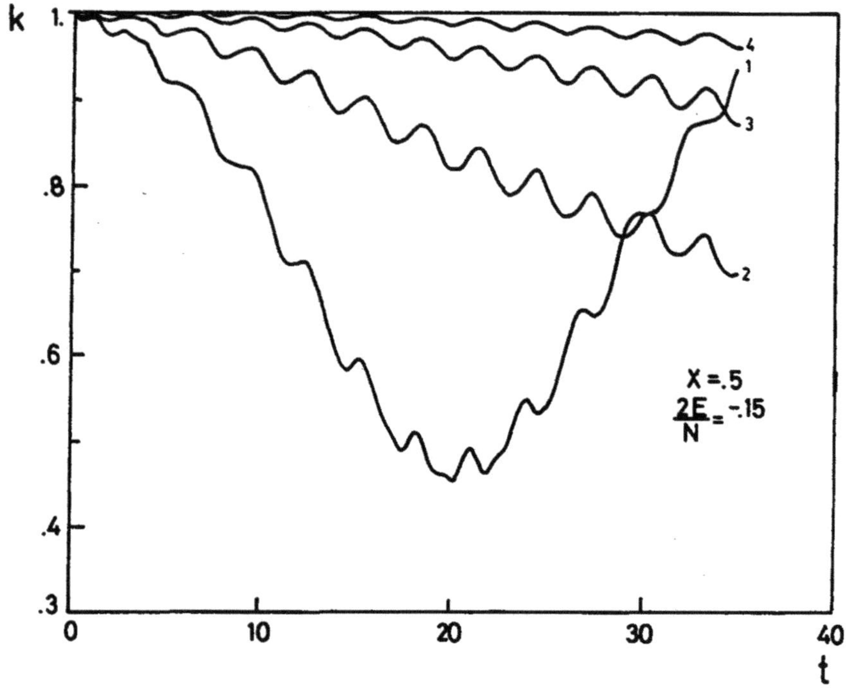

- Fig.6 -

Quantity k as a function of time for χ=.5 and 2E/N=-.15. The numbers 1,2,3,4 correspond to N=4,8,20,50.

Notice that on calculating D_M through (8.25) one uses the initial values of α and ψ.

The results for k(t) are shown in fig.6 for $\chi = 0.5, 2E/N = -0.15$ and different N values. It is clear that the evolution of the density comes closer to the TDHF behaviour when N becomes large. Moreover, a decay is observable with a time which increases with N, actually of order $\sqrt{N}$ (see ref.34). The conclusion is that, as in the case of the action, the leading order behaviour is the one predicted by TDHF whereas for large but finite N another universal behaviour is apparent. It is interesting to remark that how to calculate this subsequent order in the expansion, both for the action and for the time evolution of ρ, is still an open question.

References

1) A.de Shalit and H.Feshbach, Theoretical Nuclear Physics, Vol.I: Nuclear Structure, ed.Wiley 1974.

2) P.Ring and P.Schuck, The Nuclear Many Body Problem, ed. Springer-Verlag 1980.

3) G.Ripka, Adv.Nucl.Phys.1 (1968) 183.

4) D.Vautherin and D.M.Brink, Phys.Rev. C5 (1972) 626.

5) M.Beiner, H.Flocard, N.Van Giai and P.Quentin, Nucl.Phys. A238 (1975) 29.

6) P.Quentin and H.Flocard, Ann.Rev.Nucl.Sci. 28 (1978) 523.

7) H.J.Lipkin, N.Meshkov and A.J.Glick, Nucl.Phys. 62 (1965) 188.

8) D.Agassi, H.J.Lipkin and N.Meshkov, Nucl.Phys. 86 (1966) 321.

9) M.C.Cambiaggio and A.Plastino, Z.Physik A291 (1979) 277.

10) A.R.Edmonds, Angular Momentum in Quantum Mechanics, ed. Princeton University Press 1957.

11) A.Kamlah, Z.Physik 216 (1968) 52.

12) R.Beck, H.J.Mang and P.Ring, Z.Physik 231 (1970) 26.

13) H.J.Mang, Phys.Rep. 18C, N°6 (1975).

14) D.R.Inglis, Phys.Rev. 96 (1954) 1059.

15) A.Faessler, K.R.Sandhya Devi, F.Grümmer, K.W.Schmid and R.R.Hilton, Nucl.Phys. A256 (1976) 106.

16) A.L.Goodman, Nucl.Phys. A265 (1976) 113.

17) J.L.Egido, P.Ring and H.J.Mang, Phys.Lett. 77B (1978) 123; Nucl.Phys. A339 (1980) 390.

18) A.Arima and F.Iachello, Ann.of Phys. 99 (1976) 253; 111 (1978) 201; 123 (1979) 468.

19) O.Scholten, F.Iachello and A.Arima, Ann. of Phys. 115 (1978) 325.

20) M.C.Cambiaggio, J.Dukelsky and G.R.Zemba, Phys.Lett 162B (1985) 203.

21) O.Scholten, XVI Annual Report, Groningen (1976) p.99.

22) E.R.Marshalek and J.Weneser, Phys.Rev. C2 (1970) 1682.

23) A.K.Kerman and S.E.Koonin, Ann.Phys. (NY) 100 (1976) 332.

24) J.W.Negele, Reviews of Modern Physics 54 (1982) 913.

25) P.Bonche, S.E.Koonin and J.W.Negele, Phys.Rev. C13 (1976) 1226.

26) R.Y.Cusson, J.A.Maruhn and H.W.Meldner, Phys.Rev. C18 (1978) 2589.

27) H.Flocard, S.E.Koonin and M.S.Weiss, Phys.Rev. C17 (1978) 1682.

28) J.W.Negele, S.E.Koonin, P.Möller, J.R.Nix and A.J.Sierk, Phys.Rev. C17 (1978) 1098.

29) K.K.Kan, P.C.Lichtner, M.Dworzecka and J.J.Griffin, Phys.Rev. C21 (1980) 1098.

30) S.J.Krieger, Nucl.Phys. A276 (1977) 12.

31) K.K.Kan, J.J.Griffin, P.C.Lichtner and Dworzecka, Nucl.Phys. A332 (1979) 109.

32) S.Levit, J.W.Negele and Z.Paltiel, Phys.Rev. C21 (1980) 1603.

33) M.C.Cambiaggio, G.G.Dussel and M.Saraceno, Phys.Lett.B135 (1984) 10.

34) M.C.Cambiaggio, G.G.Dussel and M.Saraceno, Condensed Matter Theories, Vol.1, p.57-65 (New York, Plenum, 1986).

DENSITY FUNCTIONAL THEORY: AN OVERVIEW

by
Eduardo V.Ludeña(*)
Venezuelan Institute of Scientific Investigations

(*)Instituto Venezolano de Investigaciones Cientificas, IVIC
Centro de Química, Apartado 21827, Caracas 1020-A, Venezuela

DENSITY FUNCTIONAL THEORY: AN OVERVIEW

by E. V. Ludeña

1.- Introduction

Physical theories, if they are to be successful in describing reality must, to a large extent, be cast in an adequate language so that they possess at the same time both a tractable mathematical structure and clearly defined set of concepts which may be used in the design and assessment of experiments.

Quantum theory, the physical theory of microscopic matter was heralded, from the time of its very inception as a definitive step toward the understanding of those physical system corresponding to well known interactions and possessing, therefore, well defined Hamiltonians, such as atoms, molecules and solids. However, although particularly in the last twenty years, an impressive progress has been made in obtaining accurate solutions to Schrödinger equation for systems with up to a few tens of electrons, the solution for systems of interest in biology, chemistry, material science or catalysis, which typically comprise thousands of electrons and hundreds of nuclei, still remain elusive in spite of the fact that the use of large computers and supercomputers has become rather commonplace in recent years.

Basically, the difficulty with the formulation of quantum theory based on the Schrödinger equation is that a state of a system of n electrons and N nuclei, for example, is characterized by means of a wavefunction $\psi(1, \ldots, n+N)$ which depends upon all spatial and spin coordinates of the n + N particles. Thus, the dimensionality of the Schrödinger equation grows very rapidlly with increasing n and N.

Alternative formulations which bypass the calculation of the wavefunction $\psi(1, \ldots, n+N)$ and which are based on reduced density matrices or on the density have been vigorously pursued particularly during the last decade[1-6]. The aim in these formulations has been to cast the quantum mechanical problem for large systems into a mathematically tractable form. Unfortunately, there remain still some difficult problems both at the level of the foundation of these theories as well as in relation to their applications.

We review here, succintly some aspects of these developments. In Section 2 we deal with the energy expressions in terms of reduced second order and first order density operators and briefly comment on the n-representability problem. In Section 3 we discuss the approach based on the density and describe some approximations employed in this context. In Section 4 we describe the method of point transformations and show how a whole class of approximations in Density Functional Theory correspond to point transformations of the one-particle density.

2.- The energy as a functional of reduced density operators

Physical observables are quantities related to operators which although depending in a general sense on all particle coordinates, represent single particle effects and also pair interactions among particles. This is observed, for example, in the Hamiltonian for n electrons and N nuclei interacting through Coulomb forces. This Hamiltonian is written as

$$\hat{H}(1,\ldots,n+N) = \sum_{i=1}^{n} h_e(i) + \sum_{I=1'}^{N} h_N(I) + \sum_{i=1}^{n-1} \sum_{j=i+1}^{n} v_{ee}(i,j)$$

$$+ \sum_{i=1}^{n} \sum_{I=1}^{N} v_{e_N}(i,I) + \sum_{I=1'}^{N-1} \sum_{J=I+1}^{N} v_{NN}(I,J) \quad (1)$$

where

$$h_e(i) = -\frac{h^2}{8\pi^2 m}\nabla_i^2, \quad h_N(I) = -\frac{h^2}{8\pi^2 M_I}\nabla_I^2, \quad v_{ee}(i,j) = \frac{e^2}{|\vec{r}_i - \vec{r}_j|}$$

$$v_{eN}(i,I) = \frac{Z_I\, e^2}{|\vec{r}_i - \vec{R}_I|}, \quad v_{NN}(I,J) = \frac{Z_I Z_J e^2}{|\vec{R}_I - \vec{R}_J|}. \quad (2)$$

The question which obviously emerges is whether, in view of this basic characteristics of the Hamiltonian operator, it is possible to bypass the usual formulation of quantum mechanics based on the wave function $\psi(1, \ldots, n+N)$ and to develop an alternative one in terms of a simpler mathematical object depending on only two particle coordinates such as the reduced second order density operator $\Gamma^2(1,2;\, 1',2')$ or even in terms of simpler objects, such as the reduced first order density operator $\gamma(1,1')$ or the single particle density $\rho(1) \equiv \gamma(1,1)$.

As is well known, these reformulations are, in principle, possible[1-6] and are known by the generic names of "Density Matrix Theory" and "Density Functional Theory". In order to highlight some of their important aspects, we consider below the expression of the energy as a functional of these reduced operators.

Let us assume, for simplicity, that the nuclei are clamped at given positions (Born-Oppenheimer approximation). The expectation value of the Hamiltonian becomes in this case

$$\langle\hat{H}\rangle = \int d1 \ldots \int dn\ \psi(1,\ldots,n) \sum_{i=1}^{n-1} \sum_{j=i+1}^{n} k(i,j)\ \psi(1,\ldots,n) \quad (3)$$

where the two-particle operator $k(i,j)$ is given by

$$k(i,j) = [h_e(i)+h_e(j)+v_{en}(i)+v_{eN}(j)+v_{ee}(i,j)]/(n-1) \quad (4)$$

with

$$v_{eN}(i) = \sum_{I=1}^{N} v_{eN}(i,I). \quad (5)$$

Introducing the reduced two-particle density operator

$$\Gamma^2(1,2;1',2') = \frac{n(n-1)}{2}\int d3\ldots\int dn\ \psi(1,2,3,\ldots,n)\psi(1',2',3,\ldots,n) \tag{6}$$

and taking into account that electrons are undistinguishable, we obtain

$$<\hat{H}>=\int d1\ldots\int dn \sum_{i'=1}^{n-1}\ \sum_{j'=i'+1}^{n} k(i',j')\,\psi(1,\ldots,n)\psi(1',\ldots,n')\,_{1'\to 1,\ldots,n'\to n}$$

$$= \int d1\int d2\ k(1',2')\ \Gamma^2(1,2;1',2')\,|_{1'\to 1,2'\to 2} \tag{7}$$

Since the operator k(1,2) is known, the energy is simply a functional of Γ^2 : $E = E[\Gamma^2]$. Variation of $E[\Gamma^2]$ with respect to Γ^2 subject to all necessary and sufficient conditions guaranteeing that Γ^2 is the reduced second order density operator corresponding to the n-electron system under consideration (n-representability conditions) would in principle lead to the Euler-Lagrange equation for Γ^2, which would be equivalent to the Schrödinger equation (the Euler-Lagrange equation for $\psi(1,\ldots,n)$).

There is an important stumbling block barring this approach. This is the lack of knowledge of all the necessary and sufficient n-representability conditions on Γ^2. Although much progress has been made, the general n-representability problem still remains unsolved[8-10].

We may rewrite the expectation value of the Hamiltonian as follows

$$<\hat{H}> = \int d1\ f(1')\ \gamma(1,1')|_{1'\to 1} + \int d1\int d2\ v_{ee}(1',2')\ \Gamma^2(1,2;1',2')|_{1'\to 1,2'\to 2'} \tag{8}$$

where we have defined

$$f(1) = h_e(1) + v_{eN}(1) \tag{9}$$

and where we have introduced the reduced first order density operator which is given by

$$\gamma(1,1') = n\int d2\ \ldots\int dn\,\psi(1,2,\ldots,n)\ \psi(1',2,\ldots,n) \tag{10}$$

It follows from Eq. (8) that the energy is a functional of γ and Γ^2 : $E = E[\gamma,\Gamma^2]$ since both f(1) and $v_{ee}(1,2)$ are known operators. In order to express the energy as a funcxtional of γ, with an implicit dependence on Γ^2, we may introduce a single particle potential

$$V(1';p^2)\ \gamma(1,1') \equiv \int d2\ v_{ee}(1',2)\ \Gamma^2(1,2;1',2') \tag{11}$$

Using Eq. (11), we get

$$E[\gamma(\Gamma^2)] = \int d1\ \mathcal{H}(1';\ \Gamma^2)\gamma(1,1')\,|_{1'\to 1} \tag{12}$$

where we have defined

$$\mathcal{H}(1';\Gamma^2) = f(1') + V(1';\Gamma^2) \qquad (13)$$

In the Hartree-Fock approximation, where Γ^2 can be expressed solely in terms of γ,

$$\Gamma^2(1,2;1',2') = \frac{1}{2}\left[g(1,1')\gamma(2.2')-\gamma(1,2')\gamma(2,1')\right], \qquad (14)$$

$\mathcal{H}(1';\Gamma^2)$ becomes $\mathcal{H}(1';\gamma^2)$ and is given by

$$\mathcal{H}(1';\gamma) = f(1') + J(1';\gamma) - K(1';\gamma) \qquad (15)$$

where

$$J(1;\gamma) = \frac{1}{2}\int d2\ v_{ee}\ (1,2)\ \gamma(2,2) \qquad (16)$$

and

$$K(1';\gamma)\gamma(1,1') = \frac{1}{2}\int d2\ v_{ee}(1',2)\gamma(1,2)\gamma(2.1') \qquad (17)$$

The importance of rewriting the energy as an explicit functional of γ comes about from the fact that much more is known about the n-representability conditions for γ than for Γ^2. In fact, when γ corresponds to an ensemble, the necessary and sufficient conditions for γ to be n-representable have been determined by Coleman[11]. These conditions are strikingly simple and just require that the eigenvalues $\{\lambda_i\}$ of the matrix $\gamma(1,1')$, satisfy the following requirements:

$$0 \le \lambda_i \le 1,\ \forall i \quad \text{and} \quad \sum_{i=1} \lambda_i = n \qquad (18)$$

For a pure state, conditions on γ become more complicated and in general they remain unknown[12-14]. It has been shown, however, that for this case the problem may still be tractable by means of a constrained variational method where the pure-state n-representability conditions are incorporated into the variational functional before the integration over redundant coordinates is carried out[15].

Notwithstanding the advantages which a formulation based on γ would provide in terms of n-representability conditions in contrast with another one based on Γ^2, the point is that in Eq(7), the energy operator $k(1,2)$ is known directly, whereas the operator $\mathcal{H}(1;\Gamma^2)$ in Eq(12), is known only indirectly and has a rather complex dependence on Γ^2. In those cases where Γ^2 can be reduced to, or sufficiently approximated by, a function of γ, this single particle operator becomes $\mathcal{H}(1;\gamma)$; such is the case of the Hartree-Fock method, where the functional form of $\mathcal{H}(1;\gamma)$ is particularly simple as it can be seen in Eqs(15-17).

3. - The energy as a functional of the one-electron density

A. The Theorem of Hohenberg and Kohn

The possibility of expressing the energy as a a functional of the density is given by the celebrated theorem of Hohenberg and Kohenberg and Kohn[16]. Consider the separation of the electronic Hamiltonian into an internal (electronic) part, H_o and an external (electron-nuclei interaction) part, $\hat{V}$:

$$\hat{H}=\hat{H}_o+\hat{V}=\left[\sum_{i=1}^{n}-\frac{1}{2}\nabla_i^2+\sum_{i=1}^{n-1}\sum_{j=i+1}^{n}v_{ee}(i,j)\right]+\left[\sum_{i=1}^{n}v_{eN}(i)\right] \qquad (19)$$

Let us assume that for a given external potential $v_{eN}(i) = v_{eN}(\vec{r}_i)$, the system is described by a wavefunction $\psi(1,...,n)$. When the external potential is changed to $v'_{eN}(\vec{r}_i)$, which differs from the previous one by more than a constant:

$$v_{eN}(\vec{r}_i) - v'_{eN}(\vec{r}_i) \neq \text{constant} \qquad (20)$$

and calling $\psi'(1,...,n)$ the wavefunction corresponding to this new potential, we have the following inequality

$$E'=\langle\psi'|\hat{H}_o+\hat{V}'|\psi'\rangle < \langle\psi|\hat{H}_o+\hat{V}'|\psi\rangle=\langle\psi|\hat{H}+\hat{V}'-V|\psi\rangle \qquad (21)$$

from where we obtain

$$E' < E + \int d\vec{r}\,\left(v'_{eN}(\vec{r}) - v_{eN}(\vec{r})\right)\rho(\vec{r}) \qquad (22)$$

But, on the other hand,

$$E = \langle\psi|\hat{H}_o+\hat{V}|\psi\rangle < \langle\psi'|\hat{H}'+\hat{V}-\hat{V}'|\psi'\rangle \qquad (23)$$

which implies that

$$E < E' -\int d\vec{r}\,\left(v'_{eN}(\vec{r}) - v_{eN}(\vec{r})\right)\rho'(\vec{r}) \qquad (24)$$

If we now assume that both external potentials give rise to the same density: $\rho'(\vec{r}) = \rho(\vec{r})$, then we are led to the contradiction

$$E' + E < E + E' \qquad (25)$$

It follows, therefore that these densities cannot be the same, or, put another way, that the external potential uniquely determines the density of the ground state of a system. Since $\hat{H}_o$ is the same for the systems under consideration, the total energy may be written as

$$E_V[\rho] = \int d\vec{r}\ v_{eN}(\vec{r})\rho(\vec{r}) + Q[\rho] \tag{26}$$

where Q[ρ] would have the same functional form for all systems with the same number of electrons. It is in this sense that Q[ρ] is referred to as a "universal" functional of the density. From variational considerations it is easily shown that

$$E_V[\rho_{exact}] \leq E_V[\rho_{approx.}] \tag{27}$$

This theorem, although giving the possibility of expressing the energy as a functional of ρ, does not indicate the steps one must follow for its determination and up to now, the exact form of this functional remains unknown. Hence, density functional theory, deals largely with the methods for approximating this functional. It must be mentioned that there exist some difficulties with this theorem. For example, Harriman[17] has shown that in a finite basis, the theorem does not hold.

B. <u>The Formulation of Levy and Lieb</u>

Another approach to Density Functional Theory has been advanced by Levy[18] and Lieb[19]. Consider the universal functional

$$Q[\rho] = \min \left\{ \langle\Psi_\rho|\hat{H}_o|\Psi_\rho\rangle ;\ \Psi_\rho \Rightarrow \rho ; \Psi_\rho \in \mathcal{H}_A^n \right\} \tag{28}$$

where Ψ_ρ is an antisymmetric n-particle wavefunction spanning the Hilbert space $\mathcal{H}^n{}_A$, which yields the density ρ upon reduction:

$$\rho(1) = n\int d2 \ldots \int dn\ \Psi_\rho^*(1,\ldots,n)\ \Psi_\rho(1,\ldots,n) \tag{29}$$

It follows from Eq(28) that the universal functional Q[ρ] corresponds to a particular function $\Psi_\rho{}^{min}$:

$$Q[\rho] = \langle\Psi_\rho^{min}|\hat{H}_o|\Psi_\rho^{min}\rangle ; \Psi_\rho^{min} \Rightarrow \rho ; \Psi_\rho^{min} \in \mathcal{H}_A^n \tag{30}$$

Thus, we have for the total Hamiltonian

$$\langle\Psi_\rho^{min}|\hat{H}_o + \hat{V}|\Psi_\rho^{min}\rangle = \int d\vec{r}\ v_{eN}(\vec{r})\rho(\vec{r}) + Q[\rho] \geq E[\rho_o] \tag{31}$$

where ρ_o is the exact ground state density. For the equality in Eq(31) to hold, it is necessary that

$$E[\rho_o] = \inf_\rho \left\{ \int d\vec{r}\ v_{eN}(\vec{r})\rho(\vec{r}) + Q[\rho] \right\} \tag{32}$$

where it is implied that the exact energy is obtained when the infimum of the right hand side is obtained by varying ρ over the set of admissible densities[9].

C. Approximate Forms for the Universal Functional

Let us consider a partition of the exact wavefunction $\Psi(1,\dots,n)$ for an n particle systems in terms of a single Slater determinant $\Psi_{SD}(1,\dots,n)$ plus $\Psi_C(1,\dots,n)$, where in the latter all correlation effects are included:

$$\Psi(1,\dots,n) = \Psi_{SD}(1,\dots,n) + \Psi_C(1,\dots,n) \tag{33}$$

The reduced second order density operator becomes

$$\begin{aligned}\Gamma^2(1,2;1',2') &= \frac{n(n-1)}{2}\int d3\dots\int dn\left\{\Psi^*_{SD}\Psi_{SD}+\Psi^*_{SD}\Psi_C+\Psi_{SD}\Psi^*_C+\Psi^*_C\Psi_C\right\}\\ &= \Gamma^2_{SD}(1,2;1',2') + \Gamma^2_C(1,2;1',2')\end{aligned} \tag{34}$$

where we have included in $\Gamma_C{}^2$ all products containing Ψ_C. Introducing a correlation factor by means of

$$P_C(1,2;1',2') \equiv \Gamma^2_C(1,2;1',2')/\Gamma^2_{SD}(1,2;1',2') \tag{35}$$

We have

$$\Gamma^2(1,2;1',2') = \Gamma^2_{SD}\ (1,2;1',2')\ \ (1+P_C(1,2;1'2')) \tag{36}$$

It follows from Eq(36) that the reduced first order operator may be also partitioned into γ_{SD} and γ_C :

$$\begin{aligned}\gamma(1,1') &= \int d2\left\{G^2_{SD}(1,2;1',2)+\Gamma^2_{SD}(1,2;1',2)P_C(1,2;1',2)\right\}\\ &= \gamma_{SD}(1,1')+\gamma_C(1,1')\end{aligned} \tag{37}$$

$\gamma_{SD}(1,1')$ may be expressed in terms of local and non-local contributions[19]

$$\gamma_{SD}(1,1') = \rho^{1/2}_{SD}(1)\ \ \rho^{1/2}_{SD}(1')\ \ G_{SD}(1,1') \tag{38}$$

where the non-local correlation function $G_{SD}(1,1')$, which satisfies $G_{SD}(1,1) = 1$, includes the spin correlation among electrons. Using Eqs(8), (36), (37) and (38), we may rewrite the total energy expression as follows

$$\begin{aligned}E\left[r_{SD}, G_{SD}, P_C\right] &= E_{KIN}\left[r_{SD}, G_{SD}, P_C\right]\\ &+ E_{COUL}[\rho_{SD}] + E_{XC}\left[r_{SD}, G_{SD}, P_C\right] + E_{EXT}\left[\rho_{SD}, G_{SD}, P_C\right]\end{aligned} \tag{39}$$

where the various terms stand for the kinetic, Coulomb, exchange-correlation and external contributions to the energy, respectively, and where we have included their functional dependence upon ρ_{SD} , G_{SD} and P_C. Explicitly, these terns are:

$$E_{KIN}\left[\rho_{SD}, G_{SD}, P_C\right] = \frac{1}{2}\int d1 \vec{\nabla}_1 \cdot \vec{\nabla}_{1'} \gamma(1,1')\,|_{1'\to 1}$$

$$= \frac{1}{8}\int d1 \left[\vec{\nabla}_1 \rho_{SD}(1)\right]^2 / \rho_{SD}(1)$$

$$+\frac{1}{2}\int d1 \rho_{SD}(1) \vec{\nabla}_1 \cdot \vec{\nabla}_{1'} \left[G_{SD}(1,1')G_{SD}(1',1)\right]^{1/2}_{1'\to 1} + \frac{1}{2}\int d1 \vec{\nabla}_{1'} \cdot \vec{\nabla}_{1'} \gamma_C(1,1')\,|_{1'\to 1} \quad (40)$$

$$E_{COUL}[\rho_{SD}] = \frac{1}{2}\int d1 \int d2\; \rho_{SD}(1)\rho_{SD}(2)/|\vec{r}_1 - \vec{r}_2| \quad (41)$$

$$E_{XC}\left[\rho_{SD}, G_{SD}, P_C\right] = -\frac{1}{2}\int d1 \int d2\; \rho_{SD}(1)\rho_{SD}(2)G_{SD}(1,2)G_{SD}(2,1)/|\vec{r}_1-\vec{r}_2|$$

$$+\frac{1}{2}\int d1 \int d2\; \rho_{SD}(1)\rho_{SD}(2)\; P_C\; (1,2;1,2)/|\vec{r}_1-\vec{r}_2|$$

$$\frac{1}{2}\int d1 \int d2\; \rho_{SD}(1)\rho_{SD}(2)G_{SD}(2,1)\; P_C\; (1,2';1,2)/|\vec{r}_1-\vec{r}_2| \quad (42)$$

and

$$E_{EXT}\left[r_{SD}, G_{SD}, P_C\right] = \int d1\, \rho_{SD}(1) v_{eN}(1) + \int d1\; \gamma_C(1,1) v_{eN}(1) \quad (43)$$

In order to develop a true density functional theory it is necessary to express both $G_{SD}(1,2)$ and $P_C(1,2;1,2)$ as functions of the one-electron density ρ. The different approximations which have been advanced in density functional theory correspond, therefore, to different types of functional forms for these quantities. As it is well known, no exact functional form has been obtained up to date. We discuss below of these approximations.

a) The Thomas-Fermi-Weiszacker-Dirac Model.

This model results of taking G(1,2) as the correlation function of the homogeneous electron gas and of setting the correlation factor $P_C(1,2;1,2) = 0$. Since in an orbital representation

$$\gamma_{SD}(1,2) = \sum_{i=1}^{n} \phi_i^*(1)\; \phi_i(2) \quad (44)$$

it is easily seen from Eq(38) that

$$G_{SD}(1,2)G_{SD}(2,1)=\sum_{i=1}^{n}\sum_{j=1}^{n}\phi_i^*(1)\phi_i(2)\phi_j^*(2)\phi_j(1)/\left(\rho_{SD}(1)\rho_{SD}(2)\right) \quad (45)$$

Introducing plane waves

$$\phi_i(\vec{r})=V^{-1/2}\exp(i\vec{k}_i\cdot\vec{r}) \quad (46)$$

we obtain

$$G_{SD}(1,2)G_{SD}(2,1)=\frac{1}{n}\sum_{i=1}^{n}\sum_{j=1}^{n}\exp\left(i(\vec{k}_i-\vec{k}_j)\cdot(\vec{r}_2-\vec{r}_1)\right) \quad (47)$$

Transforming the summation into an integral over momenta, we obtain $G_{SD}(1,2)$ for a homogeneous electron gas:

$$G_{SD}(1,2) = G_{SD}(\vec{r}_2-\vec{r}_1) = G_{SD}(\vec{r}) = 3(k_F r)^{-3}(\sin k_F r-k_F r\cos k_F r) \quad (48)$$

Substituting this expression into the second term of Eq(40), we obtain a contribution to the kinetic energy

$$\frac{1}{2}\int d1\rho_{SD}(1)\vec{\nabla}_1\cdot\vec{\nabla}_2\left[G_{SD}(1,2)G_{SD}(2,1)\right]^{1/2}\Big|_{2\to 1}=\frac{3}{10}(3\pi^2)\int d1\rho_{SD}(1)^{5/3}\equiv T_{TF}[\rho_{SD}] \quad (49)$$

which is the well known Thomas-Fermi term[20]. Similarly, substituting Eq(48) into the first term of the right hand side of Eq(42), we obtain Dirac's exchange correction[21].

$$\frac{1}{2}\int d1\int d2\ \rho_{SD}(1)\rho_{SD}(2)G_{SD}(1,2)G_{SD}(2,1)/|\vec{r}_2-\vec{r}_1|=-3\left(\frac{3}{8\pi}\right)\int d1\ \rho_{SD}^{4/3}(1)\equiv E_{X,Dirac}[\rho_{SD}] \quad (50)$$

As in this approximation $P_c(1,2;1,2)$ is assumed to be zero, the total energy is a functional of ρ_{SD}:

$$E[\rho_{SD}] = T_W[\rho_{SD}]+T_{TF}[\rho_{SD}]+E_{COUL}[\rho_{SD}]+E_{X,Dirac}[\rho_{SD}]+E_{EXT}[\rho_{SD}] \quad (51)$$

where $T_W[\rho_{SD}]$ is the Weiszacker term[22] given by the first RHS term of Eq(40). Unfortunately, Eq(51) is not sufficiently accurate for dealing with many-electron systems in atoms, molecules, and solids. Nevertheless, Dirac's exchange correction or modifications to it have been used as substitutes of the cumbersome exchange terms which appear in the Hartree-Fock or Kohn-Shan[23] single particle equations.

b) Equidensity Orbitals Approximations

The orbital expansion for G(1,2)G(2,1) given by Eq(45), opens the possibility of simulating these terms using different types of orbitals. Thus, as discussed above, plane-wave orbitals result in the homogeneous electron gas correlation function. Harriman's equidensity orbitals[24] are closely related to plane waves but in a way, much more elaborate, as they include implicitly and explicitly the one-electron density $\rho(\vec{r})$. A modified version of these orbitals is given by[25]

$$\phi_k(\rho) = \left[\frac{\rho(r)}{n}\right]^{1/2} \exp\left(i\left(k - \frac{n+1}{2}\right) f(r)\right) \tag{52}$$

where

$$f(r) = 2\pi \int_0^r dr' (r')^2 \rho(r') \tag{53}$$

Extensions of these orbitals to three-dimensions have been discussed by Zumbach and Maschke[26] and the incorporation of spherical harmonics has been dealt with by Gosh and Parr[27]. Using Eq(52), G(1,2) becomes[28]

$$G(1,2) = \sin\frac{n}{2}\left(f(r_2) - f(r_1)\right) \Big/ \left(n \sin\frac{(f(r_2) - f(r_1))}{2}\right) \tag{54}$$

and we obtain for the kinetic energy[25]

$$E_{KIN}[\rho_{SD}] = T_W[\rho_{SD}] + \frac{8}{3}\pi^4 \left(1 - \frac{1}{n^2}\right) \int dr\, r^4\, \rho(r)^3 \tag{55}$$

and corresponding expressions for the remaining terms[28].

Clearly, all sorts of approximations to the universal functional of the density may be obtained by using different G(1,2)'s, or equivalently, different orbital expansions. But again, the level of accuracy required for the description of shell structure in atoms is rather high and at present these approximations are not yet sufficiently developed to attain this result. Nevertheless, a systematic analysis of G(1,2) for atoms has shown that G(1,2) almost entirely describes the Fermi hole and also that its contribution to the kinetic energy is localized in the intershell regions[29].

D. Gradient Expansions

The basic idea behind the gradient expansions of the kinetic and the exchange correlation energy densities is to consider that a many electron system may be described, as a first approximation, by means of the homogeneous electron gas model, which, of course, may be subsequently modified by the progressive addition of terms which account for the inhomogeneities present in real systems.

In this context, for example, the exchange correlation energy may be written as

$$E_{XC}[\rho] = E_{XC}^{LDA}[\rho] + \text{gradient corrections} \tag{56}$$

where the local density approximation is given by

$$E_{XC}^{LDA}[\rho] = \int d\vec{r}\, \rho(\vec{r})\, \varepsilon_{XC}[\rho] \tag{57}$$

and where $\varepsilon_{XC}[\rho]$ is the exchange correlation energy density for a homogeneous electron gas. The exchange part of $\varepsilon_{XC}[\rho]$ corresponds to Dirac's term, discussed in Eq(50), so that $\varepsilon_{XC}[\rho]$ is proportional to $\rho^{1/3}(\vec{r})$. For the kinetic energy, we have

$$E_{KIN}[\rho] = T_{TF}[\rho] + \text{gradient corrections} \tag{58}$$

where $T_{TF}[\rho]$ is given by Eq(49). The first gradient correction to the Thomas-Fermi term is the Weiszacker term $T_W[\rho]$ of Eq(40).

We shall not deal here with uneven history of the gradient expansion. But for revisions of its accomplishments and shortcomings, the reader is referred to references[30,31,32]. Let us simply mention that the gradient expansion presents the following three basic difficulties: a) the only terms that satisfy the asymptotic behavior demanded by the properties of the density at large r, are $T_{TF}[\rho]$, $T_W[\rho]$, for the kinetic energy and $E_{X,Dirac}[\rho]$ for the exchange-correlation correction[31], b) the additional terms diverge when the Euler-Lagrange equations for the density are obtained from a gradient expansion of the energy density[33], and c) the approximate treatment of exchange and correlation gives rise to a spurious self-interaction term[34].

In spite of these difficulties, the gradient expansion has provided useful guidelines for the development of highly sophisticated approximations to the energy density functional.

A very successful approach has been developed by Langreth and Mehl[35], who have examined the corrections to the local decomposition of E_{XC}. According to these authors, the exchange correction energy may be expanded as follows:

$$E_{XC}[\rho] = E_{XC}^{LDA-RPA}[\rho] + a\int d\vec{r}\,\frac{[\nabla\rho(\vec{r})]^2}{\rho^{4/3}(\vec{r})}\left(2e^{-F} - \frac{7}{9}\right) \tag{59}$$

where the local density term is that corresponding to the random phase approximation. The function F is

$$F = b|\nabla\rho(\vec{r})|/\rho^{7/6}(\vec{r}) \tag{60}$$

The problems caused by the spurious presence of self-interactions have been largely eliminated by means of the Self Interaction Correction (SIC) method, where these effects are directly subtracted fron the energy expression[34].

Finally, the problems due to the improper asymptotic behavior of the energy density at large r, have been remedied by a cut-off technique which tailors the individual terms of the gradient expansion so as to make them satisfy the proper asymptotic behavior[32]. In view of all these modifications there exist, at present, very accurate approximations to the elusive energy functional. The energy result obtained from these carefully trimmed functionals in the context of the Kohn-Sham formalism, where $E_{XC}[\rho]$ defines the effective potential of a Schrödinger type orbital equation

$$\left(-\frac{1}{2}\nabla^2 + v_{EXT}(\vec{r}) + v_{COUL}(\vec{r}) + v_{XC}(\vec{r})\right)\psi_i(\vec{r}) = \lambda_i\psi_i(\vec{r}) \tag{61}$$

with

$$v_{xc}(\vec{r}) = \frac{\delta E_{xc}[\rho]}{\delta \rho} \tag{62}$$

have been rather encouraging in that they show a high degree of agreement with experimental data. It seems, however, that the density obtained from this approach by means of

$$\rho(\vec{r}) = \sum_{i=1}^{n} \psi_i^*(\vec{r})\psi_i(\vec{r}) \tag{63}$$

is not yet sufficiently accurate. Furthemore, from a more fundamental point of view it has been shown that a Kohn-Sham type of equation is just one out of a system of coupled equations[15c,36], so that perhaps a way to improve accuracy might not depend entirely upon the degree of approximation to the universal functionals, but rather on the proper definition of the relevant coupled equations which describe the pure-state quantum mechanical problem.

4.- Point transformation

A plane wave representation of a physical system is, clearly, not the most adequate one if the system presents inhomogeneities and has spatial regions where the density varies considerably. Hence, the use of plane waves for the calculation of correlation functions for inhomogeneous systems, is a bit forced in the sense that plane-waves do not mimick the actual spatial features of such systems. Obviously, any function may be represented by means of an infinite plane-wave expansion. But this is not the point as what we would like to have at our disposal is an orbital set which already incorporates the physical aspects of a given problem.

A plane-wave is given by

$$U_{\vec{k}}(\vec{r}) = \left(\frac{\rho}{n}\right)^{1/2} \exp\,(\vec{k}\,.\,\vec{r}) \tag{64}$$

where we have used the term $(\rho/n)^{1/2}$ instead of the usual normalization factor $V^{-1/2}$ (of course, both are equivalent in view of $\rho = n/V$) to emphasize their resemblance to the equidensity orbitals given by Eq(52)). An important drawback with plane-waves is that they do not incorporate directly information concerning the functional relationship between $\rho(r)$ and $\vec{r}$ as it is done in the case of equidensity orbitals. In what follows, we sketch a general procedure based on point transformations which permits us to relate these two types of functions.

The basic idea of point transformations[37,38] is to go from a cartesian coordinate system $\vec{r} = (x,y,z)$ to anotherone $f(\vec{r}) \equiv (f_x(\vec{r}), f_y(\vec{r}), f_z(\vec{r}))$ where the latter is chosen so as to closely follow the distortions of the density $\rho(\vec{r})$ in $\Re^3$. In a way, these are density-adapted coordinates.

Consider the vector transformation

$$\vec{r} \equiv (x,y,z) \overset{\Re}{\rightarrow} \vec{f} \equiv (f_x, f_y, f_z) \equiv \vec{f}(\vec{r}) \tag{65}$$

This transformation $\Re$ exists if its Jacobian is non-zero

$$D\{\vec{f}(\vec{r});\vec{r}\} \equiv \frac{D(f_x, f_y, f_z)}{D(x,y,z)} \neq \tag{66}$$

Among the many ways in which this transformation may be realized, a particularly interesting one is given by

$$\vec{f}(\vec{r}) = \vec{r}\,\frac{f(\vec{r})}{r} = \vec{r}g(\vec{r}) \tag{67}$$

where $f(\bar{r}) = f(x,y,z)$. Clearly, $\vec{f}(\vec{r})$ retains the properties of $\vec{r}$, i.e., it is a Cartesian coordinate system, where the axis have been distorted by the presence of the function $g(\bar{r})$. It has been shown[37] that for this transformation the Jacobian is

$$D\left\{\vec{f}(\vec{r});\vec{r}\right\} = \left(\frac{\partial f(\vec{r})}{r}\right)^2 \frac{\partial f(\bar{r})}{\partial r} \tag{68}$$

Consider now this type of transformation applied to an n particle wavefunction. Since the normalization must be preserved, we have

$$\begin{aligned}&\int d\vec{r}_1 \ldots \int d\vec{r}_n \Psi^*(r_1, \ldots, \vec{r}_n)\Psi(\bar{r}_1, \ldots, \bar{r}_n)\\ &= \int d\vec{f}_1 \ldots \int d\vec{f}_n \Psi^*(\vec{f}_1, \ldots \vec{f}_n)\Psi(\vec{f}_1, \ldots, \vec{f}_n) = 1\end{aligned} \tag{69}$$

We have in this case that

$$d\vec{f}_i \equiv d\vec{f}(\vec{r}_i) = d\vec{r}_i D\left\{\vec{f}(\vec{r}_i);\vec{r}_i\right\} \tag{70}$$

In particular, it is seen that the reduced first order density operator

$$\gamma(\vec{r}_1,\vec{r}_1') = n\int d\vec{r}_2 \ldots \int d\vec{r}_n \Psi^*(\vec{r}_1,\ldots,\vec{r}_n)\Psi(\vec{r}_1', \ldots, \vec{r}_n) \tag{71}$$

becomes

$$\gamma(\vec{r}_1,\vec{r}_1') = D^{1/2}\left\{\vec{f}(\vec{r}_1);\vec{r}_1\right\} D^{1/2}\left\{\vec{f}(\vec{r}_1');\vec{r}_1'\right\} g_f\left(\vec{f}(\vec{r}_1), \vec{f}(\vec{r}_1')\right) \tag{72}$$

where

$$\gamma_f\left(\vec{f}(\vec{r}_1), \vec{f}(\vec{r}_1')\right) = n\int d\vec{f}_2 \ldots \int d\vec{f}_n \Psi^*(\vec{f}_1, \ldots, f_n\Psi(\vec{f}_1', \ldots, \vec{f}_n) \tag{73}$$

Equivalently, therefore, the density $\rho(\vec{r})$ and $\rho_f(\vec{f}(\vec{r}))$ are related by means of the non-linear differential equation

$$\rho(\vec{r}) = \left[\frac{f(\vec{r})}{r}\right]^2 \frac{\partial f(\vec{r})}{\partial r} \rho_f \; (\vec{f}(\vec{r})) \tag{74}$$

which may be put in the form

$$\frac{\partial f^3(\vec{r})}{\partial r} = 3r^2 \rho(\vec{r}) / \rho_f(\vec{f}(\vec{r})) \tag{75}$$

Now, assuming spherical symmetry, a very convenient choice for $\rho_f(\vec{f}(\vec{r}))$ is to require that it be a constant function of $\vec{f}(\vec{r})$ equal to n/V. This choice defines f(r) since we obtain upon integration of Eq(75)

$$f(r) = \left[\frac{3V}{n} \int_0^r dr' (r')^2 \rho(r')\right]^{1/3} \tag{76}$$

Using Eq(76) in Eq(68) one obtains the explicit expression for the Jacobian

$$D\left\{\vec{f}(\vec{r});\vec{r}\right\} = V\frac{\rho(r)}{n} \tag{77}$$

It follows, therefore, that upon application of this point transformation to a plane-wave, we obtain

$$V^{-1/2}\exp\;(i\vec{k}_j.\vec{r}) \overset{\Re}{\rightarrow} V^{-1/2}D^{1/2}\left\{\vec{f}(\vec{r});\vec{r}\right\}\exp(i\vec{k}_j.\vec{f}(\vec{r})) \tag{78}$$

The transformed functions are precisely the equidensity orbitals discussed by Harriman[24] and others[39].

A particular point of interest which may have relevance for condensed matter theory in general is whether point transformations could be fruitfully incorporated in the description of many-body effects through cluster expansions and whether in that context, they could lead to simplifications.

- References -

(1) E.R.Davidson: "Reduced Density Matrices in Quantum Chemistry", Academic, New York (1976).

(2) J.Keller and J.L.Gázquez, eds.: "Density Functional Theory", Springer- Verlag, Berlin (1983).

(3) J.P.Dahl and J.Avery, eds.: "Local Density Approximations in Quantum Chemistry and Solid State Physics", Plenum, New York (1984).

(4) S.Lundquist and N.H.March, eds.: "Theory of the Inhomogeneous Electron Gas", Plenum, New York (1983).

(5) R.M.Dreizler and J.da Providencia, eds.: "Density Functional Methods in Physics", Plenum, New York (1985).

(6) R.M.Erdhal and V.H.Smith, eds.: "Density Matrices and Density Functionals", Reidel, Dordrecht, in press.

(7) P.-D.Löwdin: Phys.Rev. 97 (1955) 1474.

(8) A.J.Coleman: Int.J.Quantum Chem. 13, (1978) 67.

(9) E.H.Lieb: Int.J.Quantum Chem. 24 (1983) 243.

(10) Ref.1, and p.104.

(11) A.J.Coleman: Rev.Mod.Phys. 35 (1963) 668.

(12) D.W.Smith: Rev.Mod.Phys. 35 (1963) 668; ibid., Phys.Rev. 147 (1966) 896.

(13) R.E.Borland and K.J.Dennis: J.Phys. B5 (1972) 7; ibid., J.Phys.B3 (1970) 887.

(14) Ref.1, p.51.

(15) a) T.T.Nguyen, E.V.Ludeña and Y.Tal: J.Mol.Struct. (Theochem) 120 (1985) 247;
b) E.V.Ludeña: J.Mol.Struct. (Theochem) 123 (1985) 371;
c) E.V.Ludeña and A.Sierraalta: Phys.Rev. A32 (1985) 19.

(16) P.Hohenberg and W.Kohn: Phys.Rev. 136 B (1964) 864.

(17) J.E.Harriman: Phys.Rev. A27, (1983) 632.

(18) M.Levy: Proc.Netl.Acad.Sci. (USA) 76 (1979) 6062; ibid, in Ref.2, p.9.

(19) E.V.Ludeña: J.Chem.Phys. 76 (1982) 3157; ibid, Int.J.Quantum Chem. 23 (1983) 127.

(20) L.H.Thomas: Proc. Cambridge & Philos.Soc. 23 (1927) 542;
E.Fermi: Z.Phys. 48 (1928) 73.

(21) P.A.M.Dirac: Proc. Cambridge & Philos.Soc. 26 (1930) 376.

(22) C.F. von Weizsäcker: Z.Phys. 96 (1935) 431.

(23) W.Kohn and L.J.Sham: Phys.Rev. 140 (1965) A1133.

(24) J.E.Harriman: Phys.Rev. A24 (1981) 680.

(25) E.V.Ludeña: in "Condensed Matter Theories 1", F.B.Malik, ed., Plenum, New York (1986), p.183.

(26) G.Zumbach and K.Maschke: Phys.Rev. A 28 (1983) 544; 29 (1984) 1585(E).

(27) S.K.Ghos and R.G.Parr: J.Chem.Phys.82 (1985) 3307.

(28) E.V.Ludeña: J.Chem.Phys. 79 (1983) 6174.

(29) A.Sierraalta and E.V.Ludeña: Int,J.Quantum Chem., in press.

(30) W.Kohn and P.Vashishta: in "Theory of the Inhomogeneous Electron Gas", S.Lundquist and N.H.March, eds., Plenum, New York (1983), p.79.

(31) Y.Tal and R.F.W.Bader: Int.J.Quantum Chem. Symp. 12 (1978) 153.

(32) J.P.Perdew: Phys.Rev.Lett. 55 (1986) 1665;
J.P.Perdew: "What is Right and What is Wrong with the Density-Gradient Expansions for the Exchange and Correlation Energies? in "Condensed Matter Theories 2", P.Vashishta, ed., Plenum, New York (1987).

(33) D.M.Murphy and W.-P.Wang: J.Chem.Phys. 72 (1980) 429.

(34) J.P.Perdew and A.Zunger: Phys.Rev. B23 (1981) 5048.

(35) D.C.Langreth and M.J.Mehl: Phys.Rev. B28 (1983) 1809.

(36) J.Keller, C.Amador and C. de Teresa in: "Condensed Mater Theories 1", F.B.Malik,ed., Plenum, New York (1986) p.195.

(37) I.Zh.Petkov, M.V.Stoitsov and E.S.Kryachko: Int.J.Quantum Chem. 29 (1986) 149.

(38) E.S.Kryachko and E.V.Ludeña: Phys.Rev.A, in press.

(39) W.Macke and P.Rennert: Ann.Phys.(Leipzig) 12 (1963) 84;
N.H.March and W.N.Young: Proc.Phys.Soc. (London) 72 (1958) 182;
J.K.Percus: Int.J.Quantum Chem. 13 (1978) 89.

www.ingramcontent.com/pod-product-compliance
Ingram Content Group UK Ltd.
Pitfield, Milton Keynes, MK11 3LW, UK
UKHW022108170926
8624IPUK00024B/251

9 789971 507435